Dictionary Learning Algorithms and Applications

Bogdan Dumitrescu • Paul Irofti

Dictionary Learning Algorithms and Applications

Springer

Bogdan Dumitrescu
Department of Automatic Control
and Systems Engineering
Faculty of Automatic Control
and Computers
University Politehnica of Bucharest
Bucharest, Romania

Paul Irofti
Department of Computer Science
Faculty of Mathematics
and Computer Science
University of Bucharest
Bucharest, Romania

ISBN 978-3-030-08761-6 ISBN 978-3-319-78674-2 (eBook)
https://doi.org/10.1007/978-3-319-78674-2

Printed on acid-free paper

This Springer imprint is published by the registered company Springer International Publishing AG part of Springer Nature.
The registered company address is: Gewerbestrasse 11, 6330 Cham, Switzerland

To our families:
Adina, Andrei, Sebastian
Ralu, Paul, Marta

Preface

This book revolves around the question of designing a matrix $\boldsymbol{D} \in \mathbb{R}^{m \times n}$ called dictionary, such that to obtain good sparse representations $\boldsymbol{y} \approx \boldsymbol{D}\boldsymbol{x}$ for a class of signals $\boldsymbol{y} \in \mathbb{R}^m$ given through a set of samples. The (also unknown) representation vectors $\boldsymbol{x} \in \mathbb{R}^n$ are sparse, meaning that they have only a few nonzero coefficients. So, each signal is a linear combination of a few atoms, as the dictionary columns are usually named.

Sparse representations have a great appeal in signal processing and related applications due to their parsimony, which allows capturing the essentials of a signal. This has physical roots in the workings of the human visual system, for example, but is also a convenient model for signals that may lack an analytic model or have a very complicated one. The applications include image denoising or inpainting, compression, compressed sensing, remote sensing, classification, and others.

The dictionary learning problem has already 20 years of history, with a slow start followed by a boom about 10 years ago, when several of the important algorithms were designed. The pace of research is still high, focusing especially on variations of the basic design problem and on applications. Our main goal in this book is to give a systematic view of the field, insisting on the relevant ways of posing the dictionary learning problem and on the most successful solutions. Our cover of applications is purposefully restrained to a small number of examples meant to open the road for the interested reader.

In a strict sense, dictionary learning can be seen as a difficult and large optimization problem, where the sparse representation error is minimized in least squares sense. We are usually interested in finding a good local optimum and not necessarily the global optimum. This is because the design is based on an inherently imperfect set of training signals; also, the optimized dictionary has to be used most often in an application where the objective is hard to express directly in terms of the dictionary. However, in some cases, a prominent example being classification, the objective can be amended such that the final goal is also considered; in such cases, a very good solution is extremely important.

Without investigating the problem from a pure optimization viewpoint, the algorithms are the main end product of our presentation. We start from the defining idea of an algorithm, trying to emphasize its place among related ideas, then give the technical details that make it efficient and ready for implementation, and finally describe the algorithm in a pseudocode that should be sufficiently clear for reproducibility. We also discuss properties of the algorithms, particularly complexity, but in a rather light manner. This pragmatic view is justified by the reality that dictionary learning algorithms are validated practically in applications and not theoretically. We present some experimental evidence that should give a sense of how well an algorithm solves a particular flavor of the problem.

This book is written for a reader who is generally knowledgeable in applied mathematics, optimization, and programming languages, for example at masters level in computer science, electrical engineering, or related fields. Some experience with sparse representations surely helps but is not strictly necessary. The book can be the support for a graduate course, since most of the notions are gently introduced. There are also many pointers for continuing the exploration of the dictionary learning topic; hence the book is useful for those who start research in this direction. Finally, the book aims to serve also those who are more advanced and simply need reference descriptions of a few selected algorithms.

We give now an overview of the book. Chapter 1 is an introduction to the sparse representations world, focusing on the two successful computational approaches: Orthogonal Matching Pursuit as a representative of greedy algorithms and FISTA as a representative of convex relaxations.

Chapters 2 and 3 are the foundation of the book, the former discussing the basic form of the dictionary learning problem, illustrated with a quick tour of the main applications, and the latter presenting the most important algorithms for solving it. Here are introduced the well-known MOD or K-SVD algorithms. The presentation follows the logic of the optimization techniques that are employed. Coordinate descent takes the largest share, its success being caused by its simplicity, which allows tackling a large problem such as dictionary learning with sometimes surprising efficiency. Gradient descent also plays its part, but the resulting algorithms are slower. Simple variations of the algorithms, like those using linearization around the current point or parallel update of the atoms, bring sometimes unexpectedly good results.

Chapters 4–6 are dedicated to enhancements of the dictionary learning problem, tailored to solve diverse issues. Regularization, as always, can improve the numerical behavior of an algorithm. Incoherence is a special property useful in sparse representations, aiming to keep the atoms far from one another. Chapter 4 presents versions of the basic algorithms that solve the regularized problem or enforce incoherence. Chapter 5 is dedicated to other kinds of modifications of the problem, the most significant being that where sparsity is enforced globally, not for the representation of each signal. Also important is the idea of changing the optimization objective to take into account the application where the dictionary is used. Other algorithmic ideas, such as replacing optimization with selection or giving an online character to the processing of the training signals, are treated here. Chapter 6 plays with a new variable, the dictionary size (the number of atoms), previously assumed as being fixed. The discrete character of the size leads to several types of heuristics that attempt to find an ideal size, although the very definition of the concept may be problematic. The narrow sense of finding the smallest dictionary that produces a given representation error is the easiest to formalize.

Chapter 7 attacks a different topic, that of endowing the dictionary with a structure, which may come from an application requirement like shift invariance or from a desire to promote incoherence and thus to work with orthogonal blocks. A special place is occupied by dictionaries that allow the direct representation of multidimensional signals, like images, which are vectorized in the standard setup. In all cases, complexity is a significant issue and most algorithms are faster than their general counterparts.

Chapters 8 and 9 deal with an important application field of sparse representations and dictionary learning, namely, classification. The diversity of ideas is extremely interesting, starting from learning an individual dictionary for each class and going to modifications of the learning objective that almost directly consider the classification results. Chapter 9 presents a nonlinear extension of dictionary learning based on the kernel trick. One can thus solve classification problems that are impossible for the usual approach. The complexity is much increased, but approximation methods like Nyström sampling offer a middle way by inserting some of the nonlinear character in the basic linear representation.

Chapter 10, the last one, takes the dual analysis view of cosparsity, where the representation is built from orthogonality relations between atoms and signals, unlike the usual sparse synthesis view where the signal is built as a linear combination of atoms. Dictionary learning takes a different shape in this case and, although the general approach remains similar, the algorithms rely on quite different operations.

The reading path is normally sequential, but depending on reader's interest, several routes can be taken. The first three chapters are mandatory for a basic comprehension of dictionary learning; they form the core of the book and should always come first. Chapters 4 and 5 cover essential areas and we recommend their (almost) full reading; the last few sections of Chap. 5 can be postponed if necessary. The remaining chapters are more or less independent and can be read in any order with the exception of the tandem made by Chaps. 8 and 9 that are better understood in succession. Also, the sections of Chap. 7 treat distinct dictionary structures with no relation between them; Sects. 7.5 and 7.6 come naturally in this order, as the first deals with 2D signals and the second with multidimensional signals in general.

We implemented in Matlab almost all algorithms presented in this book. Together with code for the examples, they can be found at https://github.com/dl-book.

Acknowledgements This book started with a short Erasmus course on dictionary learning given by BD at Tampere University of Technology in 2016. We are indebted to Cristian Rusu, who was the first in our group to work on dictionary learning problems for his PhD thesis; many discussions along several years helped our understanding of the topic. We thank our office mate and friend Florin Stoican for his contribution to the figures related to the water network application from Chap. 8. Finally, we congratulate each other for the patience with which this writing enterprise has been undertaken, sacrificing some of our more immediate research interests.

Bucharest, Romania
January 2018

Bogdan Dumitrescu
Paul Irofti

Contents

1 **Sparse Representations** 1
1.1 The Sparse Model 1
1.2 Algorithms 8
1.3 Orthogonal Matching Pursuit 10
1.4 Algorithms for Basis Pursuit: FISTA 12
1.5 Guarantees 16
1.6 The Choice of a Dictionary: Fixed vs Learned 20

2 **Dictionary Learning Problem** 25
2.1 The Optimization Problem 25
2.2 An Analysis of the DL Problem 29
2.3 Test Problems 32
2.3.1 Representation Error 33
2.3.2 Dictionary Recovery 35
2.4 Applications: A Quick Overview 35
2.4.1 Denoising 36
2.4.2 Inpainting 40
2.4.3 Compression 40
2.4.4 Compressed Sensing 41
2.4.5 Classification 42

3 **Standard Algorithms** 45
3.1 Basic Strategy: Alternating Optimization 45
3.2 Sparse Coding 47
3.3 Simple Descent Methods 51
3.3.1 Gradient Descent 51
3.3.2 Coordinate Descent 54
3.4 Method of Optimal Directions (MOD) 57
3.5 K-SVD 58
3.6 Parallel Algorithms 62
3.7 SimCO 64
3.8 Refinements 67

3.9 Practical Issues 69
3.9.1 Initialization 70
3.9.2 Dictionary Size and Other Size Parameters 71
3.9.3 Unused or Redundant Atoms 71
3.9.4 Randomization 74
3.10 Comparisons: Theory 74
3.11 Comparisons: Some Experimental Results 78
3.11.1 Representation Error Results 79
3.11.2 Dictionary Recovery Result 82
3.11.3 Denoising Results 82
3.12 Impact of Sparse Representation Algorithm 83

4 Regularization and Incoherence 89
4.1 Learning with a Penalty 89
4.2 Regularization 91
4.2.1 Sparse Coding 91
4.2.2 Regularized K-SVD 92
4.2.3 Comparison Between Regularized K-SVD and SimCO 95
4.3 Frames 97
4.4 Joint Optimization of Error and Coherence 103
4.5 Optimizing an Orthogonal Dictionary 105
4.6 Imposing Explicit Coherence Bounds 107
4.7 Atom-by-Atom Decorrelation 111

5 Other Views on the DL Problem 115
5.1 Representations with Variable Sparsity Levels 115
5.2 A Simple Algorithm for DL with ℓ_1 Penalty 118
5.3 A Majorization Algorithm 121
5.4 Proximal Methods 122
5.5 A Gallery of Objectives 126
5.6 Task-Driven DL 129
5.7 Dictionary Selection 131
5.8 Online DL 134
5.8.1 Online Coordinate Descent 135
5.8.2 RLS DL 137
5.9 DL with Incomplete Data 140

6 Optimizing Dictionary Size 145
6.1 Introduction: DL with Imposed Error 145
6.2 A General Size-Optimizing DL Structure 147
6.3 Stagewise K-SVD 150
6.4 An Initialization Method 153
6.5 An Atom Splitting Procedure 155
6.6 Clustering as a DL Tool 157
6.7 Other Methods 161
6.8 Size-Reducing OMP 162

7 Structured Dictionaries 167
7.1 Short Introduction 167
7.2 Sparse Dictionaries 168
7.2.1 Double Sparsity 168
7.2.2 Greedy Selection 170
7.2.3 Multi-Layer Sparse DL 172
7.2.4 Multiscale Dictionaries 172
7.3 Orthogonal Blocks 173
7.3.1 Orthogonal Basis Training 174
7.3.2 Union of Orthonormal Bases 176
7.3.3 Single Block Orthogonal DL 177
7.4 Shift Invariant Dictionaries 181
7.4.1 Circulant Dictionaries 181
7.4.2 Convolutional Sparse Coding 184
7.5 Separable Dictionaries 187
7.5.1 2D-OMP 188
7.5.2 SeDiL 190
7.6 Tensor Strategies 192
7.6.1 CP Decomposition 193
7.6.2 CP Dictionary Update 195
7.6.3 Tensor Singular Valued Decomposition 199
7.6.4 t-SVD Dictionary Update 201
7.7 Composite Dictionaries 204
7.7.1 Convex Approach 204
7.7.2 Composite Dictionaries with Orthogonal Blocks 206

8 Classification 209
8.1 Classification Problem 209
8.2 Water Networks Application 211
8.3 Simple Algorithms with Many Dictionaries 214
8.4 Discriminative DL with Many Dictionaries 216
8.4.1 DL with Discriminative Penalty Functions 217
8.4.2 Adding a Shared Dictionary 219
8.5 Discriminative DL with a Single Dictionary 222
8.5.1 Classification Using the Representation Matrix 222
8.5.2 Discriminative DL 223
8.5.3 Label Consistent DL 224
8.6 Other Developments 227

9 Kernel Dictionary Learning 231
9.1 Kernels 231
9.2 Dictionary Form 234
9.3 Kernel OMP 236
9.4 Kernel DL Algorithms 238
9.4.1 Kernel MOD 239
9.4.2 Kernel SGK 239

9.4.3 Kernel AK-SVD ... 240
9.4.4 Kernel K-SVD ... 242
9.5 Size Reduction ... 244
9.5.1 Nyström Sampling ... 244
9.5.2 Changing the Target ... 246
9.6 Classification with Kernel DL ... 247
9.6.1 Kernel SRC ... 248
9.6.2 An Example ... 249
9.6.3 Kernel Discriminative DL ... 250
9.6.4 Multiple Kernels ... 253

10 Cosparse Representations ... 257
10.1 The Cosparse Model ... 257
10.2 Representation ... 259
10.2.1 Backward Greedy ... 261
10.2.2 Optimized Backward Greedy ... 263
10.2.3 Other Algorithms ... 264
10.3 Cosparse DL ... 264
10.3.1 Analysis K-SVD ... 265
10.3.2 Analysis SimCO ... 267

References ... 271

Index ... 281

Chapter 1
Sparse Representations

Abstract Sparse representations using overcomplete dictionaries have found many signal processing applications. We present the main ways of formulating sparse approximation problems and discuss their advantages over the classical orthogonal transforms. The foremost difficulty is the computation of sparse representations, since it amounts to find the sparsest among the infinite number of solutions of an underdetermined linear system, a problem that has a combinatorial character. The most successful classes of algorithms are based on greedy approaches and convex relaxation. We describe in detail a representative algorithm from each class, namely Orthogonal Matching Pursuit and FISTA. In some circumstances, the algorithms are guaranteed to find the sparsest solution and we present sets of conditions that ensure their success. In preparation for stating the dictionary learning problem, we debate the advantages and drawbacks of learned dictionaries with respect to fixed ones. Since learning is based on training signals from the application at hand, adapted dictionaries have the potential of more faithful sparse representations, an advantage that overwhelms the necessity of (mainly off-line) extra computation.

1.1 The Sparse Model

The main character in this book is a matrix $\boldsymbol{D} \in \mathbb{R}^{m \times n}$ called dictionary. Its columns are named atoms; we denote $\boldsymbol{d}_j$ the jth column of $\boldsymbol{D}$; we assume that all atoms are normalized

$$\|\boldsymbol{d}_j\| = 1, \; j = 1 : n, \tag{1.1}$$

where $\|\cdot\|$ is the Euclidean norm, which is the default norm hereafter. In most cases the dictionary is overcomplete, which means $m < n$.

The typical use of a dictionary is for sparse representation. A vector $\boldsymbol{y} \in \mathbb{R}^m$, called here signal, has a sparse representation if it can be written as a linear combination of few atoms, i.e.,

B. Dumitrescu, P. Irofti, *Dictionary Learning Algorithms and Applications*,
https://doi.org/10.1007/978-3-319-78674-2_1

$$y = Dx = \sum_{j=1}^{n} x_j d_j = \sum_{j \in \mathcal{S}} x_j d_j, \tag{1.2}$$

where most of the coefficients x_j are zero. We name $\mathcal{S} = \{j \mid x_j \neq 0\}$ the support of the signal. For example, in Fig. 1.1, the signal can be represented with $s = 3$ atoms, those for which the coefficients x_j are nonzero. We denote $\|x\|_0$ the number of nonzero elements of the vector x; this is the so-called 0-norm, although not actually a norm, since it does not satisfy the homogeneity property. The number of atoms involved in the representation (1.2) is also named *sparsity level*; we will typically denote it s and note that $s = |\mathcal{S}| = \|x\|_0$. The notion of sparsity is not exactly defined, but it assumes that $\|x\|_0 \ll n$ and, more often than not, $\|x\|_0 \ll m$.

Example 1.1 Let us see first what kind of sparsity is usually *not* interesting. Figure 1.2 shows a signal y that is sparse in the canonical basis, which means that many of its elements are equal to zero. In this case, the dictionary is trivially the unit matrix, $D = I$, and the representation coincides with the signal, $y = x$. Such sparse signals may appear in practice, but are mostly devoid of content (exactly because they are sparse). A sound with these values would be mostly silence; an image would be nearly black, with a few gray dots. ■

Example 1.2 Sinusoidal signals are not sparse in the time domain, but their Discrete Fourier Transform (DFT) is sparse. Sinusoids are easily recognizable in sounds,

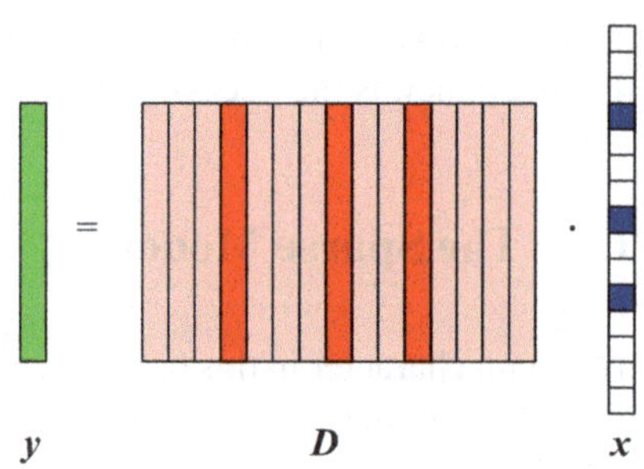

Fig. 1.1 Sparse representation of a signal. The used atoms are red and the nonzero coefficients are blue. The unused atoms are pink

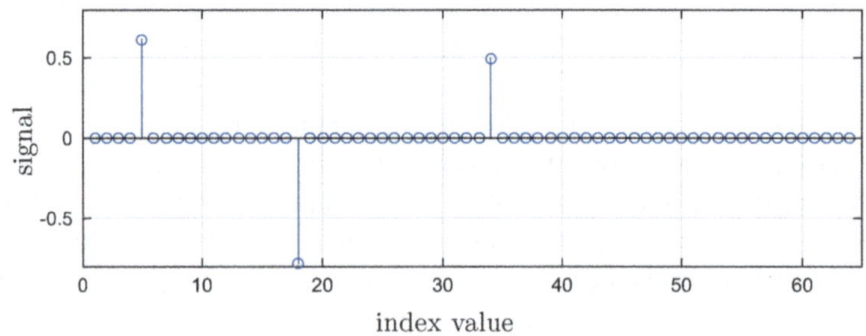

Fig. 1.2 The uninteresting case: a signal that is sparse in the canonical base

since they are pure tones, and they are important ingredients in music and in speech signals. A basis of sinusoids is easy to make; since we aim at real transforms and thus avoid the DFT, the Discrete Cosine Transform (DCT) is an immediate candidate. However, for other signals than sinusoids, the representation in the DCT base may not be even approximately sparse. To increase the family of sparsely representable signals, we can add more atoms to the DCT, obtaining an overcomplete dictionary. An example is the *overcomplete DCT*, that can be built by setting first $d_{ij} = \cos(\pi(i-1)(j-1)/n)$, $i = 1 : m$, $j = 1 : n$; the average of each column is subtracted from that column, excepting the first, which has equal elements; finally, the atoms are normalized. (Note that this construction can be different, depending on what type of DCT transform is taken as prototype; here it is the type I DCT, slightly modified.) Figure 1.3 shows the atoms of this overcomplete DCT with $m = 8$, $n = 16$; although they are discrete signals, the atoms are drawn with continuous line, to better visualize the shape of the corresponding signals. The first and third columns of subplots contain the atoms of the standard DCT (since $n = 2m$, they are obtained when $j - 1$ is even).

An example of sparse signal in the overcomplete DCT basis is $\boldsymbol{y} = 0.5\boldsymbol{d}_1 - 0.2\boldsymbol{d}_6$, since it can be represented with only two atoms of the dictionary. However, if we represent the same signal in the DCT base (the odd-numbered atoms in Fig. 1.3), then all the coefficients of the representation are nonzero. ■

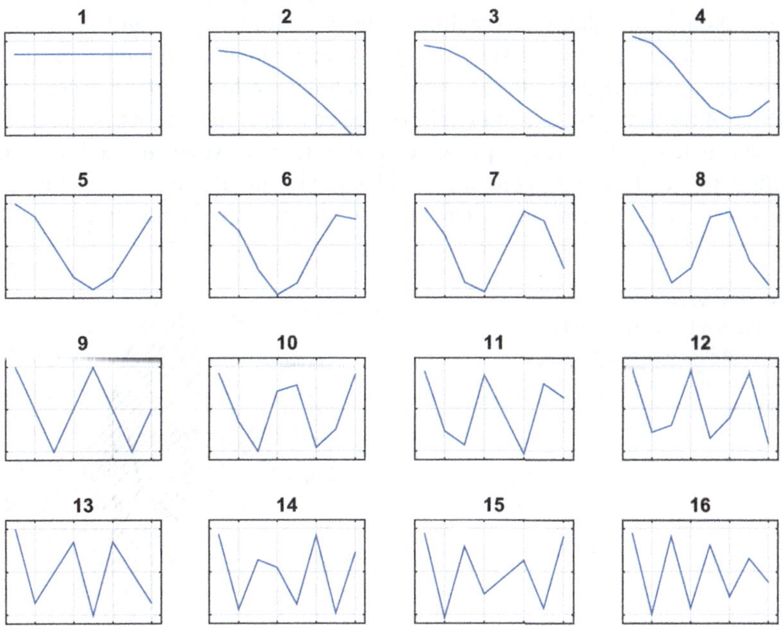

Fig. 1.3 The atoms of an 8 × 16 overcomplete DCT transform

Remark 1.3 Orthogonal transforms have found many uses in signal processing. They have many advantages, amongst which the easiness of representation. If $\boldsymbol{A} \in \mathbb{R}^{m \times m}$ is such a transform, then the transform of a signal $\boldsymbol{y}$ is $\boldsymbol{x} = \boldsymbol{A}\boldsymbol{y}$. The inverse transform is simply $\boldsymbol{A}^T$, which in our case can be viewed as the dictionary, since $\boldsymbol{y} = \boldsymbol{A}^T\boldsymbol{x}$. So, finding the representation consists of a matrix multiplication. Moreover, for many transforms, like the DFT or DCT, there are fast algorithms that reduce the complexity to $O(m \log m)$ operations.

Could overcomplete dictionaries compete with the orthogonal transforms? We have already had a glimpse of their much larger representation power. Let us develop a bit this topic. Given sparsity level s, it is clear that not all vectors in $\mathbb{R}^m$ have a sparse representation. Given a support, the signals that can be represented with the respective atoms, assumed linearly independent, lie in an s-dimensional subspace, which is "small" in $\mathbb{R}^m$. However, there are $\binom{n}{s} \approx n^s/s!$ such subspaces if all sets of s atoms are linearly independent (and distinct one from another), which creates a richness of representation already for small s. Obviously, a larger n increases the possibilities of representation with such a union of subspaces. It is thus considered that, with appropriate dictionaries, many practical signals are well approximated with sparse representations.

In any case, since for an orthogonal transform there are $\binom{m}{s} \approx m^s/s!$ possible subspaces, it is clear that the richness of sparse representations is $(n/m)^s$ higher for the overcomplete dictionary, which can amount with ease to large numbers. (Note that n/m is often called the overcompleteness factor.)

Figure 1.4 shows the subspaces that can be represented with combinations of $s = 2$ atoms taken from a dictionary with $n = m = 3$ atoms. We bound the subspaces, which are actually planes, to a sphere in the space with $m = 3$ dimensions; note that the intersection of a plane with the sphere is a great circle (the plane passes through the center of the sphere). The figure shows the case of a square (but not orthogonal) dictionary, where only $\binom{3}{2} = 3$ planes can be obtained. Atoms $\boldsymbol{d}_1$ and $\boldsymbol{d}_2$ generate the yellow plane, $\boldsymbol{d}_1$ and $\boldsymbol{d}_3$ generate the blue plane, and $\boldsymbol{d}_2$ and $\boldsymbol{d}_3$ generate the red plane. Try to imagine the case where $n = 6$, meaning an overcompleteness factor

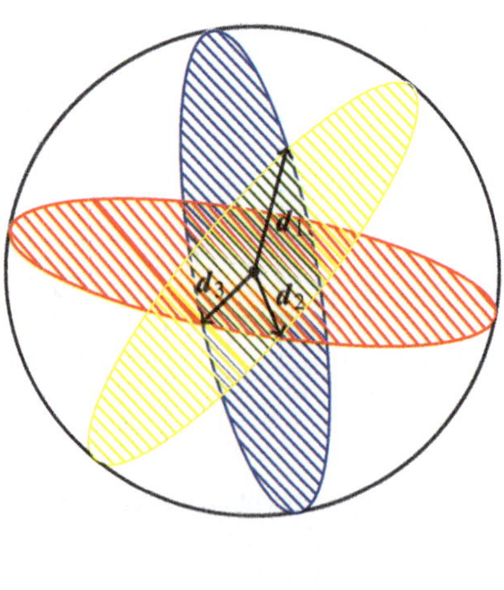

Fig. 1.4 The vectors from a sphere that can be represented as linear combinations of two atoms from a dictionary with three atoms

equal to two. There would be $\binom{6}{2} = 15$ planes, covering much better the interior of the sphere.

The clear representation advantage of the overcomplete dictionary could be shadowed by the difficulty of computing the sparse representation. We will see that, although the complexity is indeed higher than for orthogonal transforms, the computation of sparse representations is practically convenient. ■

In the remainder of this chapter we will examine several ways of posing and solving the problem of finding the sparse representation. The *exact sparse representation* problem is: given the dictionary $\boldsymbol{D}$ and the signal $\boldsymbol{y}$, we want to find the sparse vector $\boldsymbol{x}$ which is the solution of the linear system (1.2). The problem is mathematically and practically interesting when $\boldsymbol{x}$ is in fact the sparsest solution, i.e., when the problem is

$$\begin{aligned} \min_{\boldsymbol{x}} \quad & \|\boldsymbol{x}\|_0 \\ \text{s.t.} \quad & \boldsymbol{y} = \boldsymbol{D}\boldsymbol{x} \end{aligned} \tag{1.3}$$

When $m < n$, the system $\boldsymbol{y} = \boldsymbol{D}\boldsymbol{x}$ has an infinite number of solutions, from which we should pick the sparsest. The problem has a combinatorial character and is in general NP-hard: there are no polynomial time algorithms guaranteed to always solve it. However, as we will see later in this chapter, there are algorithms based on heuristics that not only have a good practical behavior, but are also guaranteed to succeed under certain conditions.

Example 1.4 Consider the dictionary

$$\boldsymbol{D} = \begin{bmatrix} 1 & 1 & 0 & -1 \\ 1 & 0 & 1 & 1 \\ 0 & 1 & 1 & -1 \end{bmatrix}. \tag{1.4}$$

Its columns are not normalized for the only purpose to have "nice" numbers in this example. The vector $\boldsymbol{y} = [2 \ \ 1 \ \ 1]^T$ can be represented as the linear combination of the first and second atoms, hence a sparse solution to $\boldsymbol{y} = \boldsymbol{D}\boldsymbol{x}$ is $\boldsymbol{x}_0 = [1 \ \ 1 \ \ 0 \ \ 0]^T$. Denoting $\boldsymbol{\xi} = [1 \ -3 \ \ 1 \ -2]^T$, we note that $\boldsymbol{D}\boldsymbol{\xi} = 0$. Since $\text{Ker}(\boldsymbol{D})$ has a single dimension and hence is generated by $\boldsymbol{\xi}$, the solutions of $\boldsymbol{y} = \boldsymbol{D}\boldsymbol{x}$ have the form $\boldsymbol{x} = \boldsymbol{x}_0 + \alpha\boldsymbol{\xi}$, with arbitrary $\alpha \in \mathbb{R}$.

In this toy example it is easy to check that $\boldsymbol{x}_0$ is the sparsest solution, since other solutions have at least three nonzeros. However, in the general case, neither finding a sparse solution nor guaranteeing that it is the sparsest are simple operations. Essentially, we should consider all combinations of m linearly independent atoms and solve the determined system with the matrix formed by these atoms for finding the sparsest solution. Such an exhaustive approach is certainly not feasible. ■

As Fig. 1.4 suggests, the signals with exact sparse representations form a set with zero size with respect to $\mathbb{R}^m$. So, the practical approach is clearly to allow a representation error and work with an approximate model, thus replacing (1.2) with

$$\boldsymbol{y} = \boldsymbol{D}\boldsymbol{x} + \boldsymbol{v}, \tag{1.5}$$

where $\boldsymbol{x}$ is sparse and $\boldsymbol{v}$ is the noise, typically Gaussian, with reasonably small variance. In this case, the signal is near from a sparse linear combinations of atoms. (Note that $\boldsymbol{y}$ can be exactly represented with m atoms, but such a representation is not useful in the current context.)

Example 1.5 To illustrate the noisy case, we take the dictionary

$$\boldsymbol{D} = \begin{bmatrix} 0 & \frac{\sqrt{3}}{2} & -\frac{\sqrt{3}}{2} \\ 1 & -\frac{1}{2} & -\frac{1}{2} \end{bmatrix}$$

and generate signals with (1.5), taking $s = 1$, and random coefficients between -1 and 1. So, the signals are just multiples of the atoms. Figure 1.5 shows the results. The left figure is obtained with $\boldsymbol{v} = 0$, hence it is the noiseless case. The right figure is obtained with noise with small variance, such that the sparse model is still obvious. It should be now easy to go to 3D and imagine Fig. 1.4 in the noisy case and then maybe go mentally to more dimensions. ■

The *approximate sparse representation* problem can then be posed in two ways. In the first, we aim to minimize the representation error with given sparsity level s:

$$\begin{aligned} \min_{\boldsymbol{x}} \quad & \|\boldsymbol{y} - \boldsymbol{D}\boldsymbol{x}\|^2 \\ \text{s.t.} \quad & \|\boldsymbol{x}\|_0 \leq s \end{aligned} \tag{1.6}$$

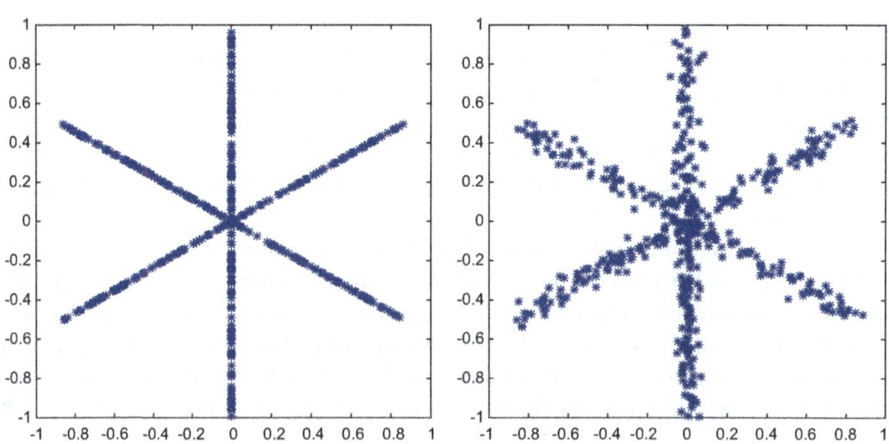

Fig. 1.5 Sparsely generated signals in 2D. Left: noiseless. Right: noisy

Although the constraint is an inequality, in most cases the solution has a sparsity level equal to s. The solution to (1.6) makes sense if we are in possession of a good guess of the sparsity level. In many applications, there is no true underlying sparsity level and the choice of s is mostly based on a try-and-error approach.

The second representation problem imposes an error bound ε and seeks the sparsest representation

$$\begin{aligned} \min_{\boldsymbol{x}} \quad & \|\boldsymbol{x}\|_0 \\ \text{s.t.} \quad & \|\boldsymbol{y} - \boldsymbol{D}\boldsymbol{x}\| \leq \varepsilon \end{aligned} \tag{1.7}$$

Of course, the bound ε should be larger than (but of the same order as) the square root of the noise variance. If ε is too small, then the problem may have no sparse solution, although the signal actually has an approximate sparse representation. If ε is too large, then the solution of (1.7) may be too sparse and possibly far from the true solution.

Like (1.3), both above problems are NP-hard, hence we are interested in fast algorithms that give suboptimal solutions.

Example 1.6 Consider the dictionary (1.4) from Example 1.4, where we have seen that the signal $\boldsymbol{y} = [2 \;\; 1 \;\; 1]^T$ can be represented as the linear combination of the first and second atoms. Perturb the signal to $\hat{\boldsymbol{y}} = [2 \;\; 1.1 \;\; 1]^T$. Solving (1.6) with $s = 2$, by trying all the combinations of two atoms (a totally impractical approach, due to the exhaustive search and the knowledge of the true sparsity level), we find the solution $\hat{\boldsymbol{x}} = [1.0667 \;\; 0.9667 \;\; 0 \;\; 0]^T$, which has the same support and is fairly close to the solution of the exact system, $\boldsymbol{x}_0 = [1 \;\; 1 \;\; 0 \;\; 0]^T$; however, note that we do not get the exact solution and we should not hope to get it in the noisy context.

If the noise is large, like in $\check{\boldsymbol{y}} = [2 \;\; 0.5 \;\; 1]^T$, then the solution is $\check{\boldsymbol{x}} = [1.25 \;\; 0 \;\; 0 \;\; -0.8333]^T$, which has a different support. So, a large noise can move the signal closer to an other sparsely representable subspace (e.g., in Fig. 1.4, by adding noise to a signal in the yellow plane we can end up nearer from the blue or red planes).

If we now attempt to solve (1.7) (using the same approach), with $0.0577 < \varepsilon < 1.1853$ we get a solution with sparsity level $s = 2$. The lower bound corresponds to $\hat{\boldsymbol{x}}$. Smaller values of ε are satisfied only by solutions with $s = 3$ (full of nonzeros). Higher values of ε bring into play solutions with $s = 1$.

So, there are several types of dangers when working with approximate sparse representations. Some of them have mostly academic flavor; for example, in many applications, there is no true sparse representation and we are happy with one having a reasonably small error. ■

1.2 Algorithms

There are many algorithms for solving the sparse representation problems (1.6) and (1.7), based on diverse heuristics. Although they cannot find the optimal solution in general, some of them are guaranteed to succeed under certain conditions; examples of such guarantees are given later in Sect. 1.5. Most algorithms have good practical behavior even if the success conditions do not hold. We review here the most important classes of algorithms.

Greedy algorithms build the support by selecting one atom at a time based on a local objective. They are usually very fast. The most popular heuristics for selecting the most promising atom are Orthogonal Matching Pursuit (OMP) [144] and Orthogonal Least Squares (OLS) [36]. OMP, discussed in detail in the next section, chooses the atom that is most aligned with the current residual. OLS chooses the atom that, added to the current support, minimizes the representation error. Both OMP and OLS only add atoms to the support. Other algorithms allow also removing atoms from the support, like Subspace Pursuit [43] or the projection-based algorithms from [34].

Convex relaxation techniques have as main representative the ℓ_1 relaxation, which consists of replacing the 0-pseudo-norm in (1.3) with an actual norm, thus obtaining the convex optimization problem

$$\begin{aligned} \min_{\boldsymbol{x}} \quad & \|\boldsymbol{x}\|_1 \\ \text{s.t.} \quad & \boldsymbol{y} = \boldsymbol{D}\boldsymbol{x} \end{aligned} \tag{1.8}$$

The 1-norm is defined by $\|\boldsymbol{x}\|_1 = \sum_{i=1}^{n} |x_i|$ and can be seen as the closest convex approximation of the 0-norm. It has the known effect of promoting sparse solutions.

In the noisy case, an immediate idea is to replace the 0-norm with the 1-norm in (1.7), thus obtaining the *Basis Pursuit* (BP) [37] problem. A more popular solution, closely related to BP, is to penalize the error with the 1-norm, thus removing all constraints. It is also called BP or BP denoising

$$\min_{\boldsymbol{x}} \quad \|\boldsymbol{y} - \boldsymbol{D}\boldsymbol{x}\|^2 + \lambda \|\boldsymbol{x}\|_1 \tag{1.9}$$

The number of nonzeros that result when solving (1.9) can be controlled by the value of the trade-off parameter λ, but a hard bound s like in (1.6) cannot be imposed.

Another approach to convex relaxation is the *lasso* problem [187], which can be seen as a convexification of (1.6) and has the form

$$\begin{aligned} \min_{\boldsymbol{x}} \quad & \|\boldsymbol{y} - \boldsymbol{D}\boldsymbol{x}\|^2 \\ \text{s.t.} \quad & \|\boldsymbol{x}\|_1 \le \theta \end{aligned} \tag{1.10}$$

Here, θ is a bound for the 1-norm of the solution. The advantage over (1.9) is that the error objective is not altered. However, there is still a parameter to be chosen and the optimization problem is slightly more difficult due to the explicit constraints.

The main advantage of (1.8)–(1.10) is the convexity, which allows reliable solutions; note, however, that the problems are not strictly convex, since the dictionary is overcomplete and so rank$\boldsymbol{D} \leq m < n$; hence, the minimum may be attained in more than one point. Since convex optimization is now a mature field, there are many algorithms that can be adapted to the BP or lasso problems and whose complexity, although higher than that of greedy algorithms, is still convenient for some practical applications. We will discuss one of these algorithms in Sect. 1.4.

A variant of convex relaxation is the *elastic net* [227], where the 1-norm is combined with the 2-norm, thus combining the sparsity penalty with standard regularization.

Other Algorithms Besides the greedy and convex relaxation approaches, there are other families of algorithms. We give here only a few pointers to some main ideas; the interested reader should go to the sources or to works that cover more widely the sparse representations algorithms, like [17, 61, 224] and their bibliographies.

A variation on the Basis Pursuit problem (1.9) is to replace the 1-norm with other penalty functions with sparsifying effect. Although these functions are not convex, and hence the optimization process is more delicate and not necessarily guaranteed to provide the optimal solution, they are better approximation of the genuine sparse representations problems (1.6) or (1.7). Among the popular penalty functions are the p-norm, with $0 < p < 1$, logarithmic functions like $\sum_{i=1}^{n} \log(x_i^2 + \alpha)$ (with $\alpha > 0$), and the minimax concave penalty [218]. Sometimes it is not even necessary to have an explicit expression of the penalty function, but only of the associated proximal (shrinkage) operator, which is used directly in optimization procedures like the alternating direction method of multipliers (ADMM), see, for example, [33]. Other algorithms using nonconvex penalties are based on iteratively reweighted least-squares (including FOCUSS [82], one of the forerunners in sparse representation computation), majorization-minimization, or homotopy methods. Iterative hard thresholding [13] can also be integrated in this category, since it implicitly works with a 0-norm penalty; a development of this algorithm, inspired from graphical models and using message passing, is [55].

Bayesian approaches [188, 205] associate a probabilistic model to the elements of the solution, with underlying random variables which essentially decide if the element is zero or not. They have higher complexity than other methods, but are especially effective in the case of group sparsity, where groups of elements are simultaneously zero; in this case, a single random variable is associated with the whole group and thus the algorithm is greatly simplified.

1.3 Orthogonal Matching Pursuit

Orthogonal Matching Pursuit (OMP) [144] is a greedy algorithm that builds the support of the sparse representation by adding one atom at a time. We denote $\mathcal{S}$ the support at some stage; it contains the already selected atoms. We assume that we know the coefficients of these atoms in the current representation. Thus, the current representation residual is

$$e = y - \sum_{j \in \mathcal{S}} x_j d_j. \tag{1.11}$$

Initially, the support is void and hence the residual is $e = y$.

The main characteristic of a greedy algorithm is the rule for selecting the next atom. OMP chooses the atom which is the most correlated with the residual; the new atom is d_k for which

$$|e^T d_k| = \max_{j \notin \mathcal{S}} |e^T d_j|. \tag{1.12}$$

Given the other *atoms and representation coefficients*, this is the choice that minimizes the next residual. The next support obviously becomes $\mathcal{S} \leftarrow \mathcal{S} \cup \{k\}$.

The second important characteristic of OMP is the (re)computation of the new coefficients. They are the solution of the least squares problem

$$\min_{\xi} \|y - D_{\mathcal{S}} \xi\|, \tag{1.13}$$

where $D_{\mathcal{S}}$ is the restriction of the matrix D to the columns with indices from $\mathcal{S}$. Note that the solution of this problem is a vector of size $|\mathcal{S}|$ containing the current nonzero elements of the sparse solution x. Analytically, the least squares solution can be written as

$$x_{\mathcal{S}} = (D_{\mathcal{S}}^T D_{\mathcal{S}})^{-1} D_{\mathcal{S}}^T y, \tag{1.14}$$

although its computation can be done in several ways. Note that all the nonzero elements of the solution are recomputed at each stage of OMP. Also, due to the optimality of the least squares solution, the residual is orthogonal to all selected atoms. (Hence, the precaution to consider only $j \notin \mathcal{S}$ in (1.12) is actually useless, since $e^T d_j = 0$ for $j \in \mathcal{S}$.)

OMP is summarized as Algorithm 1.1. Two stopping criteria can be used, corresponding to the two ways of seeing the approximate sparse representation problem, (1.6) or (1.7). The first imposes a maximum sparsity level s; once this level is attained, the algorithm stops regardless of the error. The second imposes an error bound ε and grows the support until the actual error becomes smaller than the

---

Algorithm 1.1: Orthogonal Matching Pursuit

Data: dictionary $\boldsymbol{D} \in \mathbb{R}^{m \times n}$
signal $\boldsymbol{y} \in \mathbb{R}^m$
sparsity level s
stopping error ε
Result: representation support $\mathcal{S}$, solution $\boldsymbol{x}$

1 Initialize $\mathcal{S} = \emptyset$, $\boldsymbol{e} = \boldsymbol{y}$
2 **while** $|\mathcal{S}| < s$ *and* $\|\boldsymbol{e}\| > \varepsilon$ **do**
3 Find new index: $k = \arg\max_{j \notin \mathcal{S}} |\boldsymbol{e}^T \boldsymbol{d}_j|$
4 Build new support: $\mathcal{S} \leftarrow \mathcal{S} \cup \{k\}$
5 Compute new solution: $\boldsymbol{x}_{\mathcal{S}} = (\boldsymbol{D}_{\mathcal{S}}^T \boldsymbol{D}_{\mathcal{S}})^{-1} \boldsymbol{D}_{\mathcal{S}}^T \boldsymbol{y}$
6 Compute new residual: $\boldsymbol{e} = \boldsymbol{y} - \boldsymbol{D}_{\mathcal{S}} \boldsymbol{x}_{\mathcal{S}}$

---

bound, regardless of the support size (hence, in principle, the solution could be no longer sparse).

Algorithm 1.1 implements simultaneously both criteria, stopping when the first one is met. Denoting OMP($\boldsymbol{D}$, $\boldsymbol{y}$, s, ε) a call to the algorithm, stopping with given sparsity level is obtained with OMP($\boldsymbol{D}$, $\boldsymbol{y}$, s, 0); only in the unlikely event of reaching an exact solution, the actual sparsity level may be smaller than s, but this is a perfectly acceptable situation. Stopping with given error is obtained with OMP($\boldsymbol{D}$, $\boldsymbol{y}$, m, ε); barring a degenerate case where $\operatorname{span}\boldsymbol{D} \neq \mathbb{R}^m$, a solution with m nonzero elements would be exact, hence the error condition will be fulfilled first. Throughout the book we assume that the sparsity level condition is used in OMP, if no details are given.

The direct implementation of Algorithm 1.1 can be easily analyzed. Let us count the arithmetic operations at iteration s, the last one. Line 3 needs basically the multiplication $\boldsymbol{D}^T\boldsymbol{e}$, hence $O(mn)$ operations. Line 6 is a matrix–vector multiplication requiring $O(ms)$ operations. The least squares solution from line 5 is the most demanding, needing $O(ms^2)$ operations. We can ignore the $O(m)$ operations for computing $\|\boldsymbol{e}\|$ in line 2. So, the overall operation count is roughly $O(ms(n + s^2))$, which is already acceptable for small s.

However, computing from scratch the least squares solution is a naive approach, since the current matrix $\boldsymbol{D}_{\mathcal{S}}$ is obtained by appending a column to the previous one. So, a more efficient implementation can be obtained by updating information from the previous step. More details are given in Sect. 3.2, in the context of dictionary learning algorithms, where OMP is preferred due to its good trade-off between complexity and representation abilities.

Remark 1.7 (OMP vs OLS) A related algorithm is Orthogonal Least Squares (OLS), which selects the next atom as the one that, together with the already selected atoms, minimizes the residual norm. The only difference between OLS and OMP is that OLS replaces (1.12) with

$$\min_{k,\xi} \left\| y - D_{\mathcal{S}\cup\{k\}}\xi \right\| . \tag{1.15}$$

So, the residual is minimized given only the previous atoms, not also the representation coefficients, like in OMP. Locally, OLS makes the best choice, and so it is usually slightly superior to OMP. However, OLS has a higher complexity, which makes it less practical. ■

1.4 Algorithms for Basis Pursuit: FISTA

Interior point methods are probably the most known among the plethora of algorithms for solving convex optimization problems [14]. Their generality and reliability recommends them for many applications. However, in sparse representation problems from the Basis Pursuit family (see Sect. 1.2), we are interested especially in fast algorithms. Nearly optimal solutions are acceptable, since convex relaxations are approximations of the original sparse representation problems. In this section we present a fast algorithm for solving (1.9), namely FISTA (fast iterative shrinkage-thresholding algorithm) [11].

We start by a reinterpretation of the gradient method for minimizing a differentiable function $f : \mathbb{R}^m \rightarrow \mathbb{R}$. If x_k is the current variable value at iteration k, the gradient method builds the next value by

$$x_{k+1} = x_k - t\nabla f(x_k), \tag{1.16}$$

where t is the step size. This is equivalent to solving

$$x_{k+1} = \arg\min_{x} \left\{ f(x_k) + (x - x_k)^T \nabla f(x_k) + \frac{1}{2t} \|x - x_k\|^2 \right\} . \tag{1.17}$$

Indeed, by setting to zero the gradient of the function in (1.17), relation (1.16) results immediately. The problem (1.17) may be seen as the minimization of the first order approximation of the function penalized with a quadratic term that keeps the solution not far from the current point.

The Basis Pursuit problem (1.9) is an instance of the more general problem

$$\min_{x} \quad f(x) + g(x), \tag{1.18}$$

where both $f, g : \mathbb{R}^m \rightarrow \mathbb{R}$ are convex and continuous, but only f is differentiable. For problem (1.9), we have $f(x) = \|y - Dx\|^2$ and $g(x) = \lambda\|x\|_1$.

For solving (1.18), the gradient approach (1.17) can be naturally generalized to

$$
\begin{aligned}
x_{k+1} = & \arg\min_{x} \left\{ f(x_k) + (x - x_k)^T \nabla f(x_k) + \frac{1}{2t} \|x - x_k\|^2 + g(x) \right\} \\
= & \arg\min_{x} \left\{ \frac{1}{2t} \|x - (x_k - t\nabla f(x_k))\|^2 + g(x) \right\}
\end{aligned}
\tag{1.19}
$$

This is the iterative shrinkage-thresholding algorithm (ISTA) [47] and is a proximal gradient method; a more thorough view to such methods will be presented in Sect. 5.4.

Denoting $\tilde{x}_k = x_k - t\nabla f(x_k)$, and focusing on our case, where $g(x) = \lambda\|x\|_1$, the problem (1.19) becomes

$$
x_{k+1} = \arg\min_{x} \left\{ \frac{1}{2t} \|x - \tilde{x}_k\|^2 + \lambda\|x\|_1 \right\} \tag{1.20}
$$

Proposition 1.8 *The solution to (1.20) is*

$$
x_{k+1} = Soft\_th\,(\tilde{x}_k, t\lambda)\,, \tag{1.21}
$$

where

$$
Soft\_th(\xi, \alpha) = \begin{cases} \xi + \alpha, & if\ \xi < -\alpha \\ 0, & if\ -\alpha \le \xi \le \alpha \\ \xi - \alpha, & if\ \xi > \alpha \end{cases} \tag{1.22}
$$

is the soft thresholding operator (applied elementwise on vectors).

Proof Since soft thresholding will appear again later in the book, let us dwell more on this proof. It is clear that the function from (1.20) is separable, hence we can look only at the scalar case. Assume that we have a differentiable convex function $h(\xi)$ and we want to find the minimum of $h_\alpha(\xi) = h(\xi) + \alpha|\xi|$, which is not differentiable in zero, but differentiable everywhere else. The function h and the ℓ_1 penalty are illustrated in the upper part of Fig. 1.6; the red line has a slope equal to α, and is tangent to the function graph in a point ξ where $h'(\xi) = \alpha$. The position of the minimum of h_α is decided by the value of $h'(0)$. If $h'(0) > \alpha$, then the minimum is negative and can be solved from

$$
h'_\alpha(\xi) = 0 \Leftrightarrow h'(\xi) = \alpha.
$$

If $-\alpha \le h'(0) \le \alpha$, then the minimum is in zero, since the derivative of h_α jumps from negative to positive values in zero. If $h'(0) < -\alpha$, then the minimum is positive and can be solved from

$$
h'_\alpha(\xi) = 0 \Leftrightarrow h'(\xi) = -\alpha.
$$

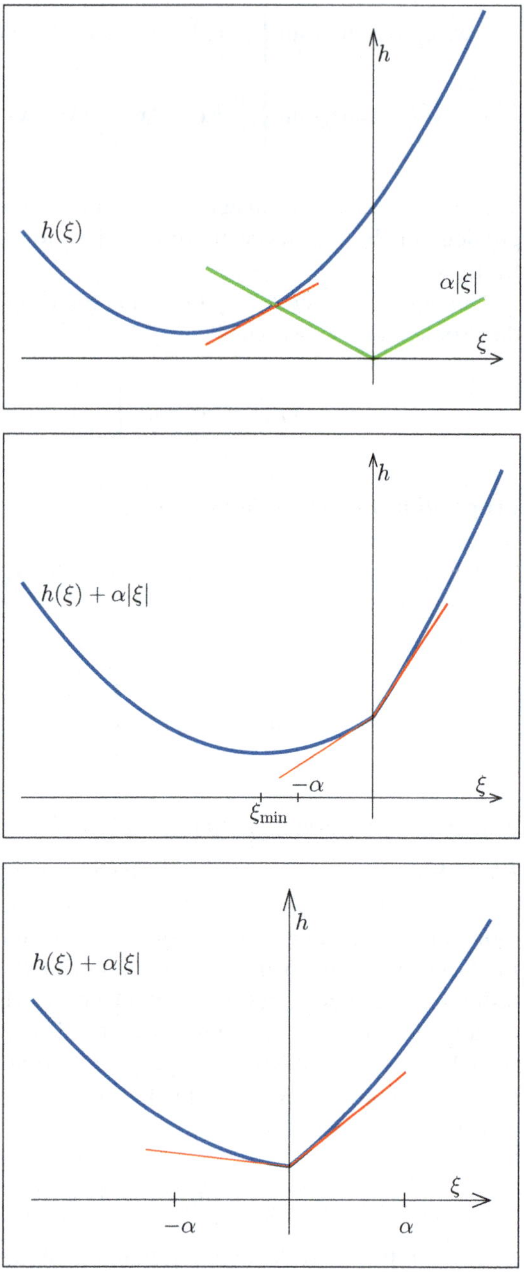

Fig. 1.6 Soft thresholding. Up: convex function $f(\xi)$ and ℓ_1 penalty $\alpha|\xi|$. Middle: their sum, when differentiable in the minimum. Down: their sum, when minimum in zero

The first and second cases are illustrated by the middle and bottom figures. For a quadratic function $h(\xi) = |\xi - \tilde{\xi}|^2 = \xi^2 - 2\xi\tilde{\xi} + \tilde{\xi}^2$, whose derivative is $h'(\xi) = 2\xi - 2\tilde{\xi}$, the above relations lead to the solution

$$\xi = \text{Soft\_th}\left(\tilde{\xi}, \frac{\alpha}{2}\right).$$

The correspondence with (1.21) is immediate. ■

Assuming that the gradient of f is Lipschitz with constant L, meaning that

$$\|\nabla f(\boldsymbol{x}) - \nabla f(\boldsymbol{\xi})\| \leq L \, \|\boldsymbol{x} - \boldsymbol{\xi}\|, \tag{1.23}$$

the ISTA iteration (1.21) is convergent if the step size is $t = 1/L$. Moreover, ISTA has a $O(1/k)$ convergence rate towards the optimal solution of (1.9).

Applying now (1.21) to the Basis Pursuit problem, the ISTA step has the expression

$$\boldsymbol{x}_{k+1} = \text{Soft\_th}\left(\boldsymbol{x}_k - 2t\boldsymbol{D}^T(\boldsymbol{D}\boldsymbol{x}_k - \boldsymbol{y}), t\lambda\right) \stackrel{\Delta}{=} \phi(\boldsymbol{x}_k). \tag{1.24}$$

FISTA uses the above relation, but constructs the next point based on an acceleration trick due to Nesterov that combines two previous points, thus improving the convergence rate to $O(1/k^2)$. Algorithm 1.2 presents the necessary operations. The sequence $\boldsymbol{x}_k$ is built with the help of an auxiliary sequence $\boldsymbol{\xi}_k$; however, the ISTA iteration (1.24) is used only once per iteration, in step 3 of the algorithm. The combination factor τ_k is updated in step 4 and the accelerated iteration is performed in step 5 (where the initialization $\boldsymbol{\xi}_{-1} = 0$ is used). The initial solution may be arbitrary, zero, or given by the user if some prior information is available. The iterations are repeated until a convergence criterion (not specified here) is satisfied.

Theoretically, the gradient step size t used in (1.24) depends on the largest eigenvalue of $\boldsymbol{D}^T\boldsymbol{D}$. Since it is hard to estimate within the computational budget of FISTA, the exact step size is replaced by a variable one, that can be adapted by simple search strategies; the same convergence rate is valid, without excessively increasing the operation count.

Among the other algorithms similar in flavor with FISTA one can cite NESTA [12], which also uses acceleration schemes and has the supplementary advantage of solving problems with constraints, like the ℓ_1 relaxation of (1.7).

Remark 1.9 No matter what algorithm we use for solving the Basis Pursuit problem (1.9), it is likely that the solution $\boldsymbol{x}$ contains many elements that are very small but not exactly zero. While in some applications this is satisfactory, in others we need a truly sparse solution. To obtain it, we set to zero the elements of $\boldsymbol{x}$ that are smaller (in absolute value) than a user-defined threshold, thus obtaining a support $\mathcal{S}$ of the representation. With this support, the solution is recomputed using the least squares formula (1.14), thus obtaining the best sparse approximation for $\mathcal{S}$. ■

Algorithm 1.2: FISTA

Data: dictionary $\boldsymbol{D} \in \mathbb{R}^{m \times n}$
signal $\boldsymbol{y} \in \mathbb{R}^m$
trade-off constant λ, step size t
initial approximation $\boldsymbol{x}_0$ of the solution
Result: solution of Basis Pursuit problem (1.9)

1 Initialize $\tau_0 = 1$
2 **for** $k = 0, 1, \ldots$ **do**
3 Compute auxiliary approximation $\boldsymbol{\xi}_k = \phi(\boldsymbol{x}_k)$ using (1.24)
4 Update stepsize: $\tau_{k+1} = \frac{1+\sqrt{1+4\tau_k^2}}{2}$
5 Compute new approximation: $\boldsymbol{x}_{k+1} = \boldsymbol{\xi}_k + \frac{\tau_k - 1}{\tau_{k+1}}(\boldsymbol{\xi}_k - \boldsymbol{\xi}_{k-1})$

1.5 Guarantees

This section presents guarantees for the existence of a sparsest solution and for OMP and other algorithms to find it.

Definition 1.10 The *mutual coherence* of a dictionary $\boldsymbol{D}$ is

$$M = \max_{1 \le i < j \le n} |\boldsymbol{d}_i^T \boldsymbol{d}_j|, \tag{1.25}$$

i.e., the maximum scalar product of two atoms (sign ignored). ■

The mutual coherence measures how close two atoms can be. The most incoherent are orthogonal matrices, for which $M = 0$. Overcomplete matrices necessarily have $M > 0$. In the worst case, when two atoms are nearly collinear, M is nearly 1.

Mutual coherence takes into account only pairs of atoms. The next definitions consider larger sets of atoms.

Definition 1.11 ([190]) The *cumulative coherence* with index s of a dictionary $\boldsymbol{D}$ is

$$M_s = \max_{|\mathcal{I}|=s} \max_{i \notin \mathcal{I}} \sum_{j \in \mathcal{I}} |\boldsymbol{d}_i^T \boldsymbol{d}_j|. \tag{1.26}$$

In particular, M_1 is the mutual coherence (1.25). ■

So, to define the cumulative coherence we split the atoms in all possible combinations of two disjoint groups, the first having s atoms and the other $n-s$, and take the largest sum of absolute scalar products of an atom from the second group with all atoms in the first group. Unlike mutual coherence, the cumulative coherence cannot be easily computed due to its inherent combinatorial character. It is clear that

$M_s \le sM$, but for most overcomplete dictionaries the actual growth of M_s with s is sublinear.

Another way of looking at groups of s atoms examines their distance from singularity (or from linear dependence).

Definition 1.12 ([23]) A dictionary $\boldsymbol{D}$ satisfies the *restricted isometry property (RIP)* with index s and constant δ_s if for any submatrix $\tilde{\boldsymbol{D}}$ made of s columns of $\boldsymbol{D}$, the constant δ_s is the smallest such that

$$(1-\delta_s)\|\boldsymbol{x}\|^2 \le \|\tilde{\boldsymbol{D}}\boldsymbol{x}\|^2 \le (1+\delta_s)\|\boldsymbol{x}\|^2 \tag{1.27}$$

for any vector $\boldsymbol{x}$ of size s. ■

Equivalently, the RIP definition can be written in terms of the extremal singular values of the submatrix:

$$\sqrt{1-\delta_s} \le \sigma_{\min}(\tilde{\boldsymbol{D}}) \le \sigma_{\max}(\tilde{\boldsymbol{D}}) \le \sqrt{1+\delta_s}.$$

Consequently, the condition number of the submatrix is bounded

$$\kappa(\tilde{\boldsymbol{D}}) = \frac{\sigma_{\max}(\tilde{\boldsymbol{D}})}{\sigma_{\min}(\tilde{\boldsymbol{D}})} \le \sqrt{\frac{1+\delta_s}{1-\delta_s}}. \tag{1.28}$$

Again, computing the RIP constant is not a feasible task, hence the use of the RIP is mostly for theoretical analysis.

Of course, Definition 1.12 is meaningful only when $\delta_s < 1$. The case $\delta_s = 1$ means that the dictionary has s columns that are linearly dependent, which is certainly not a desired property; in fact, we would like that every set of m columns is linearly independent. It is clear from Definition 1.12 that δ_s is a nondecreasing sequence (with index s), up to $\delta_{m+1} = 1$. The ideal value, true only for orthogonal matrices, is $\delta_s = 0$; overcomplete dictionaries necessarily have $\delta_s > 0$.

Finer distinctions can be made if two RIP constants are defined, one for the lower bound and one for the upper, i.e.

$$(1-\delta_{ls})\|\boldsymbol{x}\|^2 \le \|\tilde{\boldsymbol{D}}\boldsymbol{x}\|^2 \le (1+\delta_{hs})\|\boldsymbol{x}\|^2, \tag{1.29}$$

but we will present only results that use (1.27).

Finally, although historically the first, linear dependence between groups of atoms can also be useful in characterizing the sparsest solution.

Definition 1.13 The spark of a dictionary $\boldsymbol{D}$ is the smallest number of columns that are linearly dependent. ■

It is clear that $2 \le spark(\boldsymbol{D}) \le m+1$. The lower bound holds for any nonzero matrix. Since maximum rank is a generic property, the spark should be $m+1$ with probability one, although nearly linear dependence may be practically as bad as exact dependence. However, computing the spark is next to impossible, since we essentially need to check the rank for all groups of s atoms, with $s = 2 : m$.

We start our sample of guarantees with the noiseless case, when relation (1.2) holds for some support $\mathcal{S}$. The mutual coherence or the spark values can guarantee the uniqueness of the sparsest solution.

Proposition 1.14 ([52])

(a) If the linear system $\boldsymbol{y} = \boldsymbol{D}\boldsymbol{x}$ has a solution $\boldsymbol{x}$ such that

$$\|\boldsymbol{x}\|_0 \leq \frac{1}{2}\left(1 + \frac{1}{M}\right), \tag{1.30}$$

then this solution is the sparsest possible.

(b) The same conclusion holds if $\|\boldsymbol{x}\|_0 < spark(\boldsymbol{D})/2$.

■

It is thus clear that a lower mutual coherence allows solutions with higher sparsity levels to be qualified as the sparsest. The above proposition does not say how the sparsest solution can be computed; it only allows checking if this is indeed the solution, if found by some means.

Example 1.15 We compute the value of the bound from (1.30) for two simple dictionary candidates of size 64×128. For the overcomplete DCT matrix defined in Example 1.2, the mutual coherence is 0.9763. For random matrices, with elements drawn from the Gaussian distribution with zero mean and unit variance, the mutual coherence has values around 0.5. In both cases, the bound (1.30) is effectively 1, which is not informative, since it is trivial to see that a solution with a single nonzero is indeed the sparsest. ■

Remark 1.16 In contrast with this pessimistic result, the spark part of Proposition 1.14 offers more hope. In the generic case, the spark of $\boldsymbol{D}$ is $m + 1$, so, for $m = 64$, the Proposition says that if $\|\boldsymbol{x}\|_0 \leq 32$, then $\boldsymbol{x}$ is the sparsest solution. The big gap between the bound given by the spark (which is impossible to compute) and the mutual coherence (which is always at hand) leaves room for other estimations, e.g., based on RIP, not presented here. ■

We present below several results that show the ability of the main algorithms to find the sparsest solution in the noiseless case, when (1.2) holds. For all of them, we assume that the system $\boldsymbol{y} = \boldsymbol{D}\boldsymbol{x}$ has a sparsest solution with sparsity level s.

Proposition 1.17 ([190]) *Assume that one of the following two conditions holds:*

$$s \leq \frac{1}{2}\left(1 + \frac{1}{M}\right),$$

$$M_{s-1} + M_s < 1.$$

Then OMP always recovers the sparsest solution. Under the same hypothesis, the solution of the ℓ_1 relaxation problem (1.8) is the sparsest. ■

Note the similarity between the first condition and (1.30). Regarding mutual coherence, Proposition 1.17 is strong, in the sense that the guarantee of the sparsest solution is the same that the guarantee that this solution can be computed.

Proposition 1.18 ([31]) *OMP finds the sparsest solution if* $\delta_{s+1} < \dfrac{\sqrt{4s+1}-1}{2s}$. ■

The line of research proposing such bounds started in [48], where the bound was $1/(3\sqrt{s})$. One can see that

$$\frac{1}{\sqrt{s}+1} < \frac{\sqrt{4s+1}-1}{2s} < \frac{1}{\sqrt{s}}.$$

It was shown [135] that OMP may fail if $\delta_{s+1} = 1/\sqrt{s}$, hence Proposition 1.18 is quite sharp.

Proposition 1.19 *The solution of the convex relaxation problem (1.8) is the sparsest if one of the following conditions hold:*

i) $\delta_s < 0.307$ *[20]*
ii) $\delta_{2s} < \sqrt{2}-1$ *[22], improved to* $\delta_{2s} < 0.472$ *[21]* ■

Remark 1.20 All the results in this section look rather pessimistic, in the sense that the bounds seem hard to achieve and they are anyway hard to compute. However, a general lesson can be drawn: a sparse representation problem is easier and the algorithm for solving it is more likely to find the solution if the atoms of the dictionary are far away one from each other. We will see in Chap. 4 some methods to achieve this goal. ■

In the noisy case, where the model (1.5) applies, the analysis is more cumbersome. Assuming that the noise power is bounded,

$$\|\boldsymbol{y} - \boldsymbol{D}\boldsymbol{x}\| \leq \varepsilon, \tag{1.31}$$

we search a sparse representation for a signal inside a ball centered in $\boldsymbol{y}$. In principle, there may be several sparse solutions in the ball, some even sparser than the original one. However, under certain conditions, one can show that the sparsest solution of the exact problem is stable, in the sense that any sparse solution of the noisy problem is in its neighborhood.

Stability results have the following typical form. Assume that $\boldsymbol{x}$ satisfies (1.31) and also a relation depending on a dictionary constant, e.g., (1.30). Then any solution $\tilde{\boldsymbol{x}}$ of (1.7) satisfies $\|\tilde{\boldsymbol{x}} - \boldsymbol{x}\| \leq C\varepsilon$, where C is a constant depending on $\|\boldsymbol{x}\|_0$ and the dictionary. So, the actual solution of the noisy problem is within a ball centered on the sparsest solution (of the noisy problem). There are many results on this topic; one can start reading from [61, Ch.5].

Passing now to guarantees for algorithm, the results are more varied. For example, in [19], OMP (stopping at error ε) is guaranteed to find the support $\mathcal{S}$ of the sparsest solution $\boldsymbol{x}$ when (1.31) holds, if the mutual coherence is bounded, $M \leq 1/(2s-1)$, and the nonzero elements of $\boldsymbol{x}$ are sufficiently large:

$$|x_i| \geq \frac{2\varepsilon}{1-(2s-1)M}.$$

Such conditions are natural. While the mutual coherence is already an usual ingredient, the bounded value of the nonzero elements ensures that there is a large enough ball around the sparsest solution which does not contain any other solution of the noisy problem with sparsity level at most s.

1.6 The Choice of a Dictionary: Fixed vs Learned

A significant issue when using sparse representations in an application is the choice of the dictionary. There are two main possibilities, fixed vs learned dictionaries, whose advantages and drawbacks we debate here. Looking for precedents, we can think (again) at orthogonal transforms. Fixed transforms like the DFT, DCT or wavelet have clear benefits for many real-world applications, managing to compress the information for some categories of signals (e.g., speech and sound) and also to capture important traits of the signal in the transform domain; most of them have fast algorithms, hence they are even more appealing. However, they may fail in some cases. When we have a sufficiently large sample or an analytic model of the signals, the optimal Karhunen–Loève transform can be computed, which optimally compacts the energy of the signal; the drawback is that there is no fast algorithm and hence the standard matrix–vector multiplication is necessary for transforming a signal.

For overcomplete dictionaries, the discussion is similar, but more complicated due to the higher number of options. Fixed dictionaries can be made from the rows of a classic transform, like the overcomplete DCT from Example 1.2. This is an easy construction, since it has an analytic form, and needs not necessarily full storage. For computing sparse representations, the advantage of fast algorithms is not always accessible; for example, in OMP, the multiplication of $\boldsymbol{D}^T$ with the residual can be implemented as, e.g., the application of the DCT (if this is the underlying transform) of size $n \times n$ on the residual padded with zeros up to length n; if the overcompleteness factor n/m is not too high, this is faster than the $O(mn)$ matrix vector multiplication. However, in other algorithms there may be no computational savings.

Other fixed dictionaries can be simply random matrices, which have very good general properties. Since coherence is important for recovering a sparse representation, we can build incoherent dictionaries (more on this topic in Chap. 4). For all these dictionaries there is no computational shortcut.

Finally, we could build structured dictionaries, for example by concatenating orthogonal transforms in a single fat matrix. Tuning algorithms is possible, but we do not go into details.

In opposition with all these types of fixed dictionaries, chosen beforehand with only heuristic information on the data to be represented, learned dictionaries are optimized for the training signals available from a specific application. We will see in the next chapter how the optimization problem is posed, but it is clear that a good dictionary allows sparse representations with smaller error (or sparser representations with the same error) than a fixed dictionary. The optimization problem is solved before the actual use in an application, hence learning does not pose computational hardship in the application per se.

The drawbacks are not so severe. A learned dictionary has usually no structure, hence there are no fast algorithms; however, only few fixed dictionaries have this feature; moreover, we will see in Chap. 7 that harmony between structure and learning is possible. Learned dictionaries are rather coherent by default, but there are training methods that join incoherence enforcing to the learning, see Chap. 4.

As we will see in the next chapters, learning dictionaries is a hard problem. However, there are several good algorithms for it, hence learning is currently a very appealing option.

As a preview of the next results, we give below an example of a learned dictionary, before stating the problem and presenting the algorithms. Intuitively, a learned dictionary should have more atoms where the density of training signals is higher and fewer atoms where the density is lower. This is in opposition with fixed dictionaries, which are not adapted to the distribution of training signals.

Example 1.21 We consider a family of signals for which visual inspection of the dictionary confirms good adaptation to sparse representation. Using high frequency sinusoids with the form $\eta_i = \cos(2\pi i\omega/m)$, with $\omega \in [\frac{m}{4}, \frac{m}{2}]$, we build signals as linear combinations of $s = 3$ such sinusoids with random frequencies and random coefficients. Note that the frequencies lie on a whole interval, hence the number of sinusoids is infinite. Such signals could appear in the study of vibrations or direction-of-arrival problems, but our model here is a didactic simplification.

With sufficient generated signals and using one of the algorithms that will be later presented, we train a dictionary, taking $m = 8$, $n = 16$. As all dictionary learning algorithms are iterative, we initialize the optimization with the overcomplete DCT from Example 1.2. The atoms of the resulting dictionary are shown in Fig. 1.7. Compare with the overcomplete DCT atoms from Fig. 1.3. It is clear that the learned dictionary is more appropriate for high frequency content; all atoms corresponding to low frequencies in the DCT have been replaced with (or transformed into) high frequency atoms. In the same time, some high frequency atoms are only barely changed, a striking example being atom 16, which is almost identical in the fixed and learned dictionaries.

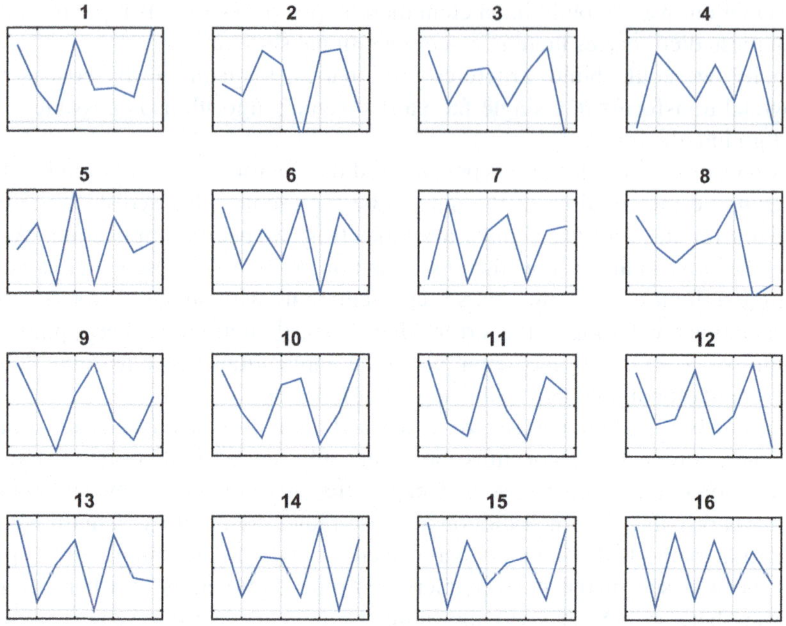

Fig. 1.7 The atoms of an 8×16 trained dictionary, for signals that are sparse combinations of high frequency sinusoids

Due to the adaptation, the learned dictionary allows sparse representations with an error more than twice smaller than when using the overcomplete DCT.

Of course, in this example we could build a very good dictionary by just taking n sinusoids with frequencies covering equidistantly the interval $[\frac{m}{4}, \frac{m}{2}]$. However, in many applications we do not have the luxury of such a neat model of the signals and learning the dictionary is the most sensible approach to obtain good sparse representations. ■

Problems

P 1.1 A norm is a function f that satisfies the following properties: i) positivity: $f(\boldsymbol{x}) > 0$, $\forall \boldsymbol{x} \neq 0$; $f(0) = 0$; ii) homogeneity: $f(\alpha \boldsymbol{x}) = |\alpha| f(\boldsymbol{x})$, $\forall \alpha \in \mathbb{R}$; iii) triangle inequality: $f(\boldsymbol{x} + \boldsymbol{y}) \leq f(\boldsymbol{x}) + f(\boldsymbol{y})$. Show that the so-called 0-norm satisfies properties i) and iii) and give a counterexample for property ii). Show also that the 0-norm is not a convex function.

P 1.2 Run OMP on the matrix $\boldsymbol{D}$ and signal $\boldsymbol{y}$ from Example 1.4. Is the sparsest representation found?

P 1.3 Show that, if a signal $\boldsymbol{y}$ can be exactly represented with sparsity level $s = 1$ (meaning that it is collinear with an atom), then OMP is able to recover the exact representation. However, when $s = 2$, this is no longer always true, as one can see fairly easily; taking $n = 3$ and $\boldsymbol{y} = (\boldsymbol{d}_1 + \boldsymbol{d}_2)/2$, build atom $\boldsymbol{d}_3$ such that it is chosen by OMP in the first step although it is not part of the representation. Note that the mutual coherence of this dictionary has relatively low value.

P 1.4 A natural property of a sparse representation algorithm is linearity, in the following sense: if the representation of $\boldsymbol{y}$ is $\boldsymbol{x}$ (meaning that $\boldsymbol{y} \approx \boldsymbol{D}\boldsymbol{x}$, with the given dictionary $\boldsymbol{D}$), then the representation of $\alpha \boldsymbol{y}$ is $\alpha \boldsymbol{x}$, $\forall \alpha \in \mathbb{R}$. So, the same support is chosen by the algorithm. Show that OMP and Basis Pursuit (1.9) have indeed this property. What about the lasso problem (1.10)?

P 1.5 It is clear that as the parameter λ grows, Basis Pursuit problem (1.9) has sparser solutions. (Note that when λ is very large, the solution is $\boldsymbol{x} = 0$.) Give an algorithm for modifying λ such that the solution has a given sparsity level s. Is there a guarantee that the number of times we solve the problem is bounded?

P 1.6 We want to solve the sparse representation problem (1.7) with error bound using a convex approximation. A possibility is to solve

$$\begin{aligned} \min_{\boldsymbol{x}} \quad & \|\boldsymbol{x}\|_1 \\ \text{s.t.} \quad & \|\boldsymbol{y} - \boldsymbol{D}\boldsymbol{x}\| \le \varepsilon \end{aligned}$$

Another option is to solve the Basis Pursuit problem (1.9) with different values of λ until reaching a satisfactory outcome. Debate the advantages and drawbacks of the two approaches.

P 1.7 The Orthogonal Least Squares is shortly described in Remark 1.7.

a) Show that OLS and OMP choose the same atom in their first step.
b) If the same support $\mathcal{S}$ was chosen by both OMP and OLS in the first t steps, then the residual norm of OLS after $t + 1$ steps is smaller. Is it, however, possible that OMP produces a smaller final residual?

P 1.8 Write a program to compute the RIP constant δ_{l2} from (1.29) (with $s = 2$) for a random matrix $\boldsymbol{D} \in \mathbb{R}^{m \times n}$. (The minimum singular values of the $m \times 2$ matrices formed by all combinations of two columns need to be computed.) Draw a graph of δ_{l2} for variable n and fixed m (taking the average over several matrices of the same size makes the results more reliable).

P 1.9 Consider a dictionary $\boldsymbol{D} = [\boldsymbol{Q}\ \boldsymbol{I}] \in \mathbb{R}^{m \times 2m}$, formed by joining the orthogonal matrix $\boldsymbol{Q}$ and the unit matrix. Show that its mutual coherence (1.25) is at least $1/\sqrt{m}$. The same result holds if the dictionary is built with two arbitrary orthogonal matrices.

Chapter 2
Dictionary Learning Problem

Abstract Dictionary learning can be formulated as an optimization problem in several ways. We present here the basic form, where the representation error is minimized under the constraint of sparsity, and discuss several views and relations with other data analysis and signal processing problems. We study some properties of the DL problem and their implications for the optimization process and explain the two subproblems that are crucial in DL algorithms: sparse coding and dictionary update. In preparation to algorithms analysis and comparisons, we present the main test problems, dealing with representation error and dictionary recovery; we give all details of test procedures, using either artificial data or images. Finally, as an appetizer for the remainder of the book, we illustrate the wide use of DL algorithms in the context of sparse representations for several applications like denoising, inpainting, compression, compressed sensing, classification.

2.1 The Optimization Problem

The interesting signals appearing in applications often lack a physical deterministic or probabilistic model; or, if such model exists, it may be quite complicated. For example, speech and music could be modeled as the outputs of dynamical systems inspired from the vocal system or from a musical instrument having as inputs sinusoids or white noise; for a single person or instrument in silent room conditions, the model could be maybe manageable, but still the inverse problem of finding the input given the output is difficult. However, acoustic scenes are often complex, and signals are the result of superposing sources and noise. In the case of images the situation is even worse: besides very restricted classes, there are no general models.

In standard operations like denoising, compression, coding, a model is however useful for structuring information, even if it has an artificial nature. Sparse representations are a natural choice, due to their generality and relative ease of manipulation. Their parsimony is also attractive; representing a signal with only few atoms is not only efficient, but follows the principle that simple models should be tried whenever possible because a simple explanation of a phenomenon, if adequate, is always better than an intricate one. In some cases sparsity may actually reflect reality: it

B. Dumitrescu, P. Irofti, *Dictionary Learning Algorithms and Applications*,
https://doi.org/10.1007/978-3-319-78674-2_2

was argued in [141] that the human vision system is intrinsically based on sparse representations, although it is not completely clear how the system works.

So, adopting (1.5) as a generative model for signals is clearly acceptable and possibly advantageous. For convenience, we write again the model,

$$\boldsymbol{y} = \boldsymbol{D}\boldsymbol{x} + \boldsymbol{v}, \tag{2.1}$$

and remind that $\boldsymbol{y} \in \mathbb{R}^m$ is our signal of interest, $\boldsymbol{D} \in \mathbb{R}^{m \times n}$ is the dictionary, $\boldsymbol{x} \in \mathbb{R}^n$ is a sparse vector, and $\boldsymbol{v} \in \mathbb{R}^m$ is noise. Such a model is appropriate if the dictionary is appropriate, meaning that sparse representations indeed exist for the signals of interest. In Sect. 1.6 we arrived at the conclusion that adapting the dictionary to the class of signals at hand is a good idea. This adaptation process is called *dictionary learning* (DL) and can be stated as an optimization problem.

In many applications we have or can generate a fairly large number of signals $\boldsymbol{y}$ that is representative for the source generating the signals; they are called training signals, since they are used to train (design, learn) the dictionary. DL means finding the best (or at least a very good) dictionary $\boldsymbol{D}$, in the sense that the model (2.1) is respected with as small as possible (or at least small) noise variance over the ensemble of signals. To give a more precise meaning to this formulation, we build a matrix $\boldsymbol{Y} \in \mathbb{R}^{m \times N}$ whose columns are the training signals; the number of training signals is N. We assume that the sparsity level s is also given. DL consists of solving the optimization problem

$$\begin{aligned} \min_{\boldsymbol{D},\boldsymbol{X}} \quad & \|\boldsymbol{Y} - \boldsymbol{D}\boldsymbol{X}\|_F^2 \\ \text{s.t.} \quad & \|\boldsymbol{x}_\ell\|_0 \le s, \ \ell = 1:N \\ & \|\boldsymbol{d}_j\| = 1, \ j = 1:n \end{aligned} \tag{2.2}$$

where the variables are the dictionary $\boldsymbol{D} \in \mathbb{R}^{m \times n}$, and the sparse representations matrix $\boldsymbol{X} \in \mathbb{R}^{n \times N}$, whose columns have at most s nonzero elements. The second constraint of (2.2) represents atom normalization; we will talk more about it in the next section, but it is a natural requirement to have atoms with equal norms; in this sense, this constraint can be seen as an inheritance from orthogonal transforms. The first constraint sets the sparsity level of the representation, which is the same for all signals; this condition is somewhat artificial and we will see later in Chap. 5 that it can be relaxed; for now, it is very convenient, although it is not yet clear how to choose s. We denote Ω the positions of the nonzero elements of $\boldsymbol{X}$, also called the support of the representation:

$$\Omega = \{(i, \ell) \mid x_{i\ell} \neq 0\}. \tag{2.3}$$

Denoting Ω^c the set complementary to Ω, we can shortly write the sparsity constraint as $\boldsymbol{X}_{\Omega^c} = 0$.

Let us talk more about the objective of (2.2). We denote

$$E = Y - DX \tag{2.4}$$

the representation error (or residual) for dictionary D and sparse representations X, not necessarily the optimal ones. The objective of (2.2) is the (squared) Frobenius norm of the error

$$\|E\|_F = \sqrt{\sum_{i=1}^{m}\sum_{\ell=1}^{N} e_{i\ell}^2}. \tag{2.5}$$

Figure 2.1 illustrates the matrices that appear in the objective of the DL problem (2.2). Since

$$\|E\|_F^2 = \sum_{\ell=1}^{N} \|e_\ell\|^2 = \sum_{\ell=1}^{N} \|y_\ell - Dx_\ell\|^2, \tag{2.6}$$

the approximation error can be explicitly split on signals. Figure 2.2 shows the part of a single signal in the DL problem; note the resemblance with Fig. 1.1. The connection with (2.1) is now obvious: we aim to have the smallest possible "noise" over all signals.

Remark 2.1 (Stochastic View) A more general view of the DL problem (2.2) has a stochastic nature. Assuming that the signals y are generated according to a probability distribution, the objective of (2.2) would become

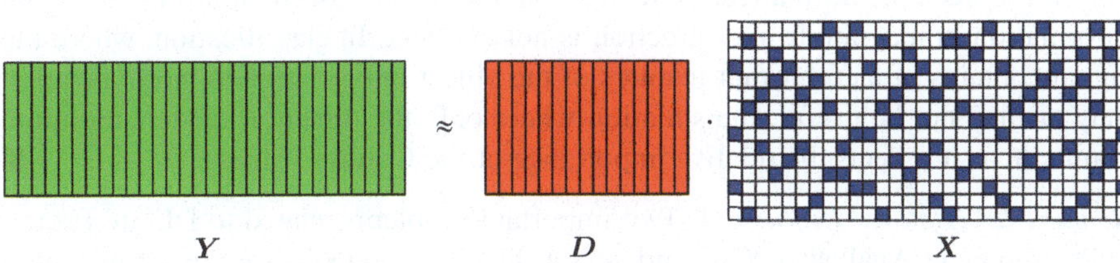

Fig. 2.1 Dictionary learning means finding the dictionary D and the sparse representations X such that the approximation $Y \approx DX$ is as good as possible

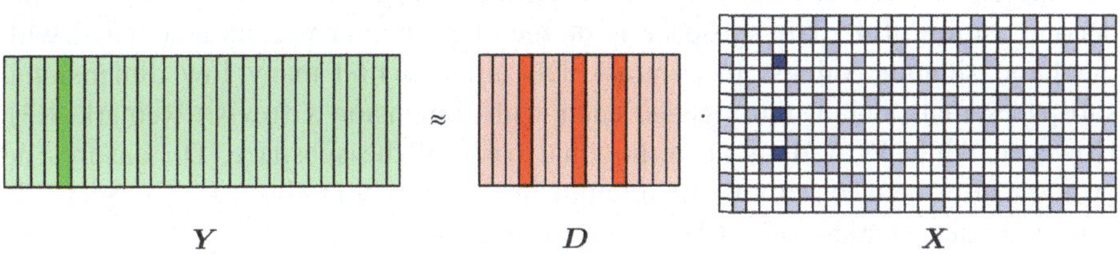

Fig. 2.2 DL problem: the contribution of a single signal

$$\mathbb{E}_y\{\|y - Dx\|^2\}, \tag{2.7}$$

meaning that we minimize the error expectation relatively to the underlying data distribution. However, since this probability distribution is usually unknown, the best we can do is to minimize the empirical average that is the objective of (2.2). The two approaches are practically equivalent when the number N of signals is large (assuming that data are indeed drawn according to the distribution, with no bias). When N is not large, near-optimality in solving (2.2) becomes less important, since the (discrete) empirical distribution may be a simplification of the (usually continuous) true distribution. Some estimations of the error for signals that are not in the training set are given in [193], for arbitrary distributions.

An implicit hypothesis when formulating the DL problem is that the noise v in (2.1) is Gaussian, hence a quadratic error is minimized. (The same approach is robust when the noise nature is unknown.) However, when the noise has a different distribution, a suited error norm is more appropriate. For example, if the noise is impulsive, the ℓ_1 norm should be minimized; see Sect. 5.5 for details. ■

Remark 2.2 (Signal Normalization) A possible dilemma at this point is whether we should normalize the signals or not. The objective of (2.2) treats the representation errors evenly for all signals. So, due to the least squares objective, the magnitude of the errors will not be necessarily linked with that of the signals. The *relative* error will be small for signals with large norms and large for signals with small norm. If we want to have similar relative errors, we should normalize the signals when building the matrix Y. If signals with large norm are actually more important, we should not normalize. For example, in (lossless) compression, there is no reason to normalize since the normalization factors should also be coded; splitting the signal information in magnitude and direction is not efficient. In classification, where the signal contains in fact relevant features of the object to be classified, normalization may or may not be useful; the situation where only the direction matters are quite common, as it reflects the relative importance of the features values. ■

Remark 2.3 (Related Problems) Two important problems related to DL are Principal Component Analysis (PCA) and vector quantization (VQ) (or clustering), with the associated K-means algorithm [122].

PCA aims to find a single low dimensional subspace that contains most of the training signals and is essentially based on the computation of the singular value decomposition of Y. The subspace is defined by singular vectors associated with the large singular values. In contrast, DL aims to find many low dimensional subspaces that contain the signals, each with dimension s (revisit Remark 1.3). However, while PCA finds an orthogonal basis of the subspace, DL ensures no orthogonality between atoms; an attempt in this direction, discussed in Sect. 7.3, is to build dictionaries with orthogonal blocks.

VQ and the similar clustering operation find vectors that are representative for a group of training signals. Each such vector is thus associated with a cluster of signals. We can see each cluster "center" (in VQ it is indeed the mass center) as an atom and thus VQ could be interpreted as DL with sparsity level $s = 1$. They are not equivalent though, since the VQ "atoms" are not normalized and there is the possibility that two such "atoms" are collinear, although with different magnitudes.

Multistage or cascaded VQ [77, Sec. 12.11] is even closer to DL. After finding the cluster centers, they are subtracted from all the signals in their cluster and the VQ algorithm is again applied to the residuals. The procedure can be repeated several times. Thus, each signal is approximated by the sum of two or several atoms. Of course, the atoms belong to distinct categories and there is a hierarchy between them, but the idea of sparse representation is certainly present. ■

Remark 2.4 (Literature) There are many research results related to the DL problem and we will try to cite all the relevant ones, where appropriate in our book. (We will certainly fail, but with the best intentions.) The DL topic in general is also covered in the review articles [158] and [189], and also in the book [221], dedicated to visual computing application. ■

2.2 An Analysis of the DL Problem

The DL problem (2.2) is obviously nonconvex. The sparsity constraint makes it even NP-hard. So, the problem is inherently difficult. Before discussing possible solutions, let us look at several characteristics of the problem, with the purpose of becoming more able to handle it.

Remark 2.5 (Indetermination) The normalization constraint has the role to remove the indetermination due to a possible multiplicative factor that can multiply $\boldsymbol{D}$ and divide $\boldsymbol{X}$ without changing the objective of (2.2); such an indetermination can hinder the optimization process by hiding actual convergence behind different values of the multiplication factor. Still, normalization leaves open the possibility of a sign flip for both the atoms (columns of $\boldsymbol{D}$) and their representation coefficients (the corresponding rows of $\boldsymbol{X}$). As we will later see, normalization can be done at different stages of an algorithm solving (2.2).

Another source of indetermination is caused by the fact that a permutation of the atoms, combined with a similar permutation of the representations, gives the same value of the objective. So, taking also the sign flip into account, if $(\boldsymbol{D}, \boldsymbol{X})$ is a solution of the DL problem, then $(\boldsymbol{DP}, \boldsymbol{P}^{-1}\boldsymbol{X})$ is also a solution, with $\boldsymbol{P}$ a permutation matrix whose nonzero elements are ± 1. Hence, there are multiple local minima with the same value. However, this combinatorial aspect is a benign issue, as finding one of these minima is perfectly satisfactory. Note also that other matrices $\boldsymbol{P}$, although preserving the value of the objective, would destroy either the unit norm of the atoms or the sparsity of $\boldsymbol{X}$, or both.

Sign flip and atom permutations are not a hurdle for the optimization process. The successful DL methods have a local character, meaning that atom changes are gradual. There is no real danger of large jumps back and forth between different sign flips or permutations. ■

Remark 2.6 (Uniqueness) Barring the above permutation issues, one may ask under what conditions the solution of (2.2) is unique. More precisely, assuming that $\boldsymbol{Y} = \boldsymbol{DX}$ (so, we assume the noiseless case) and the constraints of (2.2) hold, are $\boldsymbol{D}$ and $\boldsymbol{X}$ the unique matrices whose product is $\boldsymbol{Y}$? The answer was provided in [5] and is positive under three relatively intuitive conditions, as follows:

(i) One of the conditions of Proposition 1.14 is satisfied for each column of $\boldsymbol{X}$. For example, condition (b) is $\|\boldsymbol{x}_\ell\|_0 < spark(\boldsymbol{D})/2$, $\ell = 1 : N$. This ensures that once $\boldsymbol{D}$ is known, the matrix $\boldsymbol{X}$ is unique, due to the sparsity of the support. It is also assumed that the sparsity level $s = \|\boldsymbol{x}_\ell\|_0$ is known, although this is not crucial.

(ii) There are enough signals. Of course, if N is small (imagine the case $N = 1$), there is simply not enough information to retrieve a unique solution. Technically, it is necessary that for each group of s atoms, there are $s + 1$ signals that are linear combinations of these atoms. This would mean that $N \geq (s + 1)\binom{n}{s}$, which is indeed a huge number. However, if we could pick the combinations of atoms that produce the signals, a clever choice reduces the number of signals to $2n(s + 1)$, which would be quite convenient. Of course, in practice we do not have the luxury of choosing, but this number shows that the condition is not as daunting as it initially appeared. This condition ensures that $\boldsymbol{D}$ is unique.

(iii) There are no degeneracies, meaning that all ranks have their maximal values. For example, the $s + 1$ signals that are produced by s atoms in condition (ii) have the rank exactly s. Or, if $s + 1$ signals are generated by distinct groups of s atoms, then their rank is exactly $s + 1$. Since these conditions are generic, they pose no actual restriction.

Other developments on uniqueness conditions (including recovery algorithms) are given in [3, 6, 184].

So, the short conclusion is: if there are enough signals, their representations are indeed sparse, and the dictionary is well behaved, the solution to $\boldsymbol{Y} = \boldsymbol{DX}$ is unique. In practice, this conclusion is not very relevant: the signals are noisy. However, we should always be careful to the number of signals and to the properties of the dictionary. ■

Remark 2.7 (Factorization) The papers cited in the previous Remark and several others deal in fact with a factorization problem: given the matrix $\boldsymbol{Y}$, find $\boldsymbol{D}$ and sparse $\boldsymbol{X}$ such that $\boldsymbol{Y} = \boldsymbol{DX}$. This factorization is often also called DL, although it is only the particular case of DL without noise and where the sparse representation model holds exactly, which is rare. Although all algorithms solving (2.2) could in principle solve also the factorization problem, there are specialized algorithms for it, with termination guarantees. These algorithms will not be discussed in the book, due to their small practical impact. We will work with (2.2) and with diverse modifications of it viewed as general optimization problems. ■

Two subproblems of the basic DL problem (2.2) have major importance in algorithms. One is the *sparse coding* (or representation) problem

$$\begin{aligned} \min_{X} \quad & \|Y - DX\|_F^2 \\ \text{s.t.} \quad & \|x_\ell\|_0 \le s, \ \ell = 1 : N \end{aligned} \tag{2.8}$$

where the dictionary is fixed and the sparse representations have to be computed. In view of (2.6), sparse coding can be decoupled on signals: the representation problem (1.6) is solved for each signal y_ℓ. As we have seen, this is an NP-hard problem, whose solution can be approximated with greedy algorithms, like OMP, see Sect. 1.3 or other algorithms from those listed in Sect. 1.2. OMP is preferred due mostly to its speed.

The sparse coding subproblem appears during learning and also after learning, when the designed dictionary is used in applications. An important remark is that the same representation algorithm has to be used in learning and after it, since the learning result depends on the choice of the representation algorithm.

The second typical subproblem is that of *dictionary update*

$$\begin{aligned} \min_{D,(X)} \quad & \|Y - DX\|_F^2 \\ \text{s.t.} \quad & X_{\Omega^c} = 0 \\ & \|d_j\| = 1, \ j = 1 : n \end{aligned} \tag{2.9}$$

Here, the sparsity pattern is fixed and the dictionary is always a variable. Depending on the approach, one can try to improve also the nonzero coefficients of the representations.

With the sparsity pattern Ω fixed, the problem (2.9) is bi-quadratic, hence not convex. When the representation matrix X is fixed and only the dictionary is variable, the problem is still not convex, although the objective is a convex function, due to the normalization constraints. Hence, finding an optimal solution may be hard, even though the combinatorial aspects have disappeared. Only when the dictionary D and the sparsity pattern Ω are fixed, the problem is convex; in fact, it is a least squares problem, for which the optimal representations can be computed independently for each signal via (1.14).

Remark 2.8 (Local Minima) It is thus a natural conclusion that the DL problem (2.2) has multiple local minima. In particular, it is easy to see that each sparsity pattern Ω leads to a distinct solution, even though this solution does not necessarily correspond to a local minimum. However, if we consider only dictionaries with atoms that have exactly s nonzero elements, it is easy to conclude that, for each sparsity pattern, the DL problem has (at least) a distinct local minimum. ■

Remark 2.9 (Conditioning) Let D, X be the solution of the DL problem (2.2) for a given signal matrix Y. Loosely speaking, we say that the problem is well conditioned if, given a slightly altered input matrix $\hat{\boldsymbol{Y}}$, the solution $\hat{\boldsymbol{D}}$, $\hat{\boldsymbol{X}}$ is also only slightly different from the "true" one.

When either $\boldsymbol{D}$ (and the sparsity pattern) or $\boldsymbol{X}$ is fixed, the sparse coding and dictionary update problems assume a least squares form. In these cases, the conditioning of the problem is strictly related to the condition numbers $\kappa(\boldsymbol{D})$, $\kappa(\boldsymbol{X})$, that act as amplifiers of the input errors. Remind that, in the spectral norm, the condition number of a matrix is the ratio between the maximum and the minimum singular values of the matrix. So, if the "true" dictionary is nearly singular, then the problem is ill-conditioned and we may expect the computed dictionary to be affected by errors. (Note that we completely ignore the effect of the algorithm for solving the problem, which may introduce errors by itself.)

Although not always, conditioning may pose practical threats to finding reliable solutions to the DL problem. A way to alleviate this issue is to regularize the problem, as done in Sect. 4.2, or to constrain the mutual coherence of the dictionary, as done in other sections of Chap. 4. ■

Remark 2.10 (Number of Variables) We remark that the size of the DL problem (2.2) may be quite significant. Only the end product, the dictionary $\boldsymbol{D}$, may easily contain thousands of entries. The sheer size is an important obstacle that optimization algorithms have to overcome. ■

Example 2.11 A DL problem of usual size, as typically encountered in image processing, may have signals of size $m = 64$, obtained from the vectorization of 8×8 patches. The dictionary may easily have $n = 128$ atoms, meaning that it is (only) twice overcomplete. So, $\boldsymbol{D}$ is a 64×128 full matrix, thus containing 8192 variables, which already makes DL a large optimization problem. Assuming that the training set contains $N = 10{,}000$ signals, which is a moderate value, and taking a sparsity level $s = 6$, it results that the representation matrix $\boldsymbol{X}$ has 60,000 nonzeros in 1,280,000 possible positions. Solving the DL problem starts to appear as a daunting task. ■

2.3 Test Problems

Dictionary learning can have many applications; some of them are discussed later in this chapter, others will follow throughout the book. Since there are many DL algorithms, as it will be clear starting with the next chapter, a significant issue is how to decide which algorithm is the most appropriate, in general or for a target application. Assessing the performance of a DL algorithm may be difficult, since the role of the sparse representations obtained with the designed dictionary might be more or less important in a specific application. The benefit of the DL algorithm may be obscured by that of other processing blocks. So, when we compare DL algorithms, we try to look at "pure" problems, where all secondary factors are hopefully absent.

This section presents the most popular test problems that are used for comparing DL algorithms. Although they have a dose of artificiality, they are good indicators of how good is a DL algorithm. If an algorithm is systematically better than another in many such tests, then it is almost surely better in a real application. Conversely, if an algorithm clearly fails a test, then we should be ready to remove it from the list of our candidates.

2.3.1 Representation Error

The first performance indicator is of course the representation error, since it was chosen as objective in the DL problem. Root Mean Square Error (RMSE)

$$\text{RMSE} = \frac{1}{\sqrt{mN}} \|\boldsymbol{Y} - \boldsymbol{DX}\|_F \tag{2.10}$$

is more appropriate for reporting than the total error that is the objective of (2.2), since it scales the error to the number N of signals and the size m of a signal. Thus, (2.10) gives the mean error per signal element and is thus more relevant and intuitive.

It is maybe the moment to stress that the DL problem contains a degree of abstraction that aims to make it general. Like in all modeling problems, the danger of overfitting the dictionary to the available data exists and is more likely when the number N of signals is small, meaning that it is not significantly larger than the number n of atoms. So, we should not rely exclusively on representation error when assessing the worth of a DL algorithm.

The available data are customarily split into two parts: the training data and the test data. The N training signals are those denoted with $\boldsymbol{Y}$ and used in the DL process. The test data are used only for measuring the error with the designed dictionary, in order to estimate its behavior on the yet unavailable data given by a further application. So, when reporting errors, we should distinguish between the RMSE obtained when solving (2.2), i.e., computed on the training signals, and the RMSE computed with the test signals. They have different significances. The former reflects the quality of the optimization process that solves (2.2); a better error simply means a better optimization algorithm. The latter shows the general qualities of the dictionary, as designed from the training data, and depends not only on the algorithm, but also on the representativity of the training data and on the adequacy of chosen dictionary parameters like the number of atoms or the sparsity level. When training data are plentiful and cover well the underlying distribution of the process generating them, the difference between the errors on training and test data should not be significant. For this reason and since our focus is on optimization algorithms, we will mostly compare the DL algorithms based on the error for training data, i.e., using the objective function (2.2).

An important issue is what data could be used for assessing the DL algorithm. Traditionally, there are two main types of data that are employed.

Artificial Data The simplest choice is to generate the training signals by assuming that they are indeed sparse combinations of the atoms of a given dictionary. We choose first the sizes m and n and generate a dictionary $\boldsymbol{D}_{\text{true}} \in \mathbb{R}^{m\times n}$, e.g., with each element drawn from a Gaussian distribution with zero mean and unit variance, taking care to normalize the atoms. Then, we generate signals via

$$\boldsymbol{y} = \boldsymbol{D}_{\text{true}}\boldsymbol{x} + \boldsymbol{v}, \tag{2.11}$$

where $\boldsymbol{x} \in \mathbb{R}^m$ has exactly s nonzero values (the sparsity level s is also chosen), in random positions, drawn from a random distribution. Finally, the noise $\boldsymbol{v}$ is added to the sparse representation; Gaussian distribution is typically used, with zero mean and variance σ^2. In numerical simulations we usually want to set the signal to noise ratio (SNR) representing the ratio of signal power $\bar{p}_y^2$ to the noise power $\bar{p}_v^2$

$$\text{SNR} = \frac{\bar{p}_y^2}{\bar{p}_v^2}. \tag{2.12}$$

When the signal and the noise both have zero mean, the ratio is expressed directly through their variances. If the SNR is set in dB, then the variance is

$$\sigma^2 = \bar{p}^2 \cdot 10^{-\frac{\text{SNR}}{10}}, \tag{2.13}$$

where $\bar{p}^2$ is the power of the "true" signals, $\bar{p}^2 = \|\boldsymbol{D}_{\text{true}}\boldsymbol{X}\|_F^2 / N$, the matrix $\boldsymbol{X}$ having as columns the representation vectors $\boldsymbol{x}$ from (2.11).

Data from Images Another easy way of obtaining training data is to simply take them from images. In this case, an exact sparse representation does not exist, but it is known that certain transforms, like the wavelets, often have sparse results when applied on images; it is natural to assume that an overcomplete dictionary can have similar properties. The images can be taken from a widely accepted database like [202]. To generate the signals, we take $\sqrt{m} \times \sqrt{m}$ patches (8×8 is the most popular choice, but 12×12 and even 16×16 are possible, if enough computing power is available) from random locations in a set of images, whose columns are stacked to obtain the signals $\boldsymbol{y}$ (the operator for this matrix-to-vector transformation is denoted *vec*). Note that the patches may overlap; this is actually beneficial, since overlapping stimulates the training of shift invariant dictionaries; this issue will be discussed in detail in Sect. 7.4. Grayscale images are made of pixels with values from 0 to 255; in most of our tests, we subtract the (overall) average of the chosen patches from all the patches and then normalize the patches. Centering the signals introduces direction diversity in the resulting vectors, while the initial signals lie only in the positive orthant. Normalization leads to similar relative representation errors for all patches, see Remark 2.2; it also allows comparisons between RMSEs resulted from different sets of training images, since all transformed patches have the same norm, no matter the original values.

2.3.2 *Dictionary Recovery*

A different way to see the success in the artificial data case is to check if the dictionary optimized by the DL algorithm coincides with the "true" one. So, we generate the data exactly as described above, more precisely in (2.11), with a dictionary $\boldsymbol{D}_{\text{true}}$ and no noise. Let $\boldsymbol{D}$ be the dictionary resulted from the optimization process. Of course, the error $\|\boldsymbol{D} - \boldsymbol{D}_{\text{true}}\|$ makes not much sense since, as we have seen, the DL problem is insensitive to atom permutations; such an error would be almost always large. We should, in fact, search the atoms permutation matrix $\boldsymbol{P}$ for which $\|\boldsymbol{D}\boldsymbol{P} - \boldsymbol{D}_{\text{true}}\|$ is minimum; this is a matching problem that is combinatorial in nature. Even if a solution is found, the error value would be significant only when very small, showing almost perfect recovery; otherwise, a single wrong atom can make the error quite large.

A more relevant and easier to compute recovery measure is obtained by the same atom matching idea, but in a simplified manner. We say that an optimized atom $\boldsymbol{d}$ is almost identical with a true atom $\boldsymbol{d}_{\text{true}}$ if

$$\left|\boldsymbol{d}^T \boldsymbol{d}_{\text{true}}\right| \geq \theta, \tag{2.14}$$

where θ is a large enough threshold, allowing for inherent recovery errors due to the imperfections of the numerical optimization; for example, we can take $\theta = 0.99$; the atoms are normalized, hence the maximum value of the scalar product from (2.14) is 1.

As a measure of the recovery performance we take the ratio between the number of true atoms that can be found in the optimized dictionary $\boldsymbol{D}$ with the test (2.14) and the total number of atoms, n. (Note that we do not count how many optimized atoms can be found in the true dictionary, since two optimized atoms could be almost identical with a single true atom.) We name this *recovery ratio* (or *recovery percentage*, if multiplied with 100); if it is equal to 1, then the true dictionary has been recovered perfectly. We note that the recovery ratio can be computed by examining the elements of $\boldsymbol{D}^T \boldsymbol{D}_{\text{true}}$ and counting how many columns contain (at least) an element larger than θ in absolute value.

Measuring the recovery ratio makes sense in view of the uniqueness of the solution, as discussed in Remark 2.6; remind that the number N of signals must be sufficiently large to guarantee uniqueness. The threshold θ leaves space for errors due to the conditioning of the problem or to (inherently) stopping the DL algorithms before full convergence.

2.4 Applications: A Quick Overview

We present here several applications where sparse representations and dictionary learning play an important role. We describe the problem and only sketch the solution, often pointing to other places in the book where more details will appear, and also give a few references that cover significant advances or review the field.

2.4.1 Denoising

The denoising problem starts with a given set of noisy signals $\boldsymbol{Y}$ and sets out to find dictionary $\boldsymbol{D}$ and sparse representations $\boldsymbol{X}$ such that the cleaned signals $\boldsymbol{Y}_c = \boldsymbol{DX}$ are an exact, or good enough, replica of the originals.

In most applications we have to deal with additive white Gaussian noise, as described around (1.5), where each signal from $\boldsymbol{Y}$ was generated following our initial model (2.1). Ideally, if the noisy set was built from the unknown original signals $\boldsymbol{Y}_0$ perturbed with noise $\boldsymbol{V}$

$$\boldsymbol{Y} = \boldsymbol{Y}_0 + \boldsymbol{V}, \tag{2.15}$$

we expect to approximate $\boldsymbol{Y}$ through dictionary learning

$$\boldsymbol{Y} \approx \boldsymbol{DX} = \boldsymbol{Y}_c, \tag{2.16}$$

such that the residual $\boldsymbol{E}$ (2.4) matches the added noise $\boldsymbol{V}$

$$\boldsymbol{E} = \boldsymbol{Y} - \boldsymbol{Y}_c \approx \boldsymbol{V} \tag{2.17}$$

implying that we recovered $\boldsymbol{Y}_0$ through the clean signals $\boldsymbol{Y}_c$.

The denoising model (2.17) presumes that the DL process identifies the common patterns among the signals in $\boldsymbol{Y}$, patterns that are in fact the building blocks of the original set $\boldsymbol{Y}_0$, and a possible solution would be to have each pattern correspond to one atom in the dictionary so that each cleaned signal is built from a linear combination of patterns. In practice, due to the limited sample size (see Remark 2.6 (ii)) or an increased SNR (2.12), some atoms might end up also modeling the present signal noise. There are many techniques that attempt to minimize this defect by modifying the standard DL process starting from data preprocessing, to specializing the sparse representation stage, up to imposing a certain dictionary structure.

In a scenario where the noisy signals set $\boldsymbol{Y}$ is rather small, but its originating signal class is known, we can separate the learning stage from the actual denoising. Like this, we train $\boldsymbol{D}$ on a sufficiently large clean signal set of the same class, thus eliminating the risk of noise modeling. Afterwards we can denoise $\boldsymbol{Y}$ by performing sparse representation with the new class-specific dictionary. Of course, we can now reuse the dictionary whenever we have to deal with any other noisy signals from this class.

The disadvantage of this approach is that $\boldsymbol{D}$ is missing signal specific patterns, which prevents us from completely recovering the original noiseless signal. Even more, if we deal with a large class of signals, there will be too many class-specific patterns that will not fit in a decently sized dictionary $\boldsymbol{D}$, the DL process thus having to compromise by only choosing the most representative ones; this leads, again, to imperfect denoising. To overcome such difficulties, one might structure the denoising dictionary as the composition of two smaller ones (see Sect. 7.7), the first learned on the large clean set and the second on the current noisy signal.

When the noise variance σ^2 (2.13) is known, it can be used to set the error stopping criterion (1.31) during sparse coding. While this provides better approximations, it also prevents, or at least attenuates, noise modeling during the DL phase. Extensive OMP-based image denoising experiments in [62] made the most common threshold used in the literature to be

$$\epsilon = C\sqrt{m}\sigma, \tag{2.18}$$

where m is the signal size and C is the constant (also called gain) described around (1.31). Empirical results showed that a good choice is to set $C = 1.15$. In order to prevent the degenerate case when the desired error threshold is never reached, which in the case of OMP would lead to a representation $\boldsymbol{x}$ with $s = m$ entries, an extra maximum sparsity constraint of $s = m/2$ is often used in tandem. So, unless otherwise specified, the image denoising experiments in this book will assume the standard noise deviation known and will use the following OMP call to perform DL and denoising:

$$\boldsymbol{x} = \text{OMP}\left(\boldsymbol{D}, \boldsymbol{y}, \frac{m}{2}, 1.15\sqrt{m}\sigma\right). \tag{2.19}$$

One last trick that helps the denoising process is to have overlapping regions amongst the noisy data-items from $\boldsymbol{Y}$. This concept was briefly mentioned as beneficial at the end of Sect. 2.3.1 in the context of shift-invariant dictionaries, but it can also prove useful in our case.

Take, for example, image denoising, where the N columns in the noisy set $\boldsymbol{Y}$ represent vectorized $\sqrt{m} \times \sqrt{m}$ image patches (or blocks), and let us first investigate the case where we extract all distinct patches. Without any loss of generality, assume that we have to deal with a square image whose dimensions are equal to $\sqrt{P}$ and that the patch size evenly divides the image dimensions. It follows that we split the image in $\sqrt{P/m}$ by $\sqrt{P/m}$ blocks and that most patches have 4 neighboring patches, except for the sides that have 3 and the corners that have only 2. Now, we take each block in column (or row) order and vectorize it producing $N = P/m$ signals arranged as columns in matrix $\boldsymbol{Y} \in \mathbb{R}^{m \times N}$. During denoising, each column is sparsely represented using OMP (2.19) resulting the cleaned signal $\boldsymbol{Y}_c \in \mathbb{R}^{m \times N}$ (2.16). To complete the process we reconstruct the image, this time from matrix $\boldsymbol{Y}_c$, by reversing the vectorization and placing the resulting patches at their corresponding positions. Note that the cleaned neighboring blocks have been computed by different calls to OMP which can lead to very different supports and coefficients in their representations. This creates visible artifacts at the crossing between two patches whose neighboring pixels might have been, for example, very similar or even identical. The phenomenon is called *blocking effect* and is annoying to the human viewer, maybe even more than the Gaussian noise.

These artifacts disappear, or at least become much less visible, through the use of overlapping patches. Let us once again take the square image case and start producing $\sqrt{m} \times \sqrt{m}$ patches from the upper-left corner. When we built the distinct blocks in column order we moved down the column in steps of $p = \sqrt{m}$ pixels until we reached the last block. Then we proceeded to do the same with the next column that was, again, $p = \sqrt{m}$ pixels to the right. When extracting overlapping blocks, we do the same only with a smaller step size ($p < \sqrt{m}$). This process produces

$$N = \left(\left\lfloor \frac{\sqrt{P} - \sqrt{m}}{p} \right\rfloor + 1\right)^2 \tag{2.20}$$

patches from our $\sqrt{P} \times \sqrt{P}$ image, meaning that the same pixel will appear at most $\lfloor \sqrt{m}/p \rfloor^2$ times in $\boldsymbol{Y}$. Single pixel stepping ($p = 1$) produces the largest number of blocks, maximizing the number of appearances of each pixel. We now proceed as usual to denoise each column from $\boldsymbol{Y}$ with OMP. When constructing the cleaned image from $\boldsymbol{Y}_c$, the final pixel is an average of its different values from the blocks it appears in. This has a smoothing effect that makes the artifacts less likely to appear. Another benefit of overlapping patches is access to a larger data set that improves dictionary learning.

Example 2.12 In Table 2.1 we illustrate pixel averaging when denoising the well-known image lena. There, the dictionary was trained on a large 8×8 patches set extracted from the noisy image and used afterwards to produce the cleaned patches through OMP-based denoising (2.19). The left-hand side of the table shows a 64×64 part of the original image. Right below is the same part to which we added zero-mean white Gaussian noise with $\sigma = 20$. Finally, the cleaned image obtained through DL and overlapping pixel averaging is depicted on the third row. On the right-hand side we depict 9 of the restored images used in the averaging process. It is interesting to see that even if the individual images are riddled with block artifacts, the boundaries between patches being clearly visible, the end result is smooth and presents large improvements with respect to its noisy origins. For more technical details about this experiment, and others, see Sect. 3.11.3. ■

In order to quantify the denoising results from our tests, we use the peak signal to noise ratio (PSNR) measure and the structural similarity (SSIM) [198] index. Both compare the original signal with the cleaned version produced through DL. Where the SNR (2.12) uses the given signal power to compute the ratio, PSNR uses the maximum possible power instead. We compute PSNR values in dB through the use of the RMSE (2.10) between the original and cleaned images

$$\text{PSNR} = 20 \log_{10} \frac{\text{DR}}{\text{RMSE}}, \tag{2.21}$$

where DR is the dynamic range representing the ratio between the largest and smallest possible signal values. For images DR is the maximum possible pixel value. In our experiments we use monochrome 8-bit images where $\text{DR} = 255$. Given a vectorized patch $\boldsymbol{y}_0$ from the original image and its corresponding cleaned patch $\boldsymbol{y}_c$, the SSIM index is computed as

Table 2.1 Denoising with overlapping patches

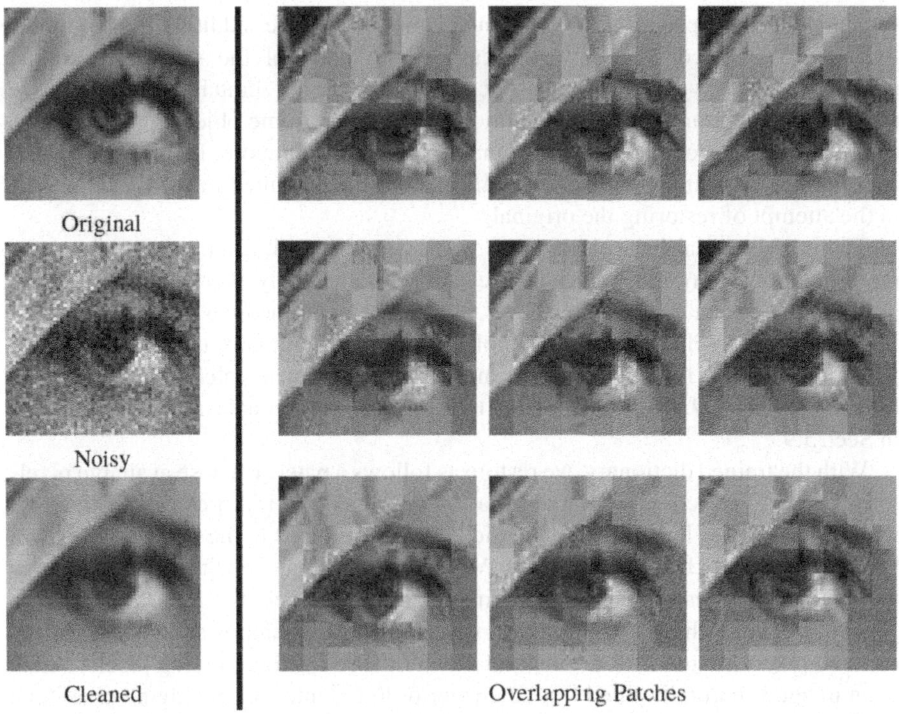

First column shows the original image, the noisy image ($\sigma = 20$) and the cleaned image obtained through overlapping pixel averaging. The columns on the right depict nine images made of the patches from which the cleaned image was built

$$\mathrm{SSIM}(\boldsymbol{y}_0, \boldsymbol{y}_c) = \frac{(2\mu_{\boldsymbol{y}_0}\mu_{\boldsymbol{y}_c} + C_1)(2\sigma_{\boldsymbol{y}_0\boldsymbol{y}_c} + C_2)}{(2\mu^2_{\boldsymbol{y}_0} + \mu^2_{\boldsymbol{y}_c} + C_1)(\sigma^2_{\boldsymbol{y}_0} + \sigma^2_{\boldsymbol{y}_c} + C_2)}, \tag{2.22}$$

where μ and σ stand for average and (co)variance, respectively, and the variables C_1 and C_2 play a stabilizing role for the division and are set based on the dynamic range.

As expected, specializing the DL problem to perform denoising changes the way we process and represent data, but also the way we structure the dictionary. Depending on the priors we have and available resources, the reader should now be able to choose the optimal denoising process for the problem at hand. The choice of DL algorithm is discussed in Sect. 3.11.3 based on experimental data. In our tests, we always used (2.19) for representation and overlapping patches with $p = 1$ (2.20) for DL and denoising (unless stated otherwise).

The first work where DL was used in denoising is [62], but sparse representations were already employed for the same purpose in [54, 147, 181]. A review of several important methods and comparisons can be found in [172].

2.4.2 *Inpainting*

Inpainting is a special case of denoising. Unlike the additive noise model from (2.15), we consider now the case where parts of the signal are simply missing. This is typical to images, for example due to scratches or superposed text, communication failures or intentional removal of some objects; the latter two causes may apply to other types of signals, e.g., speech or audio. Inpainting consists in filling the gaps in the signal with information that is suited to the local context, in the attempt of restoring the original.

There are several basic techniques, as reviewed in [88], among which sparse representations play a significant role; some of the early work in using sparse representations can be found in [65, 69, 89]. DL is mentioned the first time in [127]. In principle, the solution is quite simple. We train a dictionary using the available parts of the signal; for example, we can use all the complete patches of an image to train a dictionary $\boldsymbol{D}$. It is even possible to use incomplete signals in DL, as discussed in Sect. 5.9.

With the trained dictionary, we restore as follows a patch $\boldsymbol{y}$ whose available pixels form a set $\mathcal{I}$ of indices. The available part of the signal, $\boldsymbol{y}_{\mathcal{I}}$ is represented with OMP, with some sparsity level s, using the dictionary $\boldsymbol{D}$ reduced to the rows $\mathcal{I}$. (This will be called masked OMP in Sect. 5.9.) We then approximate $\boldsymbol{y}$ with $\boldsymbol{D}\boldsymbol{x}$, where now the full dictionary is used, thus supplying the missing pixels.

The order in which the pixels are restored plays an important role, since the restored pixels are used for the computation of the remaining missing pixels; a hole in an image is naturally filled from the exterior to the interior. Averaging pixels can be used, as explained in Sect. 2.4.1, but now the number of distinct patches that contain the same missing pixel(s) is limited by the necessity to use patches that miss only few pixels, if not only a single one. Several passes can be made.

Inpainting can have many applications, some light, such as the removal of a stranger from a holiday photo in the nature, other serious, like art restoration, and also scientific applications like cloud and cloud shadow removal [209] from multispectral satellite images.

2.4.3 *Compression*

Perhaps the immediate application that one thinks of when studying dictionary learning is compression. Indeed, the standard sparse representation algorithm can be interpreted as a lossy compression routine with m/s data compression ratio. Given a communication channel, where the dictionary is known to both sides, and a signal $\boldsymbol{y}$, the source computes its sparse representation $\boldsymbol{x}$ and transmits the nonzero coefficients and their positions to the receiver. If lossless transmission is necessary one also has to transmit the residual $\boldsymbol{e} = \boldsymbol{y} - \boldsymbol{D}\boldsymbol{x}$. A good dictionary trained on relevant data can lead to better compression rates (permitting smaller s values) and better approximations.

Let us assume we are given the task of compressing a large signal (such as an image). The first step is to split it into N blocks of size m each (see Sect. 2.4.1). Standard OMP represents each block with s atoms leading to a transmission of Ns elements. We can improve information loss, while maintaining the overall size, by switching to a global sparsity constraint, as later explained in Sect. 5.1. This allows us to juggle with the number of atoms used for representation by each block. We decrease the number of nonzero coefficients from very well represented signals and increase the number of atoms for blocks where compression leads to large residuals, thus large information losses. Note however that the size of transmitted information increases; when each block is represented with s atoms, the source transmits only the Ns atom indices; when the sparsity level is variable, the source also needs to transmit the number of atoms for each block. Hence, the best trade-off has to be sought experimentally.

Other optimizations include extra signal preprocessing steps, that usually involve block averaging and overlapping, and sparse coefficients quantization, commonly implemented through simple Huffman tables.

Among the pioneering works in image compression are [136] and [4, 18], the latter two for the particular case of facial images. In [178], a wavelet transform is applied before sparse representation, the results being better than when working directly in the image domain. Several DL algorithms are compared and the results, with a very simple compression scheme, are nearly those of JPEG-2000. At the other end of the compression process, DL is used in [32, 106] to reduce artifacts in JPEG decompression, an operation that is performed similarly with denoising.

2.4.4 Compressed Sensing

Compressed sensing or compressive sampling [24, 25, 51] is based on the intuitive idea that a signal $\boldsymbol{\eta} \in \mathbb{R}^n$ having a sparse representation in a basis $\boldsymbol{T} \in \mathbb{R}^{n \times n}$ can be recovered from much less than n measurements. The typical example is that of a signal with low spectral content (e.g., sums of few sinusoids); using a sampling rate that is twice the maximum frequency (known as Nyquist rate) ensures that the signal can be recovered from the discrete Fourier transform (DFT); however, much less information should be necessary, because the Nyquist rate ignores sparsity.

Compressed sensing proposes the use of an acquisition (or measurement) matrix $\boldsymbol{A} \in \mathbb{R}^{m \times n}$, with $m \ll n$. Instead of acquiring the full vector $\boldsymbol{\eta}$, we get $\boldsymbol{y} = \boldsymbol{A}\boldsymbol{\eta}$, i.e., m linear combinations of the signal elements. To recover $\boldsymbol{\eta}$, we take into account that $\boldsymbol{\eta} = \boldsymbol{T}\boldsymbol{x}$, with sparse $\boldsymbol{x}$. Denoting $\boldsymbol{D} = \boldsymbol{A}\boldsymbol{T}$, we are left with the sparse representation problem $\boldsymbol{y} = \boldsymbol{D}\boldsymbol{x}$ that can be (approximatively) solved with the methods described in Chap. 1.

The acquisition process is relatively simple, as it implies a matrix–vector multiplication that is highly parallel. Recovery is more difficult, since a sparse representation needs to be computed.

A significant example is that of single-pixel camera [56], where the measurement matrix $\boldsymbol{A}$ is implemented in hardware and the scalar products between a row of $\boldsymbol{A}$ and the actual image are computed sequentially, using different orientations for the mirrors that correspond to image pixels. Instead of storing the full many-megapixels image, a much lower number of measurements is enough for visually satisfactory recovery.

Theoretically, the number of measurements necessary for recovery is $m = O(s \log(n/s))$, where s is the sparsity level of the representation $\boldsymbol{x}$. Practically, 3–4s measurements are enough. Of course, in most cases there is no exact true value for the sparsity level, but it was observed that compressed sensing gives good results with sampling rates that are tens of times sub-Nyquist.

The first theoretical results were given for random or other application-independent measurement matrices $\boldsymbol{A}$, but it was proved in [57] that DL can bring benefits in conjunction with compressed sensing. Examples of use are in magnetic resonance imaging [153], plenoptic cameras [132], and in the analysis of the vibrations of rotating machinery [39].

2.4.5 *Classification*

Sparse representation can be easily applied for performing supervised classification. Given a representative, or at least large enough, set of signals, a straightforward approach is to build the dictionary from the signals themselves. That way, when we have to classify new signals, we perform sparse representation and the coefficients correspond to known signals each belonging to a known class. Thus this approach allows us to quickly classify by simple inspection; for example we can choose the class whose coefficient is largest.

The disadvantages of this fixed dictionary solution are an increased complexity due to the large number of dictionary atoms, possible class inequality depending on the ratios of sample signals per class, and misclassification due to the commonality between classes. When we talk about commonality, we refer to the representation information contained in atoms across different classes.

Performing dictionary learning instead, shrinks the dictionary size, because a single atom can now incorporate the information of many signals, allows us to control the number of atoms per class, so that each can be fairly represented, and gives us enough flexibility in the way we formulate either the optimization objective or the dictionary structure such that atoms commonality is no longer an issue.

Existing dictionary learning algorithms perform classification following two strategies. First, split the dictionary into smaller blocks learned separately for each class (which sometimes includes a common class-agnostic block); a given signal is sparsely represented with the whole dictionary and classified based on the dictionary block with most used atoms. Second, learn a linear classification matrix together with the dictionary, such that when applied to the sparse representation of a given signal the resulting vector will have its largest entry on the position corresponding to the class that it belongs to.

Due to its large interest and application, we dedicate Chap. 8 to further discuss the implementation and the algorithmic details of classification through DL.

Problems

P 2.1 What happens with the objective of the DL problem (2.2) corresponding to the (optimal) solution when n or s grow?

What do you expect to happen with the RMSE (2.10) when, everything else being fixed, the number N of signals grows? (We assume that all signals come from the same source.)

Similarly, what do you expect to happen with the RMSE when the size m of the signals grows? (We assume that the length m signals are obtained by windowing the same long signal.)

P 2.2 Consider the weighted DL problem, where some signals are more important than others and hence a weight is associated with the error of each signal. So, in this case, the representation error can be written as $\|(\boldsymbol{Y} - \boldsymbol{D}\boldsymbol{X})\boldsymbol{W}\|_F$, where $\boldsymbol{W}$ is a given diagonal matrix with positive elements on the diagonal. Do we need special algorithms for this problem or can it be solved with the same algorithms as the usual DL problem?

P 2.3 What happens if a training signal appears several times in the matrix $\boldsymbol{Y}$? Is this equivalent to associating a larger weight to this signal? If so, which is the value of the weight?

P 2.4 The representation matrix $\boldsymbol{X}$ from (2.2) is sparse. The number of nonzeros on a column is s, the sparsity level. What is the average number of nonzeros on a row? Interpret the number of nonzeros on row j as the number of signals that use the atom $\boldsymbol{d}_j$ in their sparse representation.

P 2.5 We replace the normalization constraint from (2.2) with the convex constraint $\|\boldsymbol{d}_j\| \leq 1$. Does the new problem have the same optimal value of the objective?

P 2.6 We have discussed in Sect. 2.3.1 that tests on images can be done on patches that overlap or are disjoint. In which case do you expect the RMSE at the end of the DL process to be smaller?

P 2.7 Assume that relation (2.14) holds for all atoms of $\boldsymbol{D}_{\text{true}}$, which means that the recovery test is fully successful and the optimized dictionary $\boldsymbol{D}$ is virtually identical with the true one. Assuming the most favorable permutation of the atoms of $\boldsymbol{D}$, give a bound for the error $\|\boldsymbol{D} - \boldsymbol{D}_{\text{true}}\|_F$.

Chapter 3
Standard Algorithms

Abstract There are many dictionary learning algorithms and this chapter is devoted to the most important ones solving the basic problem of optimizing the representation error. The most successful approaches rely on alternate optimization, iteratively solving sparse coding and dictionary update problems. OMP is often the choice for sparse coding, and we present a batch version that is efficient in the DL context, when many signals have to be represented at the same time. The dictionary update step makes the main difference between algorithms. We start with the oldest and simplest method, Sparsenet, which uses gradient descent, and then describe the block coordinate descent idea. The most known DL algorithms, MOD and K-SVD, come next. Due to the DL complexity, we also explore parallel versions of the coordinate descent algorithms, where more atoms are updated simultaneously and not sequentially, as usual. Other algorithms, like SimCO and NSGK, appeal to different viewpoints or subtle modifications of the classic approaches. After acquiring information about all these algorithms, we discuss implementation issues that may have a great effect on the result. Then, we attempt comparisons of the algorithms, examining especially their numerical behavior on test problems. It results that the competition between the algorithms is quite tight and that many of them are good candidates for practical use.

3.1 Basic Strategy: Alternating Optimization

In this chapter, we discuss several prominent algorithms for solving the basic dictionary learning problem (2.2). They all use a simple strategy in the attempt to overcome the difficulties of the optimization process, due to nonconvexity and to the huge size of the problem. Alternate optimization is a generic name for algorithms that iteratively improve a subset of the variables while keeping the others fixed, taking care to alternate the subsets such that all variables are optimized; such a strategy is also called block coordinate descent. As we have seen in the previous chapter, the DL problem can be naturally split into two subproblems: sparse coding and dictionary update. Solving alternatively these two subproblems, optimally or

B. Dumitrescu, P. Irofti, *Dictionary Learning Algorithms and Applications*,
https://doi.org/10.1007/978-3-319-78674-2_3

Algorithm 3.1: DL by Alternate Optimization

Data: signals set $\boldsymbol{Y} \in \mathbb{R}^{m \times N}$
sparsity level s
initial dictionary $\boldsymbol{D} \in \mathbb{R}^{m \times n}$
number of iterations K
Result: trained dictionary $\boldsymbol{D}$

```
for k = 1 : K do
    Sparse coding: keeping D fixed, solve (2.8) to compute sparse
    representations X
    Dictionary update: keeping the nonzero pattern Ω fixed, solve (2.9) to
    compute new dictionary D; the matrix X may be changed or not
    Atoms normalization, if not already performed: d_j ← d_j/||d_j||, j = 1 : n
```

suboptimally, is the core strategy of a large family of DL algorithms, described by Algorithm 3.1.

The two main operations are iterated for a high enough number of iterations K. First, keeping fixed the current dictionary $\boldsymbol{D}$, the sparse representations $\boldsymbol{X}$ are computed; this is the sparse coding step and is typically solved with simple algorithms like OMP, as detailed in Sect. 3.2. Then, in the dictionary update step, the dictionary is improved; the nonzero pattern Ω of the representations is kept fixed, hence this is a usual nonlinear optimization problem, without a combinatorial character; some of the DL algorithms improve also the representations coefficients, precisely those elements of $\boldsymbol{X}$ that are already nonzero. Atoms normalization is done explicitly by some algorithms or left at the end by others; we have put it explicitly at the end of each loop of Algorithm 3.1 only to ensure that it is always done; in any case, this is a very inexpensive operation compared with sparse coding and dictionary update.

An initial dictionary is necessary to start the alternate optimization by performing the sparse coding step. This can be done in several ways; if no better solutions are available, the simplest initialization is either to give random values to the atoms or randomly selecting n signals to serve as initial atoms. DL practice has shown that such initializations are usually satisfactory, see also Sect. 3.9.

The decision to let the algorithm run a fixed number of iterations is mostly due to the lack of a monotonous decrease of the representation error. The dictionary update step is almost always designed such that the error is guaranteed to decrease. However, due to its NP-hard character, the sparse coding step does not necessarily succeed in this purpose. So, while the error generally decreases, one may see steps in which the error increases; examples of error evolution can be found in Sect. 3.11.1.

Using a stopping criterion that takes into account the change in the error is thus hard to design. Of course, one could stop when the error reaches a predefined threshold. The difficulty here is to choose the right value of the threshold; this is possible usually only when there are already some available dictionaries whose error can be used as a yardstick.

Since sparse coding can be done by algorithms specialized in finding sparse representations and designed independently from the DL problem, it is the dictionary update step that makes the difference between the algorithms that have the basic structure exposed in Algorithm 3.1. So, with the exception of Sects. 3.2 and 3.12 (the latter being dedicated to alternatives to OMP), the remainder of this chapter will be dedicated to the description of the most successful dictionary update algorithms.

3.2 Sparse Coding

Orthogonal Matching Pursuit, whose basic form is given in Sect. 1.3, Algorithm 1.1, is the main candidate for the sparse coding step, due to its good trade-off between quality and complexity. OMP can be implemented efficiently by taking into account i) the fact that the selected atoms are never dismissed, and ii) its use for coding many signals at a time in the sparse coding stage of DL. These implementation ideas were first adapted to DL in [156].

Cholesky Factorization The first implementation trick regards the computation of the least squares solutions $\boldsymbol{x}_{\mathcal{S}}$ from (1.14) when the support $\mathcal{S}$ grows. There are at least two approaches to this kind of problem, one updating a Cholesky factorization, the other updating a QR factorization [80, Sec. 6.5]. The latter has better numerical properties, but since our matrices have a small number of columns and so numerical issues are not very significant, the simpler Cholesky approach is preferred. A numerical comparison of several OMP implementations can be found in [183].

To lighten the notation, let us denote $\boldsymbol{A} = \boldsymbol{D}_{\mathcal{S}}$ the matrix of the currently selected atoms, $\boldsymbol{d}$ the newly selected atom (with index k), and $\tilde{\boldsymbol{A}} = \boldsymbol{D}_{\mathcal{S}\cup\{k\}}$ the next matrix of atoms. Hence, we have $\tilde{\boldsymbol{A}} = [\boldsymbol{A}\ \boldsymbol{d}]$. We assume that we have computed the Cholesky factorization $\boldsymbol{A}^T\boldsymbol{A} = \boldsymbol{L}\boldsymbol{L}^T$, with lower triangular $\boldsymbol{L}$. Our purpose is to compute the Cholesky factorization $\tilde{\boldsymbol{A}}^T\tilde{\boldsymbol{A}} = \tilde{\boldsymbol{L}}\tilde{\boldsymbol{L}}^T$. We note that

$$\tilde{\boldsymbol{A}}^T\tilde{\boldsymbol{A}} = \begin{bmatrix} \boldsymbol{A}^T\boldsymbol{A} & \boldsymbol{A}^T\boldsymbol{d} \\ \boldsymbol{d}^T\boldsymbol{A} & \boldsymbol{d}^T\boldsymbol{d} \end{bmatrix} = \begin{bmatrix} \boldsymbol{L} & 0 \\ \boldsymbol{w}^T & 1 \end{bmatrix} \begin{bmatrix} \boldsymbol{L}^T & \boldsymbol{w} \\ 0 & 1 \end{bmatrix} = \tilde{\boldsymbol{L}}\tilde{\boldsymbol{L}}^T. \tag{3.1}$$

So, only the last row $\boldsymbol{w}^T$ of the Cholesky factor $\tilde{\boldsymbol{L}}$ needs to be computed. It results immediately that

$$\boldsymbol{L}\boldsymbol{w} = \boldsymbol{A}^T\boldsymbol{d}, \tag{3.2}$$

which means that only a matrix–vector multiplication and the solution of a triangular system are needed. We note that the diagonal elements of $\boldsymbol{L}$ are equal to one, due

Algorithm 3.2: OMP with Updated Cholesky Factorization

Data: dictionary $\boldsymbol{D} \in \mathbb{R}^{m \times n}$
signals $\boldsymbol{y} \in \mathbb{R}^m$
sparsity level s
stopping error ε
Result: representation support $\mathcal{S}$, solution $\boldsymbol{x}$

1 Initialize $\mathcal{S} = \emptyset$, $\boldsymbol{e} = \boldsymbol{y}$, $\boldsymbol{L} = 1$, $\boldsymbol{b} = []$
2 **while** $|\mathcal{S}| < s$ *and* $\|\boldsymbol{e}\| > \varepsilon$ **do**
3 Find new index: $k = \arg\max_{j \notin \mathcal{S}} |\boldsymbol{e}^T \boldsymbol{d}_j|$
4 Update right hand term: $\boldsymbol{b} \leftarrow \begin{bmatrix} \boldsymbol{b} \\ \boldsymbol{d}_k^T \boldsymbol{y} \end{bmatrix}$
5 **if** $|\mathcal{S}| > 1$ **then**
6 Compute $\boldsymbol{w}$ by solving the triangular system $\boldsymbol{L}\boldsymbol{w} = \boldsymbol{D}_{\mathcal{S}}^T \boldsymbol{d}_k$
7 Update $\boldsymbol{L} = \begin{bmatrix} \boldsymbol{L} & 0 \\ \boldsymbol{w}^T & 1 \end{bmatrix}$
8 Build new support: $\mathcal{S} \leftarrow \mathcal{S} \cup \{k\}$
9 Compute new solution by solving: $\boldsymbol{L}\boldsymbol{L}^T \boldsymbol{x}_{\mathcal{S}} = \boldsymbol{b}$
10 Compute new residual: $\boldsymbol{e} = \boldsymbol{y} - \boldsymbol{D}_{\mathcal{S}} \boldsymbol{x}_{\mathcal{S}}$

to the normalization of the atoms; note, however, that this might not be true outside the DL context.

When the Cholesky factorization is available, the solution of the least squares problem at the next iteration can be computed from

$$\tilde{\boldsymbol{L}}\tilde{\boldsymbol{L}}^T \tilde{\boldsymbol{x}} = \tilde{\boldsymbol{b}}, \tag{3.3}$$

where the right-hand term

$$\tilde{\boldsymbol{b}} = \tilde{\boldsymbol{A}}^T \boldsymbol{y} = \begin{bmatrix} \boldsymbol{A}^T \boldsymbol{y} \\ \boldsymbol{d}^T \boldsymbol{y} \end{bmatrix} = \begin{bmatrix} \boldsymbol{b} \\ \boldsymbol{d}^T \boldsymbol{y} \end{bmatrix} \tag{3.4}$$

is also available from the current step, excepting its last element. The computation of $\tilde{\boldsymbol{x}}$ is thus reduced to the solution of two triangular systems.

Inserting these shortcuts in Algorithm 1.1 we obtain Algorithm 3.2. Counting the operations for step s gives $O(mn)$ for line 3, $O(m)$ for line 4, $O(s(m + s))$ for line 6, $O(s^2)$ for line 9, and $O(sm)$ for line 10. Adding up for all steps gives $O(smn + s^2 m + s^3)$ operations, which is a clear reduction compared with the count reported in Sect. 1.3.

Precomputations The second implementation trick relies on the fact that, in DL, OMP is used for coding all the training signals gathered in the matrix $\boldsymbol{Y}$. Even though a single signal is represented with only a few atoms, each atom is used many times when all signals are represented. So, it makes sense to precompute the inner products that appear in the algorithm, namely

$$G = D^T D, \quad f = D^T y. \tag{3.5}$$

Elements of these products appear explicitly, e.g., in lines 4 and 6 of Algorithm 3.2. More precisely, we have

$$b = D_{\mathcal{S}}^T y = f_{\mathcal{S}},$$

$$D_{\mathcal{S}}^T d_k = G_{\mathcal{S},k}.$$

Other operations can be modified to take advantage of the precomputation. The matrix (3.5) is also called Gram matrix associated with the dictionary. Of course, the whole product $F = D^T Y$ can be computed from the beginning, not just f for the current signal. Besides their use for reducing the operation count in OMP, the matrix multiplications that compute F and G can be implemented efficiently due to their regular structure and are also adequate for parallel architectures.

Line 3 of Algorithm 3.2 can be implemented immediately by using the residual expression for computing the vector

$$c = D^T e = D^T(y - Dx) = f - Gx = f - G_{\mathcal{S}} x_{\mathcal{S}}. \tag{3.6}$$

Here we already spare some operations, going from $O(mn)$ to $O(sn)$ due to the use of a smaller matrix–vector multiplication.

The downside seems to be the lack of an explicit error vector e. However, we only need its norm in the stopping criterion. To derive an expression for the error norm, we use a few simple relations. Since OMP computes a least squares solution, the residual is orthogonal on the span of selected atoms and hence

$$e^T Dx = 0. \tag{3.7}$$

It results that

$$y^T Dx = (e + Dx)^T Dx = x^T Gx.$$

So, the error norm has the expression

$$\begin{aligned} \|e\|^2 &= (y - Dx)^T(y - Dx) \\ &= \|y\|^2 - 2y^T Dx + x^T D^T Dx \\ &= \|y\|^2 - x^T Gx \\ &= \|y\|^2 - x^T G_{\mathcal{S}} x_{\mathcal{S}}. \end{aligned} \tag{3.8}$$

Algorithm 3.3: Batch OMP

Data: precomputed $\boldsymbol{G} = \boldsymbol{D}^T\boldsymbol{D} \in \mathbb{R}^{n\times n}$, $\boldsymbol{f} = \boldsymbol{D}^T\boldsymbol{y} \in \mathbb{R}^n$
sparsity level s
stopping error ε
Result: representation support $\mathcal{S}$, solution $\boldsymbol{x}$

1 Initialize $\mathcal{S} = \emptyset$, $\boldsymbol{h} = 0$, $\boldsymbol{L} = 1$, $\boldsymbol{b} = []$, $\varsigma = \boldsymbol{y}^T\boldsymbol{y}$
2 **while** $|\mathcal{S}| < s$ *and* $\|\boldsymbol{e}\| > \varepsilon$ **do**
3   Find new index: $k = \arg\max_{j\notin\mathcal{S}} |\boldsymbol{f} - \boldsymbol{h}|$
4   **if** $|\mathcal{S}| > 1$ **then**
5   Compute $\boldsymbol{w}$ by solving the triangular system $\boldsymbol{L}\boldsymbol{w} = \boldsymbol{G}_{\mathcal{S},k}$
6   Update $\boldsymbol{L} = \begin{bmatrix} \boldsymbol{L} & 0 \\ \boldsymbol{w}^T & 1 \end{bmatrix}$
7   Build new support: $\mathcal{S} \leftarrow \mathcal{S} \cup \{k\}$
8   Compute new solution by solving: $\boldsymbol{L}\boldsymbol{L}^T\boldsymbol{x}_{\mathcal{S}} = \boldsymbol{f}_{\mathcal{S}}$
9   Compute auxiliary vector: $\boldsymbol{h} = \boldsymbol{G}_{\mathcal{S}}\boldsymbol{x}_{\mathcal{S}}$
10   Update error norm: $\|\boldsymbol{e}\|^2 = \varsigma - \boldsymbol{x}_{\mathcal{S}}^T\boldsymbol{h}_{\mathcal{S}}$

We note that the product $\boldsymbol{h} \triangleq \boldsymbol{G}_{\mathcal{S}}\boldsymbol{x}_{\mathcal{S}}$ is also necessary in (3.6), hence the error norm can be computed via

$$\|\boldsymbol{e}\|^2 = \|\boldsymbol{y}\|^2 - \boldsymbol{x}_{\mathcal{S}}^T\boldsymbol{h}_{\mathcal{S}}. \tag{3.9}$$

The above computation of the error norm is extremely cheap: only $O(s)$ operations, since $\|\boldsymbol{y}\|^2$ is computed only once.

Algorithm 3.3, called also Batch OMP, summarizes all these developments. The only new notation is $\varsigma = \|\boldsymbol{y}\|^2$, to stress that this is a one time computation. The significant number of operations for iteration s are as follows: $O(s^2)$ for lines 5 and 8, and $O(sn)$ for line 9. The total number is thus $O(s^2n + s^3)$, an obvious reduction with respect to Algorithm 3.2. Of course, one needs to take into account the precomputation, which is made for all N signals; the quota of one signal is $mn^2/N + mn$ operations. The value of N for which Batch OMP is faster than Algorithm 3.2 depends on the actual implementation, but Batch OMP is asymptotically faster. For an example with $m = 256$, $n = 512$, it is reported in [156] that Batch OMP is almost 7 times faster for large N and is already faster for $N = 100$.

Similarly to the notation from Sect. 1.3, we will denote OMP($\boldsymbol{D}, \boldsymbol{Y}, s, \varepsilon$) the computation of sparse representations for all the signals in $\boldsymbol{Y}$. The underlying algorithm is supposed to be Batch OMP, although no such constraint is necessary for the correctness of the representation.

It is also worth remarking that all signals can be coded independently, hence the Batch OMP algorithm can be run in parallel for each column of the training signals matrix $\boldsymbol{Y}$. The simultaneous access of all Batch OMP instances to the Gram matrix $\boldsymbol{G}$ may create a bottleneck, but the potential parallelism is still significant.

Even so, one can try other ways to increase the speed. For example, a coefficient reuse strategy was proposed in [180], based on the idea that the representations support does not change radically from one DL iteration to the next and the largest coefficients are likely to remain nonzero. The idea is to start OMP not from scratch in each sparse coding step, but using a number, for example $s/3$, of the largest coefficients from the previous DL iteration. This strategy gives slightly better execution times and also improves the error. However, it was found especially productive in conjunction with more evolved sparse representation algorithms that imply augmenting the support over the desired sparsity level s and then pruning. This opens a much larger discussion on alternatives to OMP that will be continued in Sect. 3.12.

3.3 Simple Descent Methods

Simple methods are often well suited to large optimization problems, where the sheer computation load for obtaining the auxiliary information required by more sophisticated methods cannot be afforded. We present here the adaptation to dictionary update of two classical approaches: gradient descent and coordinate descent. Both methods update the atoms without touching the representations.

3.3.1 Gradient Descent

The first DL method [141, 142] used gradient descent with fixed step for the dictionary update. The method needs only the gradient of the objective function. Although not among the best methods, its presentation is a good opportunity to review a few basic results regarding matrix manipulation and the derivative of a function of a matrix, that will be later useful. Other algorithms based on gradient computation can be found in [113], but in the context of Bayesian learning. Let us denote

$$f(\boldsymbol{D}) = \|\boldsymbol{Y} - \boldsymbol{D}\boldsymbol{X}\|_F^2 \tag{3.10}$$

the DL objective viewed as a function of the dictionary $\boldsymbol{D}$, as typical in the dictionary update step. Remind also the notation (2.4) of the representation error: $\boldsymbol{E} = \boldsymbol{Y} - \boldsymbol{D}\boldsymbol{X}$.

Lemma 3.1 *The gradient of (3.10) with respect to* $\boldsymbol{D}$ *is*

$$\frac{\partial f}{\partial \boldsymbol{D}} = 2(\boldsymbol{DX} - \boldsymbol{Y})\boldsymbol{X}^T = -2\boldsymbol{EX}^T. \tag{3.11}$$

Proof We give two proofs, using different mathematical tools.

Proof 1 A first idea is to express (3.10) as a function of a vector, since the Frobenius norm of a matrix is the Euclidean norm of a vector obtained by concatenating its columns (operation denoted by vec), i.e., $\|\boldsymbol{E}\|_F = \|\text{vec}(\boldsymbol{E})\|$. To vectorize a matrix product, we use the relation

$$\text{vec}(\boldsymbol{ABC}) = (\boldsymbol{C}^T \otimes \boldsymbol{A})\text{vec}(\boldsymbol{B}), \tag{3.12}$$

where $\otimes$ denotes the Kronecker product. Thus, we get

$$f(\boldsymbol{D}) = \left\|\text{vec}(\boldsymbol{Y}) - (\boldsymbol{X}^T \otimes \boldsymbol{I})\text{vec}(\boldsymbol{D})\right\|^2.$$

The derivative with respect to a vector is easier to compute. Reminding that

$$\frac{\partial \boldsymbol{c}^T \boldsymbol{d}}{\partial \boldsymbol{d}} = \boldsymbol{c}, \quad \frac{\partial \boldsymbol{d}^T \boldsymbol{Q}\boldsymbol{d}}{\partial \boldsymbol{d}} = 2\boldsymbol{Q}\boldsymbol{d}, \tag{3.13}$$

and

$$\|\boldsymbol{b} - \boldsymbol{A}\boldsymbol{d}\|^2 = \boldsymbol{b}^T\boldsymbol{b} - 2\boldsymbol{b}^T\boldsymbol{A}\boldsymbol{d} + \boldsymbol{d}^T\boldsymbol{A}^T\boldsymbol{A}\boldsymbol{d},$$

we get

$$\begin{aligned}\text{vec}\left(\tfrac{\partial f}{\partial \boldsymbol{D}}\right) &= 2(\boldsymbol{X} \otimes \boldsymbol{I})\left[(\boldsymbol{X}^T \otimes \boldsymbol{I})\text{vec}(\boldsymbol{D}) - \text{vec}(\boldsymbol{Y})\right] \\ &= 2\left[(\boldsymbol{XX}^T \otimes \boldsymbol{I})\text{vec}(\boldsymbol{D}) - (\boldsymbol{X} \otimes \boldsymbol{I})\text{vec}(\boldsymbol{Y})\right].\end{aligned}$$

Relation (3.11) results by applying (3.12) backwards.

Proof 2 We express the error function (3.10) as

$$f(\boldsymbol{D}) = \text{tr}[(\boldsymbol{Y} - \boldsymbol{DX})^T(\boldsymbol{Y} - \boldsymbol{DX})] = \text{tr}[\boldsymbol{X}^T\boldsymbol{D}^T\boldsymbol{DX}] - 2\text{tr}[\boldsymbol{Y}^T\boldsymbol{DX}] + \text{tr}[\boldsymbol{Y}^T\boldsymbol{Y}].$$

The relevant, but probably less familiar rules for computing the derivative of the trace with respect to the matrix $\boldsymbol{D}$ are

$$\frac{\partial \text{tr}(\boldsymbol{ADB})}{\partial \boldsymbol{D}} = \boldsymbol{A}^T\boldsymbol{B}^T, \quad \frac{\partial \text{tr}(\boldsymbol{AD}^T\boldsymbol{BDC})}{\partial \boldsymbol{D}} = \boldsymbol{B}^T\boldsymbol{DA}^T\boldsymbol{C}^T + \boldsymbol{BDCA}. \tag{3.14}$$

Their application leads directly to (3.11). ∎

Algorithm 3.4: Sparsenet Dictionary Update

Data: signals set $\boldsymbol{Y} \in \mathbb{R}^{m \times N}$
current dictionary $\boldsymbol{D} \in \mathbb{R}^{m \times n}$
current representation matrix $\boldsymbol{X} \in \mathbb{R}^{n \times N}$
step size α

Result: updated dictionary $\boldsymbol{D}$

for $j = 1 : n$ **do**
  Update atom: $\boldsymbol{d}_j \leftarrow \boldsymbol{d}_j + \alpha(\boldsymbol{Y} - \boldsymbol{D}\boldsymbol{X})(\boldsymbol{x}_j^T)^T$
  Normalize atom: $\boldsymbol{d}_j \leftarrow \boldsymbol{d}_j / \|\boldsymbol{d}_j\|$

For the sake of completeness, we give also the following:

Lemma 3.2 *The gradient of (3.10) with respect to* $\boldsymbol{X}$ *is*

$$\frac{\partial f}{\partial \boldsymbol{X}} = 2\boldsymbol{D}^T(\boldsymbol{D}\boldsymbol{X} - \boldsymbol{Y}) = -2\boldsymbol{D}^T\boldsymbol{E}. \tag{3.15}$$

Proof Left to the reader. ■

Computing the gradient has a relatively high complexity, due to the multiplication of possible large size matrices implied by (3.11). More precisely, the operation count is $O(mnN)$ without taking into account the sparsity of $\boldsymbol{X}$. With a careful implementation, the complexity can decrease to $O(smN)$, since each column of $\boldsymbol{X}$ has only s nonzeros out of its n elements.

Sparsenet [141] updates the dictionary by making a fixed step gradient descent on each atom, successively. Using (3.11), the update rule for atom $\boldsymbol{d}_j$ is

$$\boldsymbol{d}_j \leftarrow \boldsymbol{d}_j + \alpha(\boldsymbol{Y} - \boldsymbol{D}\boldsymbol{X})(\boldsymbol{x}_j^T)^T, \tag{3.16}$$

where $\boldsymbol{x}_j^T$ is row j of $\boldsymbol{X}$ and α is the step size. Sparsenet dictionary update is listed as Algorithm 3.4. Only one sweep of the atoms is shown, but several sweeps can be done as well. Note that (3.16) can be computed with only matrix–vector multiplications and the sparsity of $\boldsymbol{x}_j$ can be used to speed up the computation. One can also take advantage of the parallel character of matrix multiplication. However, one sweep of all atoms in Sparsenet implies computation similar to matrix–matrix multiplication.

A problematic choice is that of the step size α. Small values are usually preferred, since they ensure convergence, but this can make the method even slower than typical gradient methods, which are already rather slow to converge; hence, the overall complexity of Sparsenet may be excessively large. Alternatively, the aggressive approach of taking a large α may make the method not convergent. Since its proposal, Sparsenet has been superseded by other methods and now has mostly historical value.

3.3.2 Coordinate Descent

The basic version of coordinate descent is to optimize a single variable at a time, while the other variables are fixed. The descent can be optimal, when the minimum value of the objective is achieved (in the above conditions), or suboptimal if the objective is only decreased. In the dictionary update method proposed in [169], it is in fact a whole atom that is optimized and the step is suboptimal. The method is viewed by the authors as a sequential generalization of the K-means algorithm and hence named SGK. (Remind Remark 2.3, that the vector quantization problem solved by Lloyd's K-means algorithm can be seen as DL with sparsity level $s = 1$; see also problem **P** 3.4.)

Let us first notice that the product $\boldsymbol{DX}$ can be expressed as a sum of outer products between atoms and the rows of $\boldsymbol{X}$:

$$\boldsymbol{DX} = \sum_{i=1}^{n} \boldsymbol{d}_i \boldsymbol{x}_i^T. \tag{3.17}$$

Since $\boldsymbol{X}$ is sparse, an atom $\boldsymbol{d}_j$ is used only for the representation of some signals; let us denote $\mathcal{I}_j$ the indices of these signals; they coincide with the nonzero positions on the jth row of $\boldsymbol{X}$, i.e., $\mathcal{I}_j = \{\ell \mid (j, \ell) \in \Omega\}$. The right side of Fig. 3.1 shows with brighter colors an atom (in red) and the nonzero coefficients (in blue); the signals that use the atom are shown in brighter green on the left side of the figure.

We aim to optimize atom $\boldsymbol{d}_j$ while all the other atoms and the representations are fixed. In view of (3.17), the representation error is

$$\|\boldsymbol{Y} - \boldsymbol{DX}\| = \left\| \boldsymbol{Y} - \sum_{i \neq j} \boldsymbol{d}_i \boldsymbol{x}_i^T - \boldsymbol{d}_j \boldsymbol{x}_j^T \right\|. \tag{3.18}$$

The fixed part in the above expression is interesting only for the signals that use atom $\boldsymbol{d}_j$ and so we denote

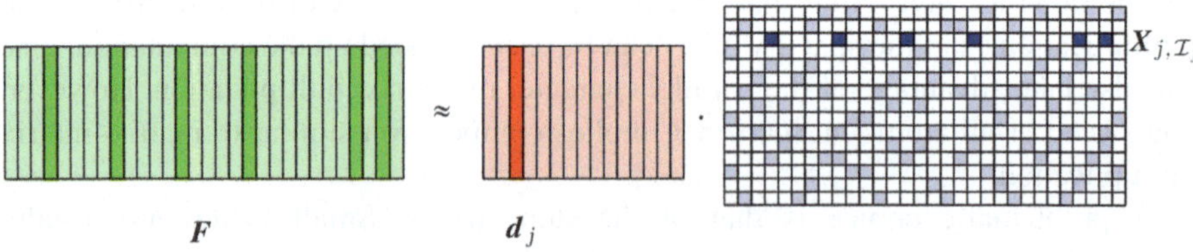

Fig. 3.1 Approximation problem (3.20), suggested by the brighter colors. The pale colors illustrate the whole DL problem

$$F = \left[Y - \sum_{i \neq j} d_i x_i^T \right]_{\mathcal{I}_j}, \tag{3.19}$$

where, as usual, the subscript $\mathcal{I}_j$ denotes the restriction of the matrix to the columns with indices in $\mathcal{I}_j$. We are thus left with the problem

$$\min_{d_j} \left\| F - d_j X_{j,\mathcal{I}_j} \right\|_F^2 \tag{3.20}$$

This is the approximation problem illustrated in Fig. 3.1. Note that we do not normalize the atom; this is only to follow the SGK derivation from [169]; we will come back later to this issue. Since several problems from the same family will appear, we lighten the notation: d is the current atom and x^T the reduced vector of representations. The problem and the solution are as follows:

Proposition 3.3 *The solution to the problem*

$$\min_{d} \left\| F - d x^T \right\|_F^2 \tag{3.21}$$

is

$$d = \frac{Fx}{\|x\|^2}. \tag{3.22}$$

Proof The objective is a quadratic function in d:

$$\begin{aligned} \| F - dx^T \|_F^2 &= \mathrm{tr}[(F^T - xd^T)(F - dx^T)] \\ &= \mathrm{tr}(F^T F) - 2\mathrm{tr}(F^T dx^T) + \mathrm{tr}(xd^T dx^T) \\ &= \|F\|_F^2 - 2x^T F^T d + \|x\|^2 d^T d. \end{aligned} \tag{3.23}$$

To obtain the last equality we have used the property of the trace operator that $\mathrm{tr}(ABC) = \mathrm{tr}(BCA)$. Computing the derivative of the last form in (3.23) according to (3.13) and equating with zero gives (3.22). ■

Coming back to the original notation, the solution to (3.20) is

$$d_j = F X_{j,\mathcal{I}_j}^T / (X_{j,\mathcal{I}_j} X_{j,\mathcal{I}_j}^T). \tag{3.24}$$

The similarity with the K-means algorithm consists in the fact that the updated atom is a weighted linear combination of the signals (after subtracting the contributions of all the other atoms) to which the atom contributes; so, it is a weighted mass center, the weights being the coefficients of the representations from $X_{j,\mathcal{I}_j}^T$.

Algorithm 3.5: SGK Dictionary Update

Data: signals set $\boldsymbol{Y} \in \mathbb{R}^{m \times N}$
current dictionary $\boldsymbol{D} \in \mathbb{R}^{m \times n}$
representation matrix $\boldsymbol{X} \in \mathbb{R}^{n \times N}$
Result: updated dictionary $\boldsymbol{D}$

1 Compute error $\boldsymbol{E} = \boldsymbol{Y} - \boldsymbol{D}\boldsymbol{X}$
2 **for** $j = 1$ **to** n **do**
3 Modify error: $\boldsymbol{F} = \boldsymbol{E}_{\mathcal{I}_j} + \boldsymbol{d}_j \boldsymbol{X}_{j,\mathcal{I}_j}$
4 Update atom: $\boldsymbol{d}_j = \boldsymbol{F}\boldsymbol{X}_{j,\mathcal{I}_j}^T / (\boldsymbol{X}_{j,\mathcal{I}_j} \boldsymbol{X}_{j,\mathcal{I}_j}^T)$
5 Optionally, normalize atom: $\boldsymbol{d}_j \leftarrow \boldsymbol{d}_j / \|\boldsymbol{d}_j\|$
6 Recompute error: $\boldsymbol{E}_{\mathcal{I}_j} = \boldsymbol{F} - \boldsymbol{d}_j \boldsymbol{X}_{j,\mathcal{I}_j}$

The modified signals (3.19) can be computed efficiently by permanently updating the error (2.4). If $\boldsymbol{E}$ is available, then (3.19) can be computed through

$$\boldsymbol{F} = \boldsymbol{E}_{\mathcal{I}_j} + \boldsymbol{d}_j \boldsymbol{X}_{j,\mathcal{I}_j}. \tag{3.25}$$

After computing the new atom via (3.24), the error matrix is recomputed via

$$\boldsymbol{E}_{\mathcal{I}_j} = \boldsymbol{F} - \boldsymbol{d}_j \boldsymbol{X}_{j,\mathcal{I}_j}. \tag{3.26}$$

The SGK dictionary update is summarized by Algorithm 3.5. The atoms are updated successively. Like in Sparsenet, a single sweep of all atoms is performed. More sweeps seem to bring only marginal improvement in the overall scheme. SGK can easily take advantage of sparsity, since it works explicitly with compacted rows of $\boldsymbol{X}$. The number of operations per dictionary update is $O(smN)$; taking into account that each row of $\boldsymbol{X}$ has in average sN/n nonzero elements, the outer products in lines 3 and 6 and the matrix–vector product in line 4 need $O(smN/n)$ operations. The potential for parallelism is low due to the inherent sequential character of an atom-by-atom optimization.

Note that the atoms are not normalized after the coordinate descent step (3.24), but only after all updates are performed (step 4 of basic DL Algorithm 3.1). So, although the problem (3.20) is solved optimally, SGK is in fact using a suboptimal coordinate descent step, since the normalization constraint is ignored, although present in the DL problem.

Introducing normalization in the optimization of an atom is in fact easy. In fact, the solution is the normalization of (3.24). With notations like in Proposition 3.3, we obtain the following result.

Proposition 3.4 *The solution to the problem*

$$\begin{aligned} \min_{d} \quad & \left\| F - dx^T \right\|_F^2 \\ s.t. \quad & \|d\| = 1 \end{aligned} \tag{3.27}$$

is

$$d = \frac{Fx}{\|Fx\|}. \tag{3.28}$$

Proof Using the norm constraint in (3.23) gives

$$\left\| F - dx^T \right\|_F^2 = \|F\|_F^2 - 2x^T F^T d + \|x\|^2. \tag{3.29}$$

Only the middle term in the right-hand part of the above relation depends on d. This scalar product is maximized when d has the same direction as Fx. With the normalization constraint, this gives (3.28). ■

Remark 3.5 Normalizing the atoms means activating line 5 of the SGK algorithm or, even better, replacing lines 4 and 5 with

$$d_j = FX_{j,\mathcal{I}_j}^T / \|FX_{j,\mathcal{I}_j}^T\|. \tag{3.30}$$

This would be indeed coordinate descent with an optimal step. Such an algorithm is barely mentioned in [169]; we have used normalization in all our experiments. Note that the two versions (with and without normalization) lead to different results, because the error $E_{\mathcal{I}_j}$ is different already after the first atom in the loop at line 2. Even though the direction given by (3.24) and (3.30) is the same, the difference in magnitude is enough to produce different error matrices and hence affect subsequent atoms. ■

3.4 Method of Optimal Directions (MOD)

The method of optimal directions (MOD) [67] is based on the simple remark that the dictionary update step is a quadratic optimization problem in the variable D. Otherwise seen, this is a linear least squares problem in the matrix variable D. No matter the interpretation, due to the convexity of the error (3.10) with respect to D, its minimization amounts to setting to zero the gradient (3.11). This gives the optimal dictionary (for the fixed X):

$$D = YX^T(XX^T)^{-1}. \tag{3.31}$$

So, the dictionary update in MOD simply consists of computing (3.31); thus, we do not give the explicit algorithm like for the other methods. (It is worth noting that in the original paper this formula was written at element level, a fact that maybe slowed down the impact of the method in DL practice.)

The numerical implementation of (3.31) can be based on the Cholesky factorization or orthogonal triangularization; the latter is more stable numerically, but more complex, since it is more difficult to take advantage of sparsity when computing a QR factorization. We analyze only the former approach.

The price for this optimal dictionary update is the complexity of the least-squares formula (3.31). Full matrix multiplication for computing $\boldsymbol{X}\boldsymbol{X}^T$ would need $O(n^2N)$ operations; taking sparsity into account reduces the number towards $O(s^2N)$, without considering the non-negligible indexing and memory management requirements. The product $\boldsymbol{Y}\boldsymbol{X}^T$ needs $O(smN)$ operations. The Cholesky factorization $\boldsymbol{X}\boldsymbol{X}^T = \boldsymbol{L}\boldsymbol{L}^T$, with lower triangular $\boldsymbol{L}$, requires $O(n^3)$ operations, since $\boldsymbol{X}\boldsymbol{X}^T$ is $n \times n$. Finally, after transforming (3.31) into

$$\boldsymbol{D}\boldsymbol{L}\boldsymbol{L}^T = \boldsymbol{Y}\boldsymbol{X}^T, \tag{3.32}$$

solving the triangular systems for $\boldsymbol{D}$ needs $O(mn^2)$. Overall, the operation count is not much higher than for SGK, especially for large number of signals N.

Parallelism is somewhat limited in MOD. Although matrix multiplication, which is genuinely very parallel, has a large weight in the algorithm, the Cholesky factorization is a clear bottleneck. Once $\boldsymbol{L}$ is available, solving (3.32) can be done in parallel on the rows of $\boldsymbol{D}$. So, MOD is potentially faster than sequential algorithms like SGK and K-SVD (to be presented in the next section) that update the atoms in succession, although not significantly.

Remark 3.6 (Relation Between SGK and MOD) Since the error (3.10) is convex in $\boldsymbol{D}$, a coordinate descent algorithm converges to the optimum (for fixed $\boldsymbol{X}$), excepting some degenerate cases. So, if the SGK dictionary update is run for sufficiently many sweeps instead of the single one shown in Algorithm 3.5, it would reach the optimum and so it would be equivalent to MOD. (Remind that neither SGK nor MOD normalizes the atoms.) Of course, doing so is not efficient. It is also not justified by the practical experience showing that optimal dictionary update is not necessarily key to success. ■

3.5 K-SVD

K-SVD [4] is probably the most popular DL method. Essentially, it is also a (block) coordinate descent algorithm but, unlike the previously presented methods, it optimizes also the representation coefficients without modifying the sparsity pattern. It still optimizes one atom at a time, but minimizes (3.20) by considering both the atom $\boldsymbol{d}_j$ and the corresponding representations $\boldsymbol{X}_{j,\mathcal{I}_j}$ as variables. The reason for such a strategy is that the approximation is better than in the case where only the atom is optimized and so it is likely that the next atoms will be more adequate to the representation; even the current atom has a different value than when it is optimized alone. Of course, the sparsity pattern (and the coefficients) are modified in the sparse coding step, hence the coefficients optimized in the dictionary update serve only to produce better atoms.

Appealing again to the notations from Proposition 3.3, the problem and the solution are as follows.

Proposition 3.7 *Consider the problem*

$$\begin{aligned} \min_{\boldsymbol{d},\boldsymbol{x}} \quad & \left\| \boldsymbol{F} - \boldsymbol{d}\boldsymbol{x}^T \right\|_F^2 \\ \text{s.t.} \quad & \|\boldsymbol{d}\| = 1 \end{aligned} \tag{3.33}$$

Let the singular value decomposition (SVD) of the matrix $\boldsymbol{F}$ *be*

$$\boldsymbol{F} = \sum_{i=1}^{r} \sigma_i \boldsymbol{u}_i \boldsymbol{v}_i^T, \tag{3.34}$$

where σ_i, $\boldsymbol{u}_i$, $\boldsymbol{v}_i$ *are the singular values, left singular vectors and right singular vectors, respectively, and* r *is the rank. Then, the solution to problem (3.33) is*

$$\boldsymbol{d} = \boldsymbol{u}_1, \quad \boldsymbol{x} = \sigma_1 \boldsymbol{v}_1. \tag{3.35}$$

Proof The problem (3.33) can be viewed as the approximation of the matrix $\boldsymbol{F}$ with a rank-1 matrix. The optimal solution is given by the truncation of the SVD to a single term, that corresponding to the largest singular value, so

$$\boldsymbol{d}\boldsymbol{x}^T = \sigma_1 \boldsymbol{u}_1 \boldsymbol{v}_1^T.$$

The multiplicative uncertainty is resolved by normalizing the atom (remind that the singular vectors have norm equal to 1) and so results (3.35). ■

The K-SVD dictionary update is detailed as Algorithm 3.6. It is clear that K-SVD is costlier than, e.g., SGK, due to the SVD computation that is significantly more expensive than a simple matrix–vector multiplication. However, K-SVD needs only the first singular vectors, which can be computed much easier than the full SVD. We will discuss later in this section how they can be computed efficiently in connection with the power method. Note that specialized implementations exist in many languages and libraries. For example, in Matlab, the function `svds` can be used for line 4 of Algorithm 3.6 instead of the general function `svd`.

Approximate K-SVD A simple solution for avoiding the (even though partial) SVD computation but keeping the idea of updating the representations is to

Algorithm 3.6: K-SVD Dictionary Update

Data: signals set $\boldsymbol{Y} \in \mathbb{R}^{m \times N}$
current dictionary $\boldsymbol{D} \in \mathbb{R}^{m \times n}$
representation matrix $\boldsymbol{X} \in \mathbb{R}^{n \times N}$
Result: updated dictionary $\boldsymbol{D}$

1 Compute error $\boldsymbol{E} = \boldsymbol{Y} - \boldsymbol{DX}$
2 **for** $j = 1$ **to** n **do**
3 Modify error: $\boldsymbol{F} = \boldsymbol{E}_{\mathcal{I}_j} + \boldsymbol{d}_j \boldsymbol{X}_{j,\mathcal{I}_j}$
4 Compute first singular value σ_1 of $\boldsymbol{F}$ and associated singular vectors $\boldsymbol{u}_1$ and $\boldsymbol{v}_1$
5 Update atom and representation: $\boldsymbol{d}_j = \boldsymbol{u}_1, \ \ \boldsymbol{X}_{j,\mathcal{I}_j} = \sigma_1 \boldsymbol{v}_1^T$
6 Recompute error: $\boldsymbol{E}_{\mathcal{I}_j} = \boldsymbol{F} - \boldsymbol{d}_j \boldsymbol{X}_{j,\mathcal{I}_j}$

optimize first the atom and *then* the associated representations, not simultaneously like in K-SVD. This amounts to splitting the block coordinate descent on both atom and representation, solved by Proposition 3.7, into two successive smaller coordinate descent blocks, on the atom first and next on the representation. For atom optimization we already have the formulation given by Proposition 3.4. The optimal representation when everything else is fixed can be obtained similarly.

Proposition 3.8 *The solution to the problem*

$$\min_{\boldsymbol{x}} \quad \left\| \boldsymbol{F} - \boldsymbol{d}\boldsymbol{x}^T \right\|_F^2 \tag{3.36}$$

is

$$\boldsymbol{x} = \boldsymbol{F}^T \boldsymbol{d}. \tag{3.37}$$

Proof Using the fact that $\|\boldsymbol{d}\| = 1$ in (3.23) gives

$$\left\| \boldsymbol{F} - \boldsymbol{d}\boldsymbol{x}^T \right\|_F^2 = \|\boldsymbol{F}\|_F^2 - 2\boldsymbol{d}^T \boldsymbol{F}\boldsymbol{x} + \boldsymbol{x}^T \boldsymbol{x}. \tag{3.38}$$

Equating with zero the derivative with respect to $\boldsymbol{x}$ of this quadratic function gives (3.37). ■

This is the idea behind Approximate K-SVD (AK-SVD) [156], whose dictionary update step is presented as Algorithm 3.7. As announced, the atom $\boldsymbol{d}_j$ and the representations $\boldsymbol{X}_{j,\mathcal{I}_j}$ are optimized successively. We can add one more twist to the algorithm: the same atom and representations can be successively optimized several times, more precisely for I iterations. The current atom is affected by the improvement of its representation coefficients only if $I > 1$, unlike the next atoms

Algorithm 3.7: AK-SVD Dictionary Update

Data: signals set $\boldsymbol{Y} \in \mathbb{R}^{m \times N}$
current dictionary $\boldsymbol{D} \in \mathbb{R}^{m \times n}$
representation matrix $\boldsymbol{X} \in \mathbb{R}^{n \times N}$
number of inner iterations $I \in \mathbb{N}$, $I \geq 1$
Result: updated dictionary $\boldsymbol{D}$

1 Compute error $\boldsymbol{E} = \boldsymbol{Y} - \boldsymbol{D}\boldsymbol{X}$
2 **for** $j = 1 : n$ **do**
3   Modify error: $\boldsymbol{F} = \boldsymbol{E}_{\mathcal{I}_j} + \boldsymbol{d}_j \boldsymbol{X}_{j,\mathcal{I}_j}$
4   **for** $i = 1 : I$ **do**
5   Update atom: $\boldsymbol{d}_j = \boldsymbol{F}\boldsymbol{X}^T_{j,\mathcal{I}_j} / \|\boldsymbol{F}\boldsymbol{X}^T_{j,\mathcal{I}_j}\|$
6   Update representation: $\boldsymbol{X}^T_{j,\mathcal{I}_j} = \boldsymbol{F}^T \boldsymbol{d}_j$
7   Recompute error: $\boldsymbol{E}_{\mathcal{I}_j} = \boldsymbol{F} - \boldsymbol{d}_j \boldsymbol{X}_{j,\mathcal{I}_j}$

which are affected even if $I = 1$. For fixed representation, atom update is done by applying (3.24); this choice is in the spirit of K-SVD and unlike SGK: the optimal atom has unit norm. When the (new) atom is fixed, the optimal representations are obtained from (3.37), i.e.,

$$\boldsymbol{X}^T_{j,\mathcal{I}_j} = \boldsymbol{F}^T \boldsymbol{d}_j. \tag{3.39}$$

The typical use of AK-SVD is with $I = 1$. In this case, the number of operations is $O(smN)$, like in SGK, but the complexity of the dictionary update is actually twice higher, since there are two matrix–vector multiplications. The fact that, ignoring the inner loop from line 4, AK-SVD (Algorithm 3.7) has a single line more than SGK (Algorithm 3.5), so it can be seen as an ulterior improvement, may be misleading: in fact AK-SVD was proposed a few years before SGK. Note also that an inner loop makes no sense for SGK.

Both K-SVD and AK-SVD guarantee the decrease of the error in the dictionary update step. While in K-SVD the decrease is maximal, AK-SVD only progresses towards the minimum. In other words, the coordinate descent (on the atom and its representation coefficients) is optimal for K-SVD and suboptimal for AK-SVD. However, in practice it was often remarked that AK-SVD needs only a slightly larger number of iterations K in Algorithm 3.1 to reach the same error as K-SVD.

Remark 3.9 (Connection with the Power Method) The left and right singular vectors of the matrix $\boldsymbol{F}$ that appear in (3.35) are eigenvectors of the positive semidefinite matrices $\boldsymbol{F}^T\boldsymbol{F}$ and $\boldsymbol{F}\boldsymbol{F}^T$, respectively. Since they are associated with the largest singular value of $\boldsymbol{F}$, they are associated with the largest eigenvalues of $\boldsymbol{F}^T\boldsymbol{F}$ and $\boldsymbol{F}\boldsymbol{F}^T$. As such, they could be computed with the power method. Moreover, especially near convergence (of the DL algorithm), the matrix $\boldsymbol{F}$ could be expected to be well approximated by a rank-1 matrix (i.e., $\boldsymbol{d}\boldsymbol{x}^T$) and hence have a dominant singular value; this would make the power method efficient.

With our usual light notation and ignoring normalization only for the sake of simpler formulas, the iteration $i = 1$ of AK-SVD starts with $\boldsymbol{d} \leftarrow \boldsymbol{F}\boldsymbol{x}$. Then, in the same iteration, we obtain $\boldsymbol{x} \leftarrow \boldsymbol{F}^T\boldsymbol{d} = \boldsymbol{F}^T\boldsymbol{F}\boldsymbol{x}$ (where the right-hand $\boldsymbol{x}$ is the current representation), which is an iteration of the power method for $\boldsymbol{F}^T\boldsymbol{F}$; so, when the process is repeated (i grows), $\boldsymbol{x}$ converges in direction to the vector $\boldsymbol{v}_1$ from (3.35). Moreover, in iteration 2 we have $\boldsymbol{d} \leftarrow \boldsymbol{F}\boldsymbol{x} = \boldsymbol{F}\boldsymbol{F}^T\boldsymbol{d}$; this is an iteration of the power method for the matrix $\boldsymbol{F}\boldsymbol{F}^T$; hence (with normalization) $\boldsymbol{d}$ converges to the vector $\boldsymbol{u}_1$ from (3.35).

We conclude that, for large enough I, AK-SVD is equivalent with K-SVD. In fact, even a few iterations can be enough, due to the good initializations that are the current values of the atom and representations in almost all stages of the DL process, excepting the first few. Even with $I = 1$, which means performing only half of a power method iteration for an atom and a full one for the representation, AK-SVD has a good behavior. ■

3.6 Parallel Algorithms

While MOD is intrinsically parallel (with the limitations discussed in Sect. 3.4), K-SVD and SGK have a clear sequential character. Updating the atoms one by one in full sweeps is a strategy that borrows the Gauss–Seidel name from the iterative method for solving linear systems. Once an atom is updated, it is immediately used for modifying subsequent atoms, since the representation error is changed according to its new value. An opposite strategy, named Jacobi, is to change all atoms in parallel [101], using the same error matrix $\boldsymbol{E}$ for all updates. The obvious computational advantage is that now all atom updates can be made in parallel. The downside is that all atoms are updated using their versions from the previous DL iteration; hence, the Jacobi strategy leads usually to somewhat slower convergence than Gauss–Seidel.

Algorithm 3.8 illustrates the idea for the SGK algorithm (extension to K-SVD or AK-SVD [98] is immediate). We name PSGK the resulting algorithm; as a general rule, we prepend the letter "P" to an algorithm name whenever atom update is made in parallel. We allow flexibility in the level of parallelism by letting $\tilde{n}$ atoms to be updated simultaneously, where $\tilde{n}$ can take a (preset) value from 1 (which is in fact the sequential strategy) to n (which is the fully parallel version). For the ease of presentation we assume that $\tilde{n}$ divides n. The overall computational burden is slightly higher, since the error has to be recomputed for each group of $\tilde{n}$ atoms that are updated together. However, computing $\boldsymbol{E}$ is a naturally parallel operation. The computation of the matrices $\boldsymbol{F}$ corresponding to the atoms from the same group can be done in parallel, as well as the update of the atom (steps 4 and 5 in Algorithm 3.8).

Algorithm 3.8: PSGK (Parallel SGK) Dictionary Update

Data: signals set $\boldsymbol{Y} \in \mathbb{R}^{m \times N}$
current dictionary $\boldsymbol{D} \in \mathbb{R}^{m \times n}$
representation matrix $\boldsymbol{X} \in \mathbb{R}^{n \times N}$
number of parallel atoms $\tilde{n}$
Result: updated dictionary $\boldsymbol{D}$

```
for ℓ = 1 to n/ñ do
    E = Y − DX
    for j = (ℓ − 1)ñ + 1 to ℓñ, in parallel do
        Modify error: F = E_{I_j} + d_j X_{j,I_j}
        Update atom: d_j = F X^T_{j,I_j} / (X_{j,I_j} X^T_{j,I_j})
        Optionally, normalize atom: d_j ← d_j / ‖d_j‖
```

So, it is clear that, on a parallel architecture, Algorithm 3.8 can run significantly faster than its sequential version from Algorithm 3.5.

Of course, the parallel and sequential version give different results; in fact, a distinct result can be obtained for each value of $\tilde{n}$. Since several atoms are updated simultaneously, the parallel algorithm gives no guarantee that the representation error decreases; although each atom is computed in an optimal manner if all other atoms are fixed, modifying several atoms at the same time in this manner may have unpredictable results, in general. However, what usually happens is that the error essentially decreases, although convergence is more erratic than for sequential algorithms; we will present later some numerical evidence, but the impatient reader can jump forward to Figs. 3.5 and 3.6. Surprisingly, for some problems, parallel algorithms are not only faster but also more effective in producing a low error value; examples will be later provided in Sect. 3.11.

Remark 3.10 (Fully Parallel SGK and AK-SVD Are Identical) When $\tilde{n} = n$, the fact that the atoms are normalized or not immediately after they are computed has no significance. Since each atom is computed using the atoms from the previous DL iteration, only their direction is important. (Normalization is performed anyway at the end of the DL iteration.) Since (3.24) and (3.30) give atoms with the same direction, the atoms produced by the fully parallel versions of SGK and AK-SVD (with $I = 1$) are identical. Moreover, the computation of the representations is useless in AK-SVD in this case, since they are not used for modifying the atoms. Of course, when $I > 1$, the above affirmations are no longer true; however, it was noticed that the parallel AK-SVD has no benefit from such a choice: the convergence is in fact slower or the algorithm may even diverge.

Similarly, the representations computed by the fully parallel K-SVD are useless. However, the atoms are different from those in SGK and AK-SVD due to the different way of computation. Hence, fully parallel K-SVD is a distinct algorithm. ■

3.7 SimCO

Simultaneous Codeword Optimization (SimCO) [44] is a DL algorithm that views the dictionary update differently than all previous algorithms by taking a different interpretation of the objective function. In its simplest version, the algorithm uses gradient search; unlike Sparsenet, the step is optimal and the search is made on the Stiefel manifold of *normalized* dictionaries.

Given the nonzero pattern Ω, fixed as usual in the dictionary update step, we define the error function

$$\begin{aligned} f_X(\boldsymbol{D}) = \min_{\boldsymbol{X}} \quad & \|\boldsymbol{Y} - \boldsymbol{D}\boldsymbol{X}\|_F^2 \\ \text{s.t.} \quad & \boldsymbol{X}_{\Omega^c} = 0 \end{aligned} \tag{3.40}$$

This function is a particular case of (3.10), with only the dictionary $\boldsymbol{D}$ as variable. For each dictionary, the representation $\boldsymbol{X}$ is the optimal one, given the sparsity pattern; it could be computed by solving the least-squares problem involved by each signal representation with fixed atoms, just as OMP computes it after selecting the relevant atoms. If the atoms used for the representation of signal $\boldsymbol{y}_\ell$ have indices in the set $\mathcal{S}_\ell$, that is

$$\boldsymbol{y}_\ell \approx \sum_{j \in \mathcal{S}_\ell} x_{j\ell} \boldsymbol{d}_j = \boldsymbol{D}_{\mathcal{S}_\ell} \boldsymbol{X}_{\mathcal{S}_\ell, \ell}, \tag{3.41}$$

then the optimal representation is

$$\boldsymbol{X}_{\mathcal{S}_\ell, \ell} = \left(\boldsymbol{D}_{\mathcal{S}_\ell}^T \boldsymbol{D}_{\mathcal{S}_\ell}\right)^{-1} \boldsymbol{D}_{\mathcal{S}_\ell}^T \boldsymbol{y}_\ell. \tag{3.42}$$

This relation holds if the matrix $\boldsymbol{D}_{\mathcal{S}_\ell}$ has maximum rank. Otherwise, there may be multiple representations $\boldsymbol{X}$ that are optimal in (3.40). We will discuss in the next chapter the difficulties raised by this non-uniqueness and show that regularization can avoid such situations.

Using definition (3.40), it is clear that the dictionary update problem (2.9) (with explicit atom normalization) is equivalent to

$$\begin{aligned} \min_{\boldsymbol{D}} \quad & f_X(\boldsymbol{D}) \\ \text{s.t.} \quad & \|\boldsymbol{d}_j\| = 1, \ j = 1 : n \end{aligned} \tag{3.43}$$

SimCO is based on classical first and second order nonlinear optimization methods to solve (3.43); note that the function (3.40) is not convex. We present only the gradient method.

The gradient of the function (3.40) is given by the usual relation (3.11), computed for the *optimal* $\boldsymbol{X}$ given by (3.42). The optimal representations are given by OMP (the nonzero pattern may be not optimal, however), so there is no need to compute them if only one gradient step is performed. For subsequent steps, the optimal $\boldsymbol{X}$ needs to be explicitly computed.

The line search is optimal (using, e.g., golden ratio) and is done on the Stiefel manifolds defined by the unit norm condition $\|\boldsymbol{d}_j\| = 1$, $j = 1 : n$. Let $\boldsymbol{G}$ be the gradient of (3.40) in the current value of $\boldsymbol{D}$ and the corresponding optimal representations $\boldsymbol{X}$. The usual line search would operate with dictionaries $\boldsymbol{D} - t\boldsymbol{G}$, trying to find a best t; then, the dictionary has to be normalized. SimCO searches only over normalized dictionaries. Let $\boldsymbol{d}_j$ be an atom and $\boldsymbol{g}_j$ the corresponding gradient. The component of the gradient that is orthogonal on the atom has the expression

$$\bar{\boldsymbol{g}}_j = \boldsymbol{g}_j - \boldsymbol{d}_j\boldsymbol{d}_j^T\boldsymbol{g}_j.$$

(The component on the direction of the atom is useless, because the norm of the atom cannot change.) The "line" search is on dictionaries with atoms of the form

$$\boldsymbol{d}_j(t) = \boldsymbol{d}_j \cos\left(\|\bar{\boldsymbol{g}}_j\|\, t\right) - \frac{\bar{\boldsymbol{g}}_j}{\|\bar{\boldsymbol{g}}_j\|} \sin\left(\|\bar{\boldsymbol{g}}_j\|\, t\right), \tag{3.44}$$

where $t > 0$ is the step size parameter. It can be checked immediately that $\|\boldsymbol{d}_j(t)\| = 1$. So, the search is actually performed on a big circle of a hypersphere. Note that the step size is approximately proportional with $\|\bar{\boldsymbol{g}}_j\|$ for small t, since $\sin(\|\bar{\boldsymbol{g}}_j\|t) \approx \|\bar{\boldsymbol{g}}_j\|t$. However, the proportionality (always holding for regular line search) is lost for larger t. (So, the advance on different atoms is not proportional with the magnitude of their respective gradients, as it would be for a standard gradient method.)

The search can be made on one atom or on more, simultaneously. SimCO can be as well a sequential or a parallel algorithm. However, unlike the parallel algorithms given in the previous section, SimCO guarantees the decrease of the objective by working with a single descent step t for all the atoms that are updated. Since the step size is optimal, the error always decreases.

For the sake of simplicity, we present in Algorithm 3.9 the version of SimCO that updates all atoms simultaneously. (Doing the same operations on groups of $\tilde{n}$ atoms, iteratively, like in Algorithm 3.8, should be a simple exercise for the reader.) The algorithm performs I inner iterations consisting each of a gradient descent with optimal step, as described above. Note that the optimal representations for the given dictionary are computed in lines 13–14 (and available in the beginning from the sparse coding step), hence the function (3.40) is indeed computed. The line search algorithm is not specified; the simple and robust *golden section* is used in [44], but other algorithms could be used as well.

The quest for optimal step size comes at a relatively large cost: the operation count for the new dictionary computation in lines 9–11 is only $O(mn)$, but rises to $O(s^2mN)$ for the optimal representations (lines 12–14) and is $O(smN)$ for the error norm value (line 15). This is already more than for the whole dictionary

Algorithm 3.9: SimCO Dictionary Update

Data: signals set $\boldsymbol{Y} \in \mathbb{R}^{m \times N}$
current dictionary $\boldsymbol{D} \in \mathbb{R}^{m \times n}$
representation matrix $\boldsymbol{X} \in \mathbb{R}^{n \times N}$ (we denote $\mathcal{S}_\ell$ the set of row indices of the nonzeros on column $\ell = 1 : N$)
number of inner iterations $I \in \mathbb{N}$, $I \geq 1$
Result: updated dictionary $\boldsymbol{D}$

1 **for** $i = 1$ **to** I **do**
2 Compute error for current dictionary (and optimal representations): $\boldsymbol{E} = \boldsymbol{Y} - \boldsymbol{D}\boldsymbol{X}$
3 Compute gradient: $\boldsymbol{G} = -2\boldsymbol{E}\boldsymbol{X}^T$
4 **for** $j = 1$ **to** n **do**
5 Compute component of gradient orthogonal on atom: $\boldsymbol{g}_j \leftarrow \boldsymbol{g}_j - \boldsymbol{d}_j \boldsymbol{d}_j^T \boldsymbol{g}_j$
6 Compute error norm: $f = \|\boldsymbol{E}\|_F$
7 Set step size t to initial value
8 **while** *t is not optimal* **do**
9 Compute tentative new dictionary:
10 **for** $j = 1$ **to** n **do**
11 $\tilde{\boldsymbol{d}}_j(t) = \boldsymbol{d}_j \cos(\|\bar{\boldsymbol{g}}_j\| t) - \frac{\bar{\boldsymbol{g}}_j}{\|\bar{\boldsymbol{g}}_j\|} \sin(\|\bar{\boldsymbol{g}}_j\| t)$
12 Compute optimal representations:
13 **for** $\ell = 1$ **to** N **do**
14 $\boldsymbol{X}_{\mathcal{S}_\ell,\ell} = \left(\tilde{\boldsymbol{D}}_{\mathcal{S}_\ell}^T \tilde{\boldsymbol{D}}_{\mathcal{S}_\ell}\right)^{-1} \tilde{\boldsymbol{D}}_{\mathcal{S}_\ell}^T \boldsymbol{y}_\ell$
15 Compute error norm: $\tilde{f} = \left\|\boldsymbol{Y} - \tilde{\boldsymbol{D}}\boldsymbol{X}\right\|_F$
16 **if** $\tilde{f} < f$ **then**
17 Save dictionary $\tilde{\boldsymbol{D}}$, representations $\boldsymbol{X}$ and put $f = \tilde{f}$
18 Set step size t according to line search algorithm
19 Restore optimal dictionary and representations into $\boldsymbol{D}$, $\boldsymbol{X}$

update in SGK or AK-SVD. This count should be multiplied with typically tens of evaluations for the line search (the while loop, line 8) and with the number I of inner iterations (which may go up to 50 in the examples from [44]). So, SimCO can be used especially for small datasets.

3.8 Refinements

We have seen in Sect. 3.6 that the idea of parallel atom updates can be applied similarly to several algorithms. We present here other refinements applicable to the algorithms presented in this chapter.

Objective Approximation Using Differences A particularly successful idea [167] is to change the optimization objective by using a first order approximation obtained by expressing the next dictionary and representations in terms of differences with respect to their current values. Let us denote $(\boldsymbol{D}_0, \boldsymbol{X}_0)$ the current values of the dictionary and representations and consider a point $(\boldsymbol{D}, \boldsymbol{X})$ in its neighborhood. Expressing this point in terms of differences

$$\begin{aligned} \boldsymbol{D} &= \boldsymbol{D}_0 + (\boldsymbol{D} - \boldsymbol{D}_0), \\ \boldsymbol{X} &= \boldsymbol{X}_0 + (\boldsymbol{X} - \boldsymbol{X}_0), \end{aligned}$$

and neglecting the product of the differences, assumed to be small, we obtain

$$\begin{aligned} \boldsymbol{D}\boldsymbol{X} &= \boldsymbol{D}_0\boldsymbol{X}_0 + \boldsymbol{D}_0(\boldsymbol{X} - \boldsymbol{X}_0) + (\boldsymbol{D} - \boldsymbol{D}_0)\boldsymbol{X}_0 + (\boldsymbol{D} - \boldsymbol{D}_0)(\boldsymbol{X} - \boldsymbol{X}_0) \\ &\approx \boldsymbol{D}\boldsymbol{X}_0 + \boldsymbol{D}_0\boldsymbol{X} - \boldsymbol{D}_0\boldsymbol{X}_0. \end{aligned} \tag{3.45}$$

Thus, ignoring for the moment all constraints, the DL problem (2.2) becomes

$$\min_{\boldsymbol{D},\boldsymbol{X}} \quad \|\boldsymbol{Y} + \boldsymbol{D}_0\boldsymbol{X}_0 - \boldsymbol{D}\boldsymbol{X}_0 - \boldsymbol{D}_0\boldsymbol{X}\|_F^2 \tag{3.46}$$

The important distinction between this and the original formulation is that now the problem is convex with respect to both $\boldsymbol{D}$ and $\boldsymbol{X}$ (when the support is fixed). Although the new objective is an approximation of the original one, the hypothesis that the variables are not changed very much in DL iterations holds especially in the later stages of the DL process and so the approximation is sufficiently good. With this formulation, the basic Algorithm 3.1 is changed as follows.

In sparse coding, the fact that the dictionary is fixed, and so we have $\boldsymbol{D} = \boldsymbol{D}_0$, makes (3.46) become identical with (2.2), so there is actually no change in the algorithm: OMP can be run as usual.

However, the dictionary update step is different. Putting

$$\boldsymbol{Z} = \boldsymbol{Y} + \boldsymbol{D}_0\boldsymbol{X}_0 - \boldsymbol{D}_0\boldsymbol{X}, \tag{3.47}$$

where $\boldsymbol{X}$ is the representation matrix that has been just computed, the problem to be solved is

$$\min_{\boldsymbol{D}} \quad \|\boldsymbol{Z} - \boldsymbol{D}\boldsymbol{X}_0\|_F^2 \tag{3.48}$$

Algorithm 3.10: NSGK Iteration

Data: signals set $\boldsymbol{Y} \in \mathbb{R}^{m \times N}$
current dictionary $\boldsymbol{D} \in \mathbb{R}^{m \times n}$
representation matrix $\boldsymbol{X} \in \mathbb{R}^{n \times N}$
Result: updated dictionary $\boldsymbol{D}$

1 Save current representation matrix: $\boldsymbol{X}_0 = \boldsymbol{X}$
2 Sparse coding: compute new representations $\boldsymbol{X}$ with OMP
3 Compute error $\boldsymbol{E} = \boldsymbol{Y} + \boldsymbol{D}\boldsymbol{X}_0 - 2\boldsymbol{D}\boldsymbol{X}$
4 **for** $j = 1$ **to** n **do**
5   Modify error: $\boldsymbol{F} = \boldsymbol{E}_{\mathcal{I}_j} + \boldsymbol{d}_j \boldsymbol{X}_{j,\mathcal{I}_j}$
6   Update atom: $\boldsymbol{d}_j = \boldsymbol{F}\boldsymbol{X}_{j,\mathcal{I}_j}^T / (\boldsymbol{X}_{j,\mathcal{I}_j} \boldsymbol{X}_{j,\mathcal{I}_j}^T)$
7   Optionally, normalize atom: $\boldsymbol{d}_j \leftarrow \boldsymbol{d}_j / \|\boldsymbol{d}_j\|$
8   Recompute error: $\boldsymbol{E}_{\mathcal{I}_j} = \boldsymbol{F} - \boldsymbol{d}_j \boldsymbol{X}_{j,\mathcal{I}_j}$

This is the usual dictionary update problem, only with different constant matrices, so all the algorithms that optimize only the dictionary in the update phase, like SGK and MOD can be adapted. We keep the terminology from [167], and prepend the word *new* to the name of an algorithm modified as described above (although the rules of scientific writing formally forbid the use of such a word in a name, for obvious reasons). The new SGK, abbreviated NSGK (same rule applies to the other algorithms), appears to be particularly successful. It was observed that better results are obtained when the (now) previous representation $\boldsymbol{X}_0$ is replaced by the latest one in (3.48). This is yet another approximation, justified by the good practical behavior.

A full iteration of NSGK is shown as Algorithm 3.10. Since we need explicitly the representations $\boldsymbol{X}_0$ from the previous iteration, we have to save it before the sparse coding step, which normally destroys the previous representations. Only line 3 deserves more explanations; note first that the current dictionary is $\boldsymbol{D}$, hence replacing $\boldsymbol{D}_0$ in (3.47); then, using $\boldsymbol{X}$ instead of $\boldsymbol{X}_0$ (as argued above), makes $\boldsymbol{E} = \boldsymbol{Z} - \boldsymbol{D}\boldsymbol{X}$, which gives line 3. The remainder of the algorithm is identical to SGK. Note that atom normalization is used in [167], unlike in the original SGK; we adopt the same option in our experiments. The number of operations for computing $\boldsymbol{E}$ appears to be doubled. However, the matrices $\boldsymbol{X}_0$ and $\boldsymbol{X}$ usually have sparsity patterns that are not very different; so, it is more efficient to compute first $\boldsymbol{X}_0 - 2\boldsymbol{X}$ (on columns) and then multiply with $\boldsymbol{D}$ (each column). Typically, the computation load is only slightly larger in the new algorithm.

Of course, one can combine the modification described above with the parallel (Jacobi) update of the atoms, obtaining an algorithm called PNSGK.

Dictionary Update Cycles A refinement that was already mentioned is to make several dictionary update steps for the same support in a single DL iteration. For a sequential DL algorithm, such as K-SVD, the dictionary update has the form listed

Algorithm 3.11: Dictionary Update in Several Cycles

Data: signals set $\boldsymbol{Y} \in \mathbb{R}^{m \times N}$
current dictionary $\boldsymbol{D} \in \mathbb{R}^{m \times n}$
representation matrix $\boldsymbol{X} \in \mathbb{R}^{n \times N}$
number of dictionary update cycles C
Result: updated dictionary $\boldsymbol{D}$

```
for c = 1 to C do
    for j = 1 to n do
        Update atom d_j and possibly representation X_{j,I_j}
```

in Algorithm 3.11. There are C sweeps, or dictionary update cycles, of the atoms; until now, all the algorithms have been written with $C = 1$. This approach can be seen as an attempt to better solve the dictionary update problem (2.9), maybe even find the optimal error value for the given support. While a single sweep guarantees only a decrease of the objective for most algorithms (for some, like the parallel ones, there is no guarantee at all), a sufficiently high C may almost reach the optimum.

It was reported in [168, 180] that a small value like $C = 2$ or $C = 3$ can increase the convergence speed, although often leading to the same final DL error value as the single sweep version. An explanation why a larger C is not useful, besides the associated high complexity, may be as follows. DL is a problem with many local minima. An optimal dictionary update decreases the objective more than what a single sweep can achieve, but may also bring the variables into a local minimum from which the sparse coding stage cannot extract them, in the sense that it cannot find a better sparsity pattern. So, the quest for too much local improvement, i.e., a large C, is not necessarily a winning strategy.

Without changing the above remarks, we need to stress, however, that Algorithm 3.11 is not necessarily optimal whatever the value of C is, due to the nonconvexity of (2.9).

The dictionary update cycles idea can be adapted to all algorithms, including MOD, the NSGK family, and the parallel versions. The reader is invited to explore this topic. For MOD, see problem **P** 3.6.

3.9 Practical Issues

We discuss here several issues that arise when implementing a DL algorithm. They are common to all algorithms presented in this chapter and refer to initialization, the selection of dictionary size and the supplementary operations that we make on atoms, for example when they are not used or are almost collinear.

3.9.1 Initialization

As already mentioned, the typical initialization methods are either to select randomly n signals that will serve as first atoms or to simply generate the atoms randomly; of course, in both cases we take care to normalize the atoms. Since the DL problem has multiple local minima and the algorithms are based on local optimization methods, it is expected that different initializations lead to different results. Like in all nonlinear optimization problems, if one can afford the time, running the same algorithm with several initializations, or more algorithms with several initializations, can only improve the final dictionary.

Figure 3.2 shows the evolution of the RMSE (2.10) for AK-SVD (chosen here only because of its speed), run with $m = 64$, $n = 128$, $s = 6$, for 50 iterations. The signals are $N = 4000$ image patches, obtained as described in Sect. 2.3.1. Each figure shows the RMSE for five different initializations, with random vectors at the left and with random signals at the right. In both cases one can see a certain expectable variability of the results, although the errors have similar values. The differences can be larger, depending on the signals and the dimensions of the problem; larger dimensions usually lead to a higher impact of the initialization. The error evolution is dictated not only by the initialization; we can see that the ranking of the errors may change during the DL process. Also, the initialization with random signals seems to lead, in this case, to a slightly faster convergence.

Since DL is a highly complex optimization problem, one can spend more time for the initialization than to quickly pick some signals or random vectors as atoms. In fact, initialization can be regarded as a mini-training process; we will present in Sect. 6.4 an initialization algorithm that builds an initial dictionary by taking advantage of more refined techniques than presented until now. The underlying idea is to initialize already with atoms that are more representative for the training set than mere samples.

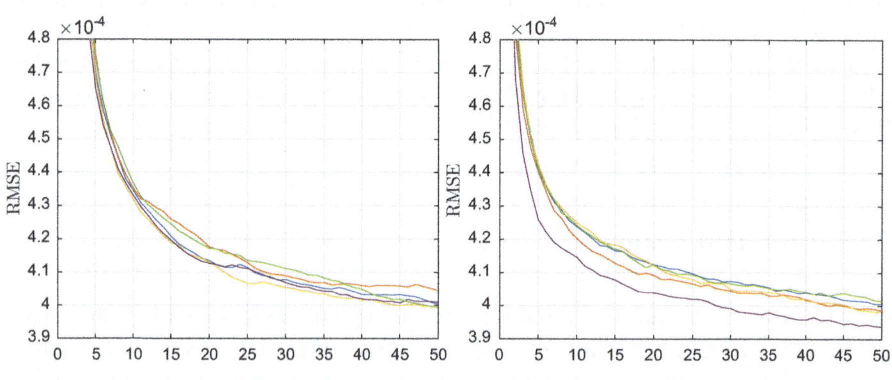

Fig. 3.2 Evolution of the RMSE in AK-SVD for several initializations with random vectors (left) and random signals (right)

3.9.2 Dictionary Size and Other Size Parameters

The size m of the signals is either given in some applications or, more often, is at designer's choice; the typical example for the latter case is when a long signal or a large image is windowed for processing. Usually, m is limited by the complexity of the signal processing operations, DL included. The number N of training samples is also easy to set: as large as we can afford in the DL process (where complexity is roughly proportional with N), if we are at liberty to choose; if not, we content ourselves with as many samples as available from the process at hand; note that N is the only size parameter that affects only the DL process and has no complexity influence in testing (or any other stage of the actual application). Finally, the last matrix size in the DL problem is the number n of atoms. This is actually an important parameter of the problem that seems totally at the designer's discretion. However, its selection is so significant that there are methods that automatically choose n; we postpone the discussion until Chap. 6, where several such methods are presented.

Another important parameter for whose choice there is no standard recipe is the sparsity level s. In some applications like compression, the value of s directly determines the result, since the compression rate depends on the number of coefficients to be stored and coded. There is also an indirect effect, due to the magnitude of the representation error; larger s means better error, hence the possibility to compress it with less bits (if we want lossless compression). So, there is a trade-off value s for which the compression rate is the best. The indirect effect is the only one in other applications like denoising, where the trade-off is harder to achieve. In some applications, keeping a fixed value of s for all signal representations may be impractical and more flexibility leads to better results. Methods with this flavor are presented in the first sections of Chap. 5. In general, the choice of s has a very empiric character.

3.9.3 Unused or Redundant Atoms

An event that occurs relatively often, but especially when there are few training signals (N is small), is that an atom is not used at all: no signal representation contains it. This may happen from the beginning (only for random initialization, since choosing signals as initial atoms ensures their use at least once) or during the DL algorithm. If the atom is $\boldsymbol{d}_j$, that means that $\mathcal{I}_j = \emptyset$; reminding Fig. 3.1, this means that the full jth row of $\boldsymbol{X}$ is zero. Of course, atom update formulas like (3.24) have no longer a meaning: the atom cannot be changed. Moreover, the MOD formula (3.31) cannot be used ad litteram, because $\boldsymbol{X}\boldsymbol{X}^T$ is now singular, but only after removing the zero row and thus computing only the other atoms. So, at the implementation of any DL algorithm, we need to provide code for detecting unused atoms and taking specific actions.

There are several possible actions in such a case, each with its merits.

- *Do nothing.* We can simply ignore the situation and hope that in next iterations the atom will become useful. However, this seldom happens and it is more likely that the atom stays unused.
- *Eliminate.* The completely opposite approach is to remove the atom. This outcomes to decreasing n with 1 and thus is an indirect and passive method to obtain an efficient number n of atoms.
- *Replace.* If we stick to the dictionary size chosen in the beginning, then we can simply replace the atom with a random vector, a signal or another vector. The first option is simple and usually effective. If we replace with a signal, then it makes sense to use one of the signals that is badly represented (in the sense that its representation error is large), since in the next DL iteration it will be perfectly represented. More sophisticated methods are possible: for example, the principal direction of a group of signals; this idea will be detailed later, in a different context; see Sect. 6.3.

Similar actions can be considered if $\|\boldsymbol{X}_{j,\mathcal{I}_j}\|$ is small, which means that, although the atom contributes to some representations, the corresponding coefficients are few or small and hence the atom is not significant. Also, a small $|\mathcal{I}_j|$ may be an alarm trigger, meaning that the atom is used by only few signals.

We present here a short experiment to see the effect of the atom replacement strategies. Figure 3.3 shows the evolution of the relative representation error

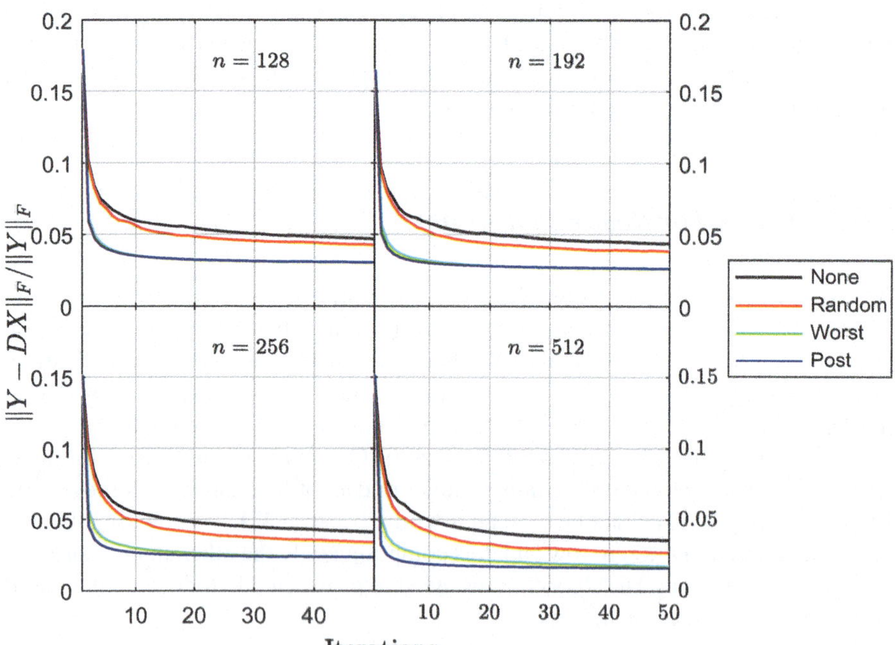

Fig. 3.3 Evolution of the relative representation error in AK-SVD for several atom replacement strategy

$$\frac{\|Y - DX\|_F}{\|Y\|_F} \tag{3.49}$$

averaged over ten runs of AK-SVD (similar results are obtained for other algorithms). The test signals are vectorized 8×8 patches from images, as described in Sect. 2.3.1. The patches are unprocessed, i.e., all their elements are positive. AK-SVD is initialized with random vectors and run with $N = 2048$, $s = 8$, and several dictionary sizes n. The replacement strategies are:

- "None": no replacement (which gives practically the same error as eliminating the atom).
- "Random": the unused atom is replaced with a random vector.
- "Worst": the unused atom is replaced with the worst represented signal, immediately after trying to update the atom (i.e., instead of step 4 in the AK-SVD Algorithm 3.7).
- "Post": after the whole dictionary update iteration, the unused atoms are replaced with the worst represented signals, the representation being computed with the updated atoms; so, an extra sparse coding step is performed.

The figure shows that, no matter the dictionary size, the "Random" strategy, which is the cheapest, brings an improvement over no replacement. The "Worst" strategy is even better and appears to be the most convenient, since, after a number of iterations, it gives the same results as "Post" but without any significant extra computation.

Interestingly, when the patches are centered (the overall average is subtracted from them) and hence the diversity of directions increases very much, there are no unused atoms and hence all replacement strategies give the same result, as shown in Fig. 3.4. This happens also in the experiments with random synthetic data. It is also notable that the relative error (3.49) is larger in this case, which suggests that it may be better to not preprocess the signals.

Another unpleasant event is when two atoms become very similar and both are used although only one could replace them without significant change in the representation error. In this case, the absolute inner product $|d_i^T d_j|$ is nearly equal to 1. The same actions as for unused atoms can be taken. Besides the lack of usefulness for one of the atoms, in this situation the mutual coherence (1.25) is high, which may impede on the quality of representation, in the sense that OMP is less able to find the sparsest representation. More generally, it is possible that a low number of atoms are linearly dependent. This limits the number of subspaces that a dictionary can cover and also destroys the RIP condition (1.27), which again may diminish the quality of representation. There is no cheap solution to this issue; some methods are discussed in [154].

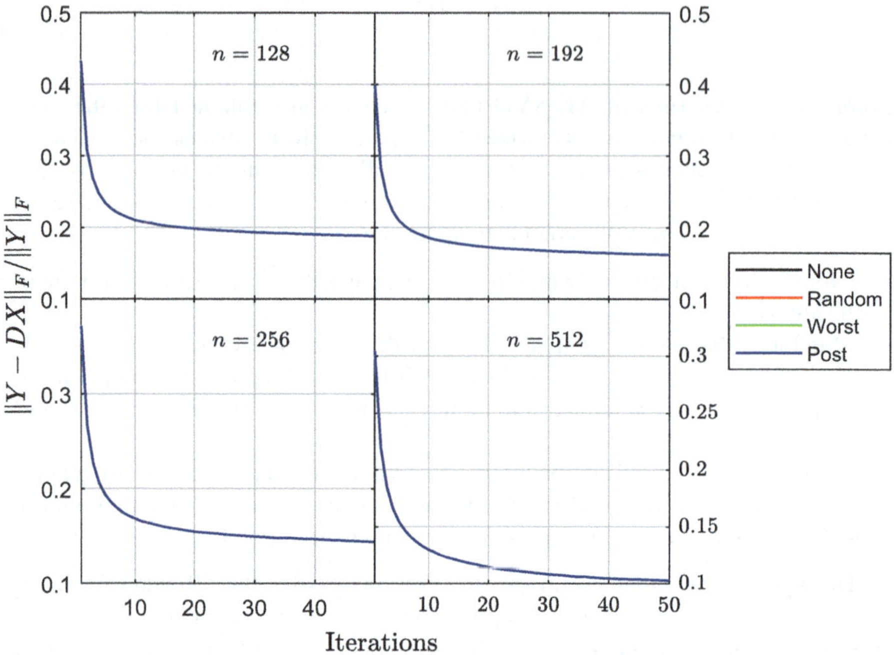

Fig. 3.4 Same as in Fig. 3.3, but with centered patches

3.9.4 Randomization

Although the algorithms presented in this chapter are generally efficient, a possibility of improvement of the algorithms that update the atoms sequentially is to sweep the atoms in a different order in each iteration. For example, this order can be generated by taking a random permutation of the set $1 : n$. At least in principle, changing the order may help convergence in coordinate descent algorithms, by removing the possibilities of stalls in error decrease due to an unfavorable order of coordinates.

3.10 Comparisons: Theory

Although numerical experiments are the most important in evaluating the ability of DL algorithms to produce low representation errors in a reasonable running time, we start with a theoretical analysis that covers an important convergence issue. The source is [124], which analyzes Sparsenet, MOD and K-SVD.

We analyze the dictionary update problem in its general form (2.9), which is the DL problem (2.2) when the support Ω of the representations $\boldsymbol{X}$ is fixed. The discrete nature of the support is not amenable to an analysis of the local behavior of an algorithm. Reminding the expressions (3.11) and (3.15) of the gradients of the error function f defined in (3.10) with respect to the dictionary $\boldsymbol{D}$ and the representations $\boldsymbol{X}$, the *critical point* of the function f is $(\boldsymbol{D}, \boldsymbol{X})$ where both gradients are zero. The gradient (3.15) should be zero only on the support Ω, not everywhere. So, a critical point satisfies the conditions

$$\boldsymbol{E}\boldsymbol{X}^T = 0, \;\; [\boldsymbol{D}^T\boldsymbol{E}]_\Omega = 0, \tag{3.50}$$

where $\boldsymbol{E} = \boldsymbol{Y} - \boldsymbol{D}\boldsymbol{X}$ is the representation error (2.4).

We investigate *fixed points* $(\boldsymbol{D}, \boldsymbol{X})$ of dictionary update algorithms; if an algorithm reaches such a point, it cannot move from it. (Of course, the sparse coding step can find a different representation, and thus move from the fixed point, but only by changing the support Ω.) If the point is a local minimum, this may be a positive outcome (unless the error value is poor, i.e., much higher than the globally optimal one); otherwise, this is not a desired feature of the algorithm; ideally, all fixed points should be local minima. We look first at the sparse coefficients update (with no support change). For a signal $\boldsymbol{y}_\ell$, the optimal coefficients are given by (3.42), which can be written

$$\boldsymbol{D}_{\mathcal{S}_\ell}^T(\boldsymbol{y}_\ell - \boldsymbol{D}_{\mathcal{S}_\ell}\boldsymbol{X}_{\mathcal{S}_\ell,\ell}) = 0, \tag{3.51}$$

which, for all signals, is equivalent to the second critical point condition from (3.50). Since OMP uses (3.42), all the DL algorithms from this chapter have fixed points that coincide with the critical points of f, when only representation is considered.

When considering also the dictionary update, the situation is different.

Proposition 3.11 *All fixed points of Sparsenet are critical points and conversely.*

Proof If $(\boldsymbol{D}, \boldsymbol{X})$ is a critical point, then the gradient is zero and hence the Sparsenet update (3.16) does not modify the atom.

If $(\boldsymbol{D}, \boldsymbol{X})$ is a fixed point, then (3.16) reads

$$\boldsymbol{d}_j + \alpha(\boldsymbol{Y} - \boldsymbol{D}\boldsymbol{X})(\boldsymbol{x}_j^T)^T = \beta\boldsymbol{d}_j,$$

for some scalar $\beta \neq 0$. The updated atom has the same direction as the current one, so they are identical after normalization. By multiplying the above relation with $\boldsymbol{d}_j^T$ and taking into account that $\boldsymbol{d}_j^T\boldsymbol{E} = 0$ (due to the representation property of fixed points), it results that $\beta = 1$. So, considering all $j = 1 : n$, it results that $\boldsymbol{E}\boldsymbol{X}^T = 0$, hence $(\boldsymbol{D}, \boldsymbol{X})$ is a critical point. ■

Proposition 3.12 *Assuming that the representation matrix $\boldsymbol{X}$ has full rank, all fixed points of MOD are critical points and conversely.*

Proof Let us notice first that the full rank assumption is very mild and it is extremely likely to be satisfied in practice.

If $(\boldsymbol{D}, \boldsymbol{X})$ is a critical point, then the gradient is zero, which is exactly the condition that gives the MOD optimal dictionary (3.31). Hence, the dictionary cannot change if the gradient is already zero.

If $(\boldsymbol{D}, \boldsymbol{X})$ is a fixed point, then it should satisfy

$$\boldsymbol{Y}\boldsymbol{X}^T(\boldsymbol{X}\boldsymbol{X}^T)^{-1} = \boldsymbol{D}\boldsymbol{\Lambda},$$

where $\boldsymbol{\Lambda} = \text{diag}(\lambda_j)$ is a diagonal matrix. That means that, after normalization, the new dictionary is identical with the current one. We need to show that $\boldsymbol{\Lambda} = \boldsymbol{I}$ and hence the gradient is zero. A short reasoning is at follows. Replacing $\boldsymbol{Y} = \boldsymbol{E} + \boldsymbol{D}\boldsymbol{X}$ in the above relation we obtain

$$\boldsymbol{E}\boldsymbol{X}^T + \boldsymbol{D}\boldsymbol{X}\boldsymbol{X}^T = \boldsymbol{D}\boldsymbol{\Lambda}\boldsymbol{X}\boldsymbol{X}^T.$$

Multiplying with $\boldsymbol{D}^T$ and taking into account that the second gradient equality from (3.50) holds, it results that

$$\boldsymbol{D}^T\boldsymbol{D}\boldsymbol{X}\boldsymbol{X}^T = \boldsymbol{D}^T\boldsymbol{D}\boldsymbol{\Lambda}\boldsymbol{X}\boldsymbol{X}^T,$$

which, due to the full rank of $\boldsymbol{X}$, simplifies to

$$\boldsymbol{D}^T\boldsymbol{D} = \boldsymbol{D}^T\boldsymbol{D}\boldsymbol{\Lambda}.$$

Multiplying with the unit vector $\boldsymbol{e}_j$, we get

$$\boldsymbol{D}^T\boldsymbol{d}_j = \lambda_j\boldsymbol{D}^T\boldsymbol{d}_j,$$

which has the unique solution $\lambda_j = 1$, because the vector $\boldsymbol{D}^T\boldsymbol{d}_j$ is nonzero. ■

For K-SVD the situation is different.

Proposition 3.13

i) A fixed point $(\boldsymbol{D}, \boldsymbol{X})$ of K-SVD satisfies the condition (3.51) and each atom $\boldsymbol{d}_j$ is the left singular vector associated with the largest singular value of the corresponding matrix (3.19).

ii) A critical point of the function f defined in (3.10) satisfies the condition (3.51) and each atom $\boldsymbol{d}_j$ is a left singular vector of the corresponding matrix (3.19).

Proof

Part i) is relatively clear, as the stated conditions define the new atom and representation. If they already satisfy the conditions, they will not change.

Part ii) is proved in [124]. ■

Corollary 3.14 *All fixed points of K-SVD are fixed points of Sparsenet and MOD, but the opposite is not true.*

Proof Since all fixed points of Sparsenet and MOD are critical points of f, it results that the atoms may be any singular vector of (3.19). For K-SVD, this is true only for the atoms that are singular vectors associated with the largest eigenvalue. So, the fixed points of K-SVD are a subset of the fixed points of Sparsenet and MOD. ■

Example 3.15 Consider $m = 2$, $N = 2$, $n = 1$ and

$$\boldsymbol{Y} = \begin{bmatrix} 2 & 2 \\ -1 & 1 \end{bmatrix}, \quad \boldsymbol{D} = \begin{bmatrix} 0 \\ 1 \end{bmatrix}, \quad \boldsymbol{X} = \begin{bmatrix} -1 & 1 \end{bmatrix}.$$

The representation error is

$$\boldsymbol{E} = \boldsymbol{Y} - \boldsymbol{DX} = \begin{bmatrix} 2 & 2 \\ 0 & 0 \end{bmatrix}, \quad \|\boldsymbol{E}\|_F = \sqrt{8}.$$

One can check immediately that the criticality conditions (3.50) are met, hence this is a fixed point for Sparsenet and MOD. The matrix (3.19) is $\boldsymbol{F} = \boldsymbol{E}$, since there is a single atom (which enters all representations). Its left singular vectors are the eigenvectors of

$$\boldsymbol{F}\boldsymbol{F}^T = \begin{bmatrix} 8 & 0 \\ 0 & 0 \end{bmatrix}.$$

Taking the one associated with the maximum eigenvalue and then updating the representation via (3.39) gives

$$\boldsymbol{D} = \begin{bmatrix} 1 \\ 0 \end{bmatrix}, \quad \boldsymbol{X} = \begin{bmatrix} 2 & 2 \end{bmatrix}, \quad \boldsymbol{E} = \begin{bmatrix} 0 & 0 \\ -1 & 1 \end{bmatrix}, \quad \|\boldsymbol{E}\|_F = \sqrt{2}.$$

So, K-SVD finds a better dictionary (which is also optimal, in this case).

We note that also SGK and AK-SVD are in a tight spot, as the new atoms computed through (3.24) or (3.30) are zero. In this case, not treated in the principle versions of the algorithms, it is natural to leave the atom unchanged. So, this is a fixed point also for SGK and AK-SVD.

Of course, the example is somewhat artificial, in the sense that the initial atom is in fact the worst that could be chosen: the representation error is maximum. However, it illustrates Corollary 3.14, showing that indeed K-SVD has fewer fixed points than the other algorithms. ■

A practical conclusion of the above discussion is that one can try to run K-SVD after the other algorithms in the hope of achieving better results. It was indeed observed in [124] that Sparsenet followed by K-SVD gives better results than any other combination of algorithms and also better than K-SVD alone.

We note that the results and conclusions presented in this section stand also for the parallel versions of the algorithms: a fixed point means that no individual atom update can succeed, hence it does not matter if the updates are made sequentially or in parallel.

3.11 Comparisons: Some Experimental Results

We present here a few results obtained by running some of the algorithms presented in this chapter on the test problems described in Sect. 2.3 and also on denoising (Sect. 2.4.1).

To be relevant, the tests must be conducted on problems as diverse as possible, with different sets of parameters m, n, s, and N, and with different types of data. Since the data have a more or less random nature, in each case we should run the algorithms on several instances of data of the same type and report averages. We present only a limited amount of results, hence our conclusions must be taken with a grain of salt.

When comparing DL algorithms we must follow a few simple fairness rules. We ensure that, for all algorithms, we use

- the same training signals
- the same initialization
- the same number of iterations
- the same rule of dealing with the unused atoms (see Sect. 3.9).

In this section, we also assume that all algorithms use the same sparse coding method, namely OMP, in order to focus the comparison on the dictionary update. Next section will examine the effect of sparse coding methods.

The algorithms that enter the comparison are: MOD, SGK, K-SVD, AK-SVD, NSGK, PAK-SVD, PNSGK. These are the main algorithms studied in this chapter and two of their parallel version (fully parallel atom updates are used), those with the most promising behavior. PK-SVD was eliminated due to its high complexity; from our experience, its behavior relatively to K-SVD (which shares the same complexity) is similar to that of PAK-SVD relatively to AK-SVD. SimCO will be examined in the context of regularized DL, in Chap. 4.

All algorithms are initialized with a random subset of the training signals. During the iterative process, the unused atoms are replaced with random vectors. The number of iterations is $K = 500$, although we will present the results for fewer iterations, when they are relevant; note that 500 is a very large number, used here only to have a full image of the optimization process; in practice, the number of iterations is often in the tens and typically less than 100. The results are averaged over ten runs of each algorithm for the same set of input parameters.

3.11.1 Representation Error Results

The first experiments measure the representation error, using random patches from images, as described in Sect. 2.3.1. The relevant dimensions appearing in the DL problem are: atom size $m = 64$ and number of signals $N = 4000$. The number of atoms, n, and the sparsity level s are varied on broad ranges: n goes from 128 to 512 and s from 4 to 12. The RMSE (2.10) is computed in each iteration of an algorithm.

Figures 3.5 and 3.6 show the RMSE of the seven considered algorithms for the four combinations of $n \in \{128, 320\}$ and $s \in \{6, 12\}$. The results are more or less representative for the studied set of values of n and s. The first figure shows the RMSE for the full 500 iterations, focusing on final values and thus allowing to infer the asymptotic behavior of the algorithms, despite the clear fact that the error would continue to decrease if the algorithms would continue to run. The second figure focuses on the initial behavior of the algorithms, up to 50 iterations, less visible in the first figure.

Noting the scale on the vertical axis, we can say that the differences between algorithms are minor. However, the best RMSE is consistently achieved by NSGK; its parallel version has also a good performance; hence, it should be considered for large problems, where execution time is critical, since it can be efficiently

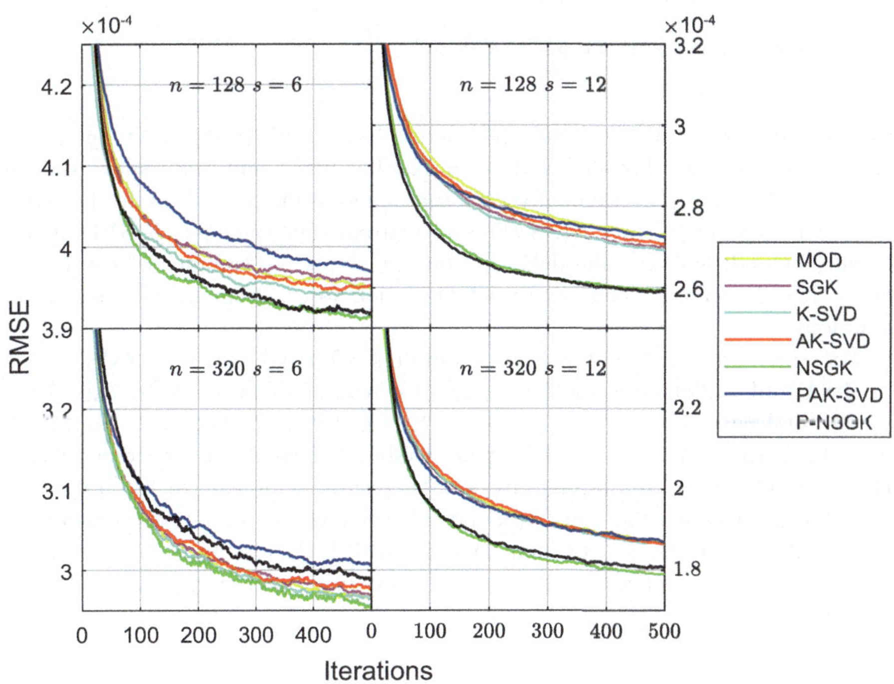

Fig. 3.5 RMSE evolution over 500 iterations

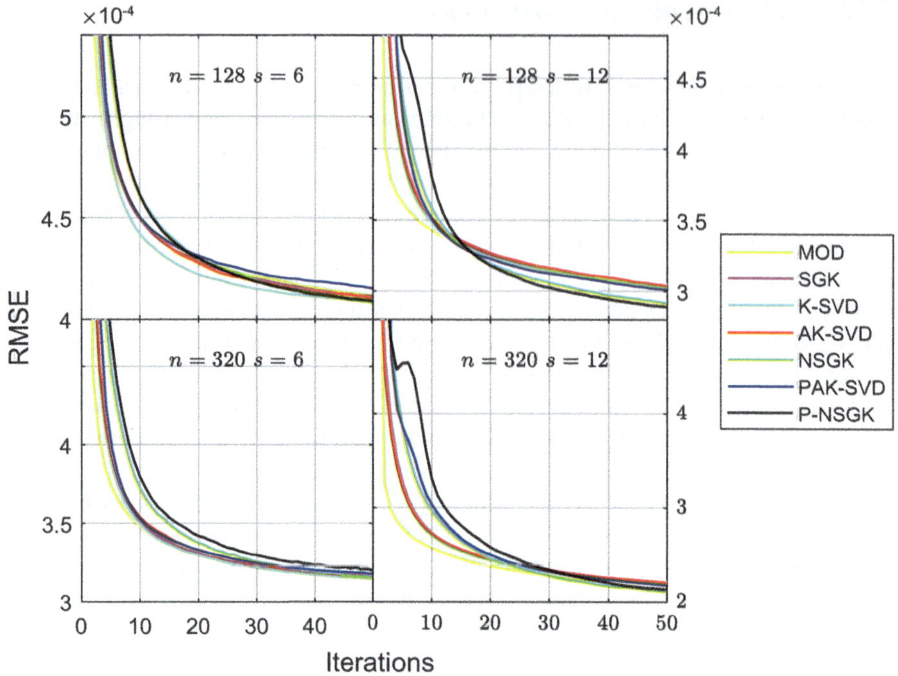

Fig. 3.6 RMSE evolution over the first 50 iterations

implemented on a multiprocessor architecture like a GPU [101]. Both algorithms have a slower start. PNSGK and other parallel algorithms may show sudden increases of the error; averaging over ten runs smooths the curves, but this behavior is typical. Expectedly, since AK-SVD is an approximated version of K-SVD, it lags slightly behind; however, the delay in convergence is not significant. The fastest initial RMSE decrease belongs to MOD, which loses steam in the subsequent iterations.

For a more thorough view, we present in Figs. 3.7 and 3.8 the final RMSE given by the DL algorithms after 500 iterations, for several values of the sparsity level s and number of atoms n, respectively. The superiority of NSGK is confirmed, while the performance of P-NSGK appears to slightly degrade for large dictionaries. However, all studied algorithms give similar results, with at most 8% difference between the best and the worst (for the same n and s), in all cases. Otherwise, all algorithms perform as expected, the final RMSE decreasing as n and s grow, due to the higher number of variables in the dictionary and representation matrix, respectively.

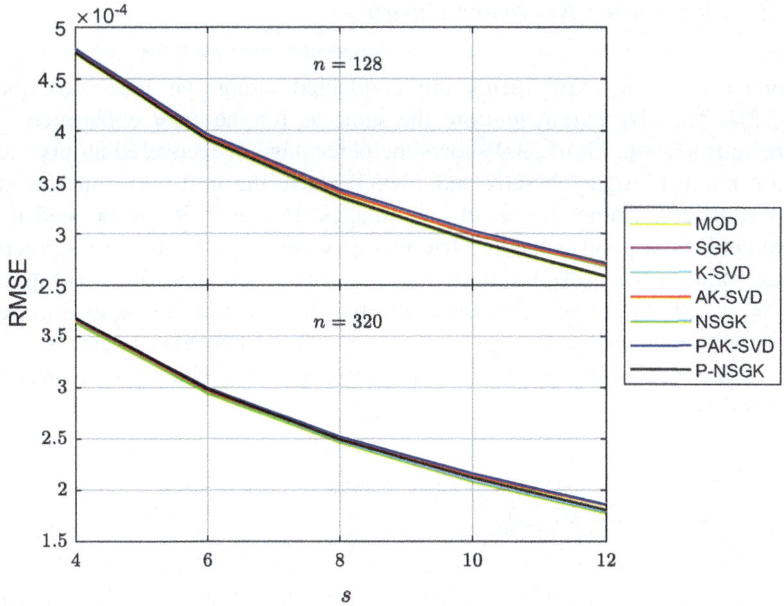

Fig. 3.7 Final RMSE (after 500 iterations) for various sparsity levels

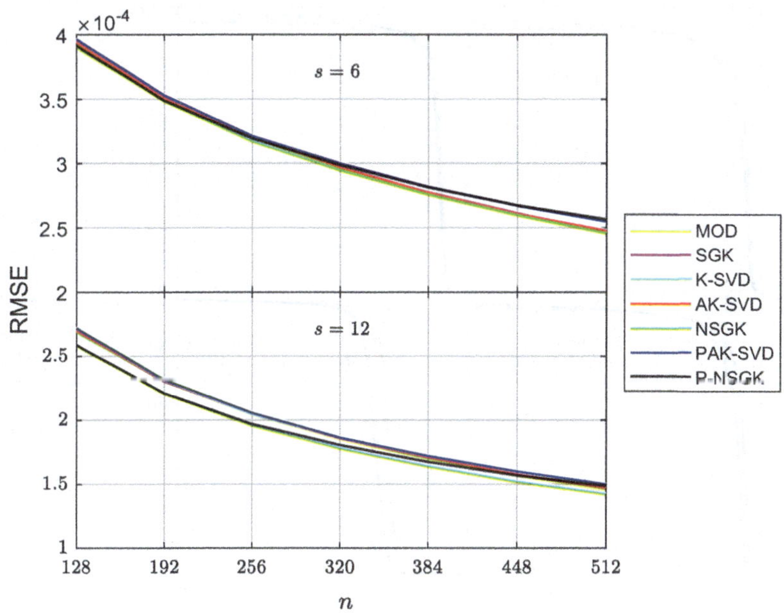

Fig. 3.8 Final RMSE (after 500 iterations) for various dictionary sizes

3.11.2 Dictionary Recovery Result

Dictionary recovery experiments are conducted along the lines described in Sect. 2.3.2. The size parameters are the same as for the error experiments from the previous section. Figure 3.9 shows the percentage of recovered atoms over the iteration number. Again, NSGK and PNSGK have the best performance, after a slower start, confirming the results for images. However, it can be said that all algorithms have a good behavior. The recovery percentage settles to a practically constant value before 150 iterations for $s = 6$ and only at around 300 iterations for $s = 12$; this kind of behavior is normal; however, it is interesting that the final recovery percentage is larger when $s = 12$; an explanation can be the fact that each atom is present in more sparse combinations, which might increase its identifiability.

3.11.3 Denoising Results

In this section we showcase DL-based image denoising. We use overlapping patches from the noisy image, as described in Sect. 2.4.1, to train the dictionary and to

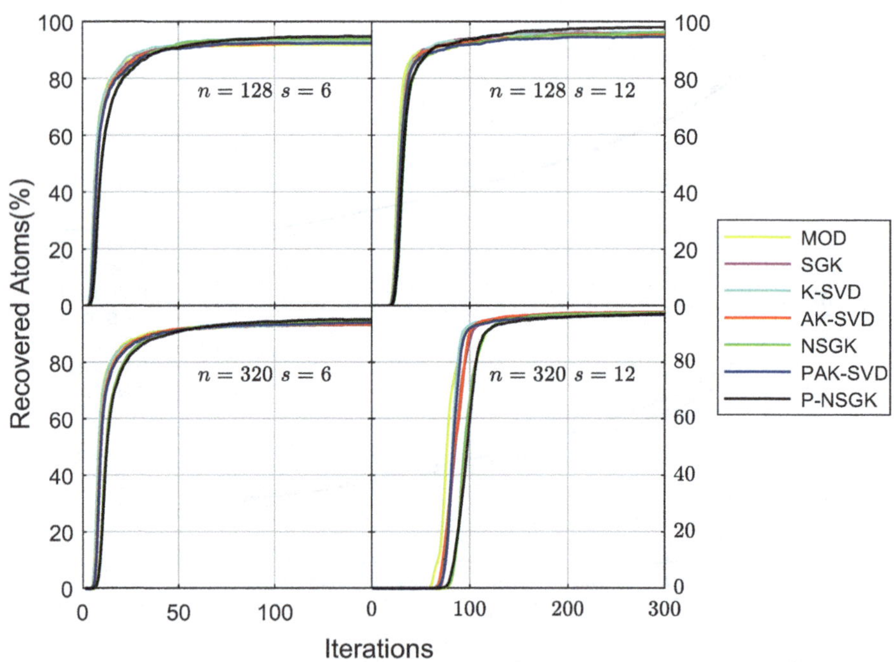

Fig. 3.9 Evolution of the atom recovery percentage during the DL process

Table 3.1 PSNR and SSIM results when denoising Lena with standard DL algorithms

σ	10		20		30	
Method	PSNR	SSIM	PSNR	SSIM	PSNR	SSIM
MOD	35.465384	0.907525	32.188647	0.858538	30.231635	0.821542
SGK	35.482473	0.907856	32.179603	0.858119	30.223639	0.821572
K-SVD	35.467575	0.907660	32.198808	0.858442	30.223013	0.821463
AK-SVD	35.478279	0.907740	32.178533	0.858107	30.202745	0.821136
NSGK	35.464369	0.907454	32.185298	0.858246	30.181760	0.821143
PAK-SVD	35.470827	0.907592	32.185756	0.858203	30.213522	0.821322
P-NSGK	35.475794	0.907553	32.180088	0.858273	30.217976	0.821557
BM3D	35.875969	0.914762	33.011568	0.876265	31.220110	0.844209

clean the image. In the following experiments we added white Gaussian noise of varied standard deviations $\sigma \in \{5, 10, 20, 30, 40, 50\}$ to the well-known USC-SIPI [202] images barbara, boat, house, lena, and peppers. For each noisy image, we randomly extracted $N = 40{,}000$ patches (with a stepping of $p = 1$ pixels) and trained a dictionary of $n = 256$ atoms of size $m = 64$. The sparsity target was set to $s = 6$ and the DL process was iterated $K = 50$ times. Our experiments did not show denoising improvements when using dictionaries trained with larger parameters values for s or K. With the resulting dictionary, we performed sparse representation on all overlapping patches using OMP as described around (2.19). At the end, we averaged the pixel values from the different patches they appeared in. For reference we also include results on the same data with the well-known BM3D [42] denoising algorithm.

In Table 3.1 we show the numerical results of the standard DL algorithms when denoising Lena from noises of various deviations. The PSNR (2.21) and SSIM (2.22) differences among DL algorithms are minimal for both indicators. BM3D is constantly ahead, improving PSNR by 0.5 dB up to 1 dB. We found this to be the case in all our experiments.

Figure 3.10 depicts the AK-SVD denoising results and compares them to BM3D. We omit the rest of the DL results as they are nearly identical. While the two methods perform similarly at first, DL-based results degrade much faster in low SNR environments. This phenomenon is reflected in the top PSNR panels, but it is made even more apparent through the SSIM curves.

3.12 Impact of Sparse Representation Algorithm

Improvements in any step of a DL algorithm should have a beneficial effect on the learned dictionary and on the representation error. We examine here the sparse coding step and study what happens when OMP is replaced by a better algorithm. Although there are many candidates, we select a single one for experimental

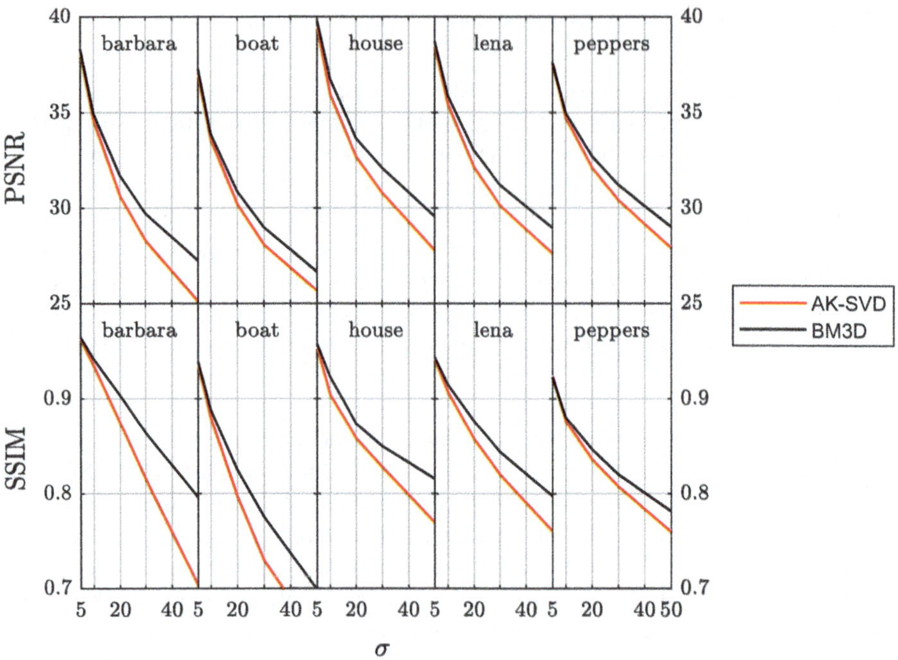

Fig. 3.10 Denoising performance of AK-SVD for various images and noise levels

comparison, namely the Look-Ahead Orthogonal Least Squares (LAOLS) algorithm [34]. LAOLS has the desired feature that it gives a representation error that is never worse than that of OMP. The behavior of other greedy sparse representation algorithms in the DL context was studied in [99].

We stress that, in terms of complexity, LAOLS is much costlier than OMP. Reminding that the sparse representation algorithm is necessary not only in DL, but also in the applications that use the learned dictionary, it is clear that low complexity is a crucial issue. So, the main purpose of our comparison between OMP and LAOLS is only to see how much the DL result could be improved by a "close-to-perfection" sparse coding algorithm.

LAOLS is presented in Algorithm 3.12. Essentially, LAOLS is an enhanced OMP that builds more supports instead of a single one. Like in OMP, the candidates are the atoms that are best correlated with the current residual. However, instead of choosing only the most correlated one, like in (1.12), LAOLS tries the L best atoms, where L is a small integer. Each of these atoms is appended to the current support and OMP is run to completion (i.e., until the support has s atoms). Then, the atom that produces the smallest *final* error is selected. This look-ahead search thus goes up to the full support and not a single step forward like OMP. When $L = 1$, this is equivalent to OMP. However, when $L > 1$, LAOLS checks more supports than OMP, including that generated by OMP, and hence gives a better result; of course, optimality is in no way guaranteed.

Algorithm 3.12: Look-Ahead Orthogonal Least Squares (LAOLS)

Data: dictionary $\boldsymbol{D} \in \mathbb{R}^{m \times n}$
signal $\boldsymbol{y} \in \mathbb{R}^m$
sparsity level s
number of look ahead attempts L
Result: representation support $\mathcal{S}$, solution $\boldsymbol{x}$

Initialize: $\boldsymbol{e} = \boldsymbol{y}$, $\mathcal{S} = \emptyset$
for $k = 1 : s$ **do**
  Compute correlations with residual: $\boldsymbol{z} = |\boldsymbol{D}^T \boldsymbol{e}|$
  Select indices $\mathcal{J}$ of the L largest elements of $\boldsymbol{z}$
  for $j \in \mathcal{J}$ **do**
    Run OMP starting from $\mathcal{S} \cup \{j\}$, obtaining $\boldsymbol{x}$ with sparsity level s
    Compute residual norm $\rho_j = \|\boldsymbol{y} - \boldsymbol{D}\boldsymbol{x}\|^2$
  Select new column: $i = \arg\min_j \rho_j$
  Increase support: $\mathcal{S} \leftarrow \mathcal{S} \cup \{i\}$
  Compute new solution: $\boldsymbol{x}_{\mathcal{S}} = (\boldsymbol{D}_{\mathcal{S}}^T \boldsymbol{D}_{\mathcal{S}})^{-1} \boldsymbol{D}_{\mathcal{S}}^T \boldsymbol{y}$
  Compute new residual: $\boldsymbol{e} = \boldsymbol{y} - \boldsymbol{D}_{\mathcal{S}} \boldsymbol{x}_{\mathcal{S}}$

At each step of LAOLS, L OMP algorithms are run; ignoring that they have actually a warm start, we can roughly assume that the complexity of LAOLS is $O(Ls)$ times larger than that of OMP. The complexity could be reduced by keeping track of the already checked supports, but this pursue is besides our purposes.

Figure 3.11 shows the evolution of three DL algorithms—NSGK and P-NSGK, as the best in the previous experiments, and AK-SVD as a representative of the faster algorithms—run with LAOLS as sparse coder. The results with OMP are also shown; they are taken from the previous section. The number of iterations is only 100 to reduce the necessary total time. All the other parameters have the same values as in the previous section. LAOLS is run with $L = 3$, i.e., a relatively low value. As expected, the LAOLS versions of the DL algorithms produce lower RMSE than the OMP versions; this is true for each algorithm, but also on the ensemble of the algorithms. Somewhat surprisingly, the order is not the same. AK-SVD is the worst of the three algorithms with OMP, but the best with LAOLS. If this fact would be always true (which is basically impossible to check), it would imply that, in the presence of nearly-ideal sparse coding, AK-SVD is the best of the three examined dictionary update strategies.

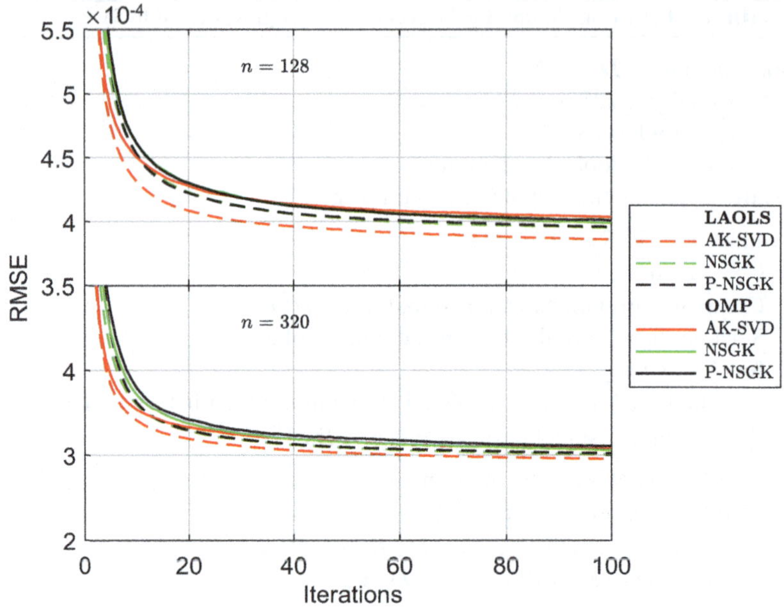

Fig. 3.11 RMSE evolution for DL algorithms with OMP (solid lines) and LAOLS (dashed lines) in the sparse coding stage

Problems

P 3.1 Prove the gradient expression (3.15) using both vectorization and matrix gradients.

P 3.2 A gradient descent algorithm for dictionary update could have the form

$$\boldsymbol{D} \leftarrow \boldsymbol{D} + \alpha(\boldsymbol{Y} - \boldsymbol{D}\boldsymbol{X})\boldsymbol{X}^T.$$

Discuss the differences with respect to relation (3.16) defining Sparsenet and list possible advantages and drawbacks. Is this a parallel algorithm?

P 3.3 Write an efficient implementation of Sparsenet using an error update trick similar with (3.25) and (3.26).

P 3.4 Show that the Lloyd's (or K-means) algorithm for vector quantization (or clustering) is essentially equivalent with SGK for sparsity level $s = 1$, when the signals $\boldsymbol{y}_\ell$, $\ell = 1 : N$, have unit norm and we consider only their direction (not changed by a sign flip). The algorithm iterates two basic operations for designing a set of n atoms, whose meaning is that of cluster centers.

(a) Assignment: with fixed atoms $\boldsymbol{d}_j$, associate each signal $\boldsymbol{y}_\ell$ with the nearest atom, thus forming clusters around the atoms. Show that the same operation is done by OMP in the sparse coding step of DL; the difference is that OMP computes also a coefficient, which is the scalar product between the atom and the signal.
(b) Update: recompute the atoms as the centroids (i.e., the average) of the signals belonging to their cluster. Show that, ignoring normalization, this is the SGK dictionary update (3.22) when the representation coefficients are all equal to 1.

P 3.5 Taking algorithm NSGK as model, write algorithm NAK-SVD; then, give the parallel (Jacobi) versions of both algorithms. Are the fully parallel versions of NSGK and NAK-SVD identical, like it is the case with SGK and AK-SVD (remind Remark 3.10)?

P 3.6 A possible dictionary update algorithm that also modifies the representations may look as follows. First, compute an optimal dictionary like in MOD, with (3.31); then, keeping the dictionary and sparsity pattern fixed, compute the optimal representations like in (3.42); repeat these two operations a few times. Comment on the properties of such an algorithm and find its "relatives" among the algorithms presented in this chapter.

P 3.7 The main disadvantage of the sparse coding algorithm LAOLS presented in Sect. 3.12 is complexity. Other sparse coding algorithms share the same drawback. Debate the pros and cons of the following DL scheme. The preferred DL algorithm is run as usual, with OMP for sparse coding, for a number of iterations. Then, in the final few iterations, OMP is replaced with a more powerful (and more complex) sparse coding algorithm.

Chapter 4
Regularization and Incoherence

Abstract A dictionary should be faithful to the signals it represents, in the sense that the sparse representation error in learning is small, but also must be reliable when recovering sparse representations. A direct way to obtain good recovery guarantees is to modify the objective of the DL optimization such that the resulted dictionary is incoherent, meaning that the atoms are generally far from each other. Alternatively, we can explicitly impose mutual coherence bounds. A related modification of the objective is regularization, in the usual form encountered in least squares problems. Regularization has the additional benefit of making the DL process avoid bottlenecks generated by ill-conditioned dictionaries. We present several algorithms for regularization and promoting incoherence and illustrate their benefits. The K-SVD family can be adapted to such approaches, with good results in the case of regularization. Other methods for obtaining incoherence are based on the gradient of a combined objective including the frame potential or by including a decorrelation step in the standard algorithms. Such a step is based on successive projections on two sets whose properties are shared by the Gram matrix corresponding to the dictionary, which reduce mutual coherence, and rotations of the dictionary, which regain the adequacy to the training signals. We also give a glimpse on the currently most efficient methods that aim at the minimization of the mutual coherence of a frame, regardless of the training signals.

4.1 Learning with a Penalty

DL is faced with a contradictory requirement when designing dictionaries. In order to better represent signals that lie closely, it is natural that atom density is higher in such zones. So, inherently, some atoms are near each other, hence the mutual coherence (1.25) or the RIP constant (1.27) of the dictionary has large values. However, this implies that the recovery properties of the dictionary are guaranteed only for extremely sparse representations, while for higher values of the sparsity level s there is no theoretical recovery guarantee. So, while the dictionaries are in principle better by following the distribution of the training signals, we cannot be sure that the algorithms using them for sparse representations

B. Dumitrescu, P. Irofti, *Dictionary Learning Algorithms and Applications*,
https://doi.org/10.1007/978-3-319-78674-2_4

can indeed take advantage of these potentially better dictionaries. Of course, sparse representation algorithms have empirical good behavior, in any case better than the theoretical guarantees. Still, their performance decreases when the dictionary has poor properties. We certainly would like to obtain the best trade-off between modeling the training signals and good recovery guarantees.

We expect that any m atoms of a dictionary are linearly independent, since this a generic property. However, the DL process, due to the reasons exposed above, may produce small groups of atoms that are nearly linearly dependent; if there are s (or fewer) such atoms, the RIP constant δ_s is nearly equal to 1 and the condition number bound (1.28) may have very large values. This is important when such atoms enter in the representation of the same signal: for a signal $\boldsymbol{y}_\ell$, the matrix $\boldsymbol{D}_{\mathcal{S}_\ell}$ is ill-conditioned. In the extreme case of rank deficiency, the solution to the approximation problem (3.41) is no longer unique and (3.42) cannot be used since the matrix $\boldsymbol{D}_{\mathcal{S}_\ell}^T \boldsymbol{D}_{\mathcal{S}_\ell}$ is not invertible; in principle, the computation problem could be overcome by resorting to the generalized least squares approach based on the singular value decomposition, with the immediate price of higher complexity. However, it was argued in [44] that such small groups of nearly linearly dependent atoms may have another ill effect on DL algorithms: they slow down the progress made in dictionary update, sometimes almost to a stall. So, the DL algorithm may simply stop prematurely and provide a dictionary that not only has bad recovery properties, but also is not fully optimized for the training signals.

A cure for these phenomena is to modify the DL process to not only optimize the representation error, but to trade it off with other properties of the dictionary. Due to its simpler definition, mutual coherence is much easier to handle than the RIP constants, hence it is a first candidate for the joint optimization. This takes care directly of the recovery properties of the dictionary.

Another idea is to act indirectly on the dictionary via the representations. Least squares regularization is long known to have beneficial effects on the solutions of approximation problems and to alleviate numerical problems for ill conditioned problems.

In this context, remind also the discussion from Remark 2.9 on conditioning, essentially stating that both the dictionary and the representation matrix should be far from singularity (or rank deficiency) in order to have a well-behaved problem.

Both above approaches—improving mutual coherence and regularizing the representations—can be implemented by adding a penalty function to the error objective of the DL problem (2.2), thus transforming it into

$$f_h(\boldsymbol{D}, \boldsymbol{X}) = \|\boldsymbol{Y} - \boldsymbol{D}\boldsymbol{X}\|_F^2 + h(\boldsymbol{D}, \boldsymbol{X}). \tag{4.1}$$

Here, the function h must have large values when the dictionary or the representations have undesired properties, thus forbidding such solutions, even if the error is small. Otherwise, h should have small values, such that its interference with the error optimization is ideally negligible and thus the best acceptable solution is chosen.

This chapter starts with regularization, which penalizes high values of the representation coefficients, since their occurrence is a sign of ill conditioned small sets of atoms. It continues with several methods to jointly optimize coherence or, as an alternative to (4.1), to bound it. We will minimize (4.1) with the usual bounds on sparsity level and with atom normalization constraints, without mentioning them explicitly.

4.2 Regularization

Least squares regularization consists of optimizing

$$f_\mu(\boldsymbol{D}, \boldsymbol{X}) = \|\boldsymbol{Y} - \boldsymbol{D}\boldsymbol{X}\|_F^2 + \mu\|\boldsymbol{X}\|_F^2, \tag{4.2}$$

where $\mu > 0$ is a parameter whose value will be discussed later; however, the value cannot be too large, such that error optimization is still pursued. Such an objective encourages small values of the representation coefficients, thus indirectly producing a better conditioned dictionary. This is a typical approach for least squares problems, but not used in DL until [44]. We can also write

$$\begin{aligned} f_\mu(\boldsymbol{D}, \boldsymbol{X}) &= \sum_{\ell=1}^{N} \left(\|\boldsymbol{y}_\ell - \boldsymbol{D}_{\mathcal{S}_\ell} \boldsymbol{X}_{\mathcal{S}_\ell,\ell}\|^2 + \mu \|\boldsymbol{X}_{\mathcal{S}_\ell,\ell}\|_F^2 \right) \\ &= \sum_{\ell=1}^{N} \left\| \begin{bmatrix} \boldsymbol{y}_\ell \\ 0 \end{bmatrix} - \begin{bmatrix} \boldsymbol{D}_{\mathcal{S}_\ell} \\ \sqrt{\mu}\boldsymbol{I} \end{bmatrix} \boldsymbol{X}_{\mathcal{S}_\ell,\ell} \right\|^2 . \end{aligned} \tag{4.3}$$

In the remainder of this section we discuss the modifications of the DL algorithms presented in the previous chapter to optimize the objective (4.2).

4.2.1 Sparse Coding

Relation (4.3) says that each signal is coded independently (which was obvious) and that we should sparsely represent a signal extended with zeros using a dictionary extended (below) with the unit matrix. (Note that s extended atoms are always linearly independent if $s \leq m$.) OMP selects atoms by looking at their scalar product with the current residual, which is a linear combination of the signal and some atoms; we note that the scalar product of the atoms with the signal or of two atoms is not modified by the extensions from (4.3). So, OMP remains unchanged in what concerns the search of the nonzero pattern. However, after selecting a group of atoms, the least-squares solution changes. In particular, the final representation (3.42) has the expression (this is the classical regularized least squares formula)

$$\boldsymbol{X}_{\mathcal{S}_\ell,\ell} = \left(\boldsymbol{D}_{\mathcal{S}_\ell}^T \boldsymbol{D}_{\mathcal{S}_\ell} + \mu \boldsymbol{I} \right)^{-1} \boldsymbol{D}_{\mathcal{S}_\ell}^T \boldsymbol{y}_\ell. \tag{4.4}$$

The different value of the coefficients makes OMP (Algorithm 1.1) and this regularized OMP (somewhat unfortunately, the name *regularized OMP* was given in [138] to a different idea of regularization) having different residuals and hence possibly selecting different supports. So, it is possible that regularization changes significantly the solution, even though the representation error is only slightly modified. We also note that plain OMP was used in [44] for the regularized DL problem, possibly in order to keep the same representation algorithm in training and testing (note that in testing we have a well-behaved dictionary, hence no reason to regularize the OMP algorithm).

4.2.2 Regularized K-SVD

We focus now on the dictionary update step. When the regularized problem (4.2) is seen only with $\boldsymbol{D}$ as variable, the presence of the regularization term does not change the outcome of the optimization. So, all the methods that optimize only $\boldsymbol{D}$ in the dictionary update step are unchanged, in the sense that they have no specific regularized version; the only modification is in the sparse coding step. From the methods presented in the previous chapter, this characterization covers Sparsenet, MOD, SGK and its parallel version. Although SimCO seems to fall into the same category, we remind that it always uses the optimal representations, now given by the regularized least squares solution (4.4); hence, although the regularized version keeps the same general form, this detail makes the results different from the plain version.

Regularized K-SVD We go now to the K-SVD family, keeping the notation (3.20) and those from Proposition 3.3 for the error matrix with respect to a single atom $\boldsymbol{d}$ and its representations $\boldsymbol{x}$. When both $\boldsymbol{d}$ and $\boldsymbol{x}$ are optimized, the prototype problem (3.33) is replaced with

$$\begin{aligned} \min_{\boldsymbol{d},\boldsymbol{x}} \quad & \|\boldsymbol{F} - \boldsymbol{d}\boldsymbol{x}^T\|_F^2 + \mu\,\|\boldsymbol{x}\|_F^2 \\ \text{s.t.} \quad & \|\boldsymbol{d}\| = 1 \end{aligned} \tag{4.5}$$

The solution to this problem gives the regularized K-SVD algorithm. It turns out that the result is very similar to that of Proposition 3.33 defining K-SVD.

Proposition 4.1 ([60]) *Using the SVD decomposition (3.34), the solution of (4.5) is*

$$\boldsymbol{d} = \boldsymbol{u}_1, \quad \boldsymbol{x} = \frac{1}{1+\mu}\sigma_1 \boldsymbol{v}_1. \tag{4.6}$$

Proof We denote $\phi(\boldsymbol{d}, \boldsymbol{x})$ the objective of (4.5) and remark that

$$\phi(\boldsymbol{d}, \boldsymbol{x}) = (1 + \mu)\|\boldsymbol{x}\|^2 - 2\boldsymbol{x}^T \boldsymbol{F}^T \boldsymbol{d} + \|\boldsymbol{F}\|_F^2. \tag{4.7}$$

We express $\boldsymbol{d}$ and $\boldsymbol{x}$ in the bases formed by the singular vectors of $\boldsymbol{F}$:

$$\boldsymbol{d} = \sum_{i=1}^{m} \rho_i \boldsymbol{u}_i, \quad \boldsymbol{x} = \sum_{i=1}^{N} \theta_i \boldsymbol{v}_i.$$

We denote $t = \|\boldsymbol{x}\|$ and (the atom being always normalized) note that

$$\sum_{i=1}^{m} \rho_i^2 = 1, \quad \sum_{i=1}^{N} \theta_i^2 = t^2. \tag{4.8}$$

Since

$$\boldsymbol{F}^T \boldsymbol{d} = \sum_{i=1}^{r} \sigma_i \boldsymbol{v}_i \boldsymbol{u}_i^T \boldsymbol{d} = \sum_{i=1}^{r} \sigma_i \rho_i \boldsymbol{v}_i,$$

the function to be minimized becomes

$$\phi(\boldsymbol{d}, \boldsymbol{x}) = (1 + \mu)t^2 - 2\sum_{i=1}^{r} \sigma_i \rho_i \theta_i + \|\boldsymbol{F}\|_F^2.$$

Since σ_i form a decreasing sequence and the norm constraints (4.8) hold, the Cauchy–Schwarz inequality gives

$$\left(\sum_{i=1}^{r} \sigma_i \rho_i \theta_i\right)^2 \leq \left(\sum_{i=1}^{r} \sigma_i^2 \rho_i^2\right) \sum_{i=1}^{N} \theta_i^2 \leq \sigma_1^2 t^2.$$

Equality is attained for $\rho_i = 0$, $\theta_i = 0$, for all $i > 1$, hence the objective function becomes

$$\phi(\boldsymbol{d}, \boldsymbol{x}) = (1 + \mu)t^2 - 2\sigma_1 t + \|\boldsymbol{F}\|_F^2$$

and depends now on a single variable; its minimum is given for $t = \sigma_1/(1 + \mu)$, which gives (4.6). ■

Remark 4.2 Comparing (4.6) with the standard K-SVD solution (3.35) we see that the atom is the same, but the representation coefficients are reduced. So, the purpose of reducing the coefficients is explicitly achieved. The regularized K-SVD algorithm, denoted K-SVDr (we will always obey to the rule of appending an 'r' to the name of an algorithm for its regularized version), is simply obtained from

Algorithm 3.6 by dividing the representations with $1 + \mu$ in line 5. The complexity is basically the same: regularization comes at almost no computational effort.

Remark 4.3 In [44], similarly to (4.3), the objective of (4.5) was written in the form

$$\phi(\boldsymbol{d}, \boldsymbol{x}) = \left\| \begin{bmatrix} \boldsymbol{F} \\ \boldsymbol{0} \end{bmatrix} - \begin{bmatrix} \boldsymbol{d} \\ \sqrt{\mu} \end{bmatrix} \boldsymbol{x}^T \right\|_F^2 \tag{4.9}$$

and it was argued that the minimum cannot be found via SVD: applying SVD directly to $\boldsymbol{F}$ gives the standard K-SVD solution; applying SVD to the extended matrix from (4.9) gives an extended atom with 0 on the last position instead of $\sqrt{\mu}$. Indeed, since minimizing (4.9) is a *structured* rank-1 approximation, one cannot expect SVD to supply a solution like in the standard approximation.

However, an alternative proof of Proposition 4.1 shows the relation with a usual rank-1 approximation. Since $\|\boldsymbol{d}\| = 1$, it follows that $\|\boldsymbol{x}\| = \|\boldsymbol{d}\boldsymbol{x}^T\|_F$. Introducing the rank-1 variable $\boldsymbol{\Upsilon} = \boldsymbol{d}\boldsymbol{x}^T$, it results that

$$\min_{\|\boldsymbol{d}\|=1, \boldsymbol{x}} \phi(\boldsymbol{d}, \boldsymbol{x}) = \min_{\text{rank}\boldsymbol{\Upsilon}=1} \|\boldsymbol{F} - \boldsymbol{\Upsilon}\|_F^2 + \mu\|\boldsymbol{\Upsilon}\|_F^2. \tag{4.10}$$

A few algebraic manipulations lead to the equality

$$\|\boldsymbol{F} - \boldsymbol{\Upsilon}\|_F^2 + \mu\|\boldsymbol{\Upsilon}\|_F^2 = \\ \left\| \frac{1}{\sqrt{1+\mu}} \boldsymbol{F} - \sqrt{1+\mu}\boldsymbol{\Upsilon} \right\|_F^2 + \left(1 - \frac{1}{1+\mu}\right) \|\boldsymbol{F}\|_F^2.$$

Since the second term is constant, it results that (4.10) has the same solution as

$$\min_{\text{rank}\boldsymbol{\Upsilon}=1} \left\| \frac{1}{1+\mu} \boldsymbol{F} - \boldsymbol{\Upsilon} \right\|_F^2, \tag{4.11}$$

whose solution is (4.6). ■

Regularized AK-SVD updates successively the atom and its representations. When the representation $\boldsymbol{x}$ is fixed in (4.5), then atom optimization is identical to that from Proposition 3.4, hence the solution (3.28) is identical to that from plain AK-SVD; line 5 of Algorithm 3.7 stays unchanged. When the representation is fixed in (4.5), we obtain the following regularized version of Proposition 3.8.

Proposition 4.4 *The solution to the problem*

$$\min_{\boldsymbol{x}} \left\| \boldsymbol{F} - \boldsymbol{d}\boldsymbol{x}^T \right\|_F^2 + \mu \|\boldsymbol{x}\|^2 \tag{4.12}$$

is

$$x = \frac{1}{1+\mu} F^T d. \tag{4.13}$$

Proof Equating with zero the derivative with respect to x of (4.7) gives (4.13). ■

So, like in K-SVDr, regularization forces an explicit reduction of the representation coefficients. Regularized AK-SVD results by simply dividing with $1 + \mu$ the coefficients computed in line 6 of Algorithm 3.7.

Remark 4.5 The expression of the optimal representations leads to a third proof of Proposition 4.1. Replacing the optimal representation (4.13) (as a function of the atom) in (4.7), we obtain

$$\phi(d) = \frac{1}{1+\mu}\|F^T d\|^2 - \frac{2}{1+\mu} d^T F F^T d + \|F\|_F^2 = -\frac{1}{1+\mu}\|F^T d\|^2 + \|F\|_F^2.$$

The normalized atom d for which the minimum is attained is u_1, the left singular vector of F corresponding to the maximum singular value (identical to the eigenvector of FF^T corresponding to the maximum eigenvalue). Using again (4.13) and taking into account that $F^T u_1 = \sigma_1 v_1$ gives (4.6). ■

4.2.3 Comparison Between Regularized K-SVD and SimCO

The first algorithm for the regularized DL problem (4.2) was SimCO [44]. As mentioned above the only modification to Algorithm 3.9 is to replace the optimal representation in step 14 with the regularized version (4.4); we name SimCOr the regularized algorithm.

In this section, we try to find out via simulation whether regularization has an effect on DL and which are the best regularized DL algorithms. The candidates are K-SVDr, AK-SVDr and SimCOr.

The first round of simulations is performed on artificial data, as described in Sect. 2.3.1. Data are generated with a random dictionary of size $m = 16$, $n = 32$, with sparsity level $s = 4$ and no noise. We follow the experiment from [44] by varying the training set size N. The DL process takes $K = 50$ iterations. For SimCOr we use a single inner iteration (parameter I in Algorithm 3.9). We execute 50 runs for each parametrization and report the average RMSE.

The regularization parameter is set to $\mu = 0.1$ for the first 30 DL iterations and $\mu = 0$ for the remaining 20. Such an approach is sensible: the dictionary, in the first part of the learning process, can change a lot; regularization prevents the occurrence of ill-conditioned dictionaries. When the dictionary values have almost stabilized, regularization has a less important role; reducing the weight of the regularization term (or even making it zero) allows the optimization process to focus on the error, which is the main objective.

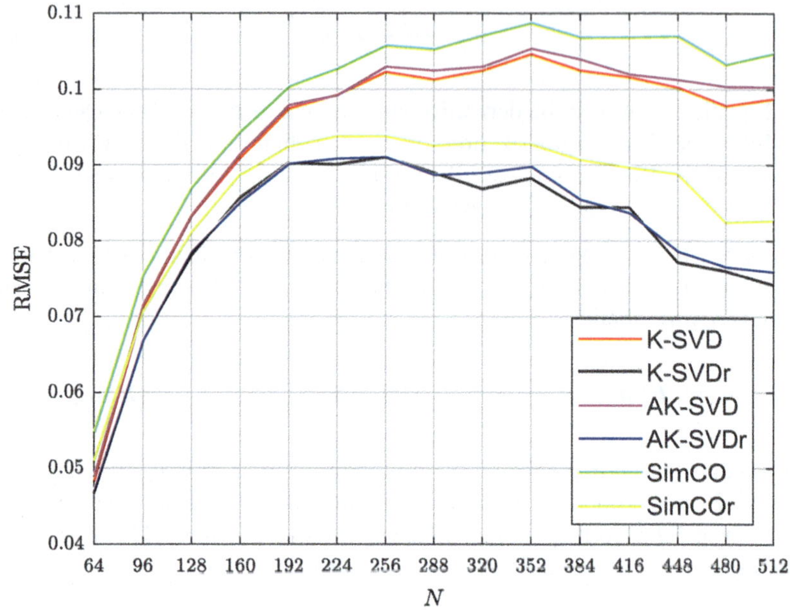

Fig. 4.1 Final RMSE for standard and regularized algorithms, for several low values of the number of signals

The results are shown in Fig. 4.1, for small values of the number N of training signals. This is where regularization shows the most benefits, as remarked in [44]. We can draw at least two conclusions from the figure. First, the regularized versions of all algorithms give a better RMSE than the standard ones; hence, regularization is useful. Second, K-SVD and AK-SVD are better than SimCO, in both the standard and the regularized versions. (We need to post here an erratum to [60], where a similar figure shows the same relative behavior of the algorithms, but with a different computation of the error, which is not correctly scaled with N; the conclusions are the same, but the correct RMSE values are here.)

The second round of simulations is on image representation, following again the description from Sect. 2.3.1, with $m = 64$, $n = 256$, $N = 4096$, $K = 50$ and several values of the sparsity level. The results shown in Table 4.1 are averages over ten runs, each starting with the same initial dictionary and training set. For SimCO we show multiple results with varied number of inner iterations. The regularization parameter is $\mu = 0.05$.

Complexity-wise, K-SVD and AK-SVD are clearly superior to SimCO. For example, with $s = 8$ in the image representation test, a single dictionary update takes about 1 s for K-SVDr and less than 0.2 s for AK-SVDr. For the same task, a single SimCO inner iteration takes about 1.5 s. Since SimCO needs many inner iterations to compete with (A)K-SVDr in terms of achieved error level, the

Table 4.1 RMSE/10^{-4} for the standard and regularized DL algorithms in image representation

Method	$s = 4$	$s = 6$	$s = 8$	$s = 10$
KSVD	3.38	2.94	2.63	2.39
KSVDr	3.44	3.02	2.69	2.44
AK-SVD	3.40	2.96	2.65	2.41
AK-SVDr	3.46	3.02	2.71	2.45
SimCO(1)	4.74	4.35	3.99	3.58
SimCOr(1)	4.66	4.21	3.85	3.43
SimCOr(10)	3.48	3.05	2.73	2.48
SimCOr(20)	3.42	2.99	2.69	2.44

latter algorithms are much faster. For the representation error, regularization seems unnecessary, giving about the same results as the standard algorithm. However, it was noticed in [60], on tests with patches that (unlike here) are not centered, regularization can bring benefits for larger values of s, where the optimization problem is more difficult.

Image denoising experiments reported in [60] show that regularization brings only marginal improvement and that AK-SVDr gives similar performance with the other algorithms but is much faster.

4.3 Frames

As matrices with more columns than rows, the dictionaries are in fact frames, hence we recall a few basic notions from frame theory, connecting also with the notion of mutual coherence. For more on frames, the reader could go to [27] and the references therein.

Definition 4.6 A matrix $\boldsymbol{D} \in \mathbb{R}^{m \times n}$ is a frame if there are positive constants A and B such that for any vector $\boldsymbol{y}$, the following inequality holds:

$$A\|\boldsymbol{y}\|^2 \leq \|\boldsymbol{D}^T\boldsymbol{y}\|^2 \leq B\|\boldsymbol{y}\|^2. \tag{4.14}$$

The lower bound A and the upper bound B are usually assumed to take the maximum and, respectively, minimum values over all possible bounds.

The columns of $\boldsymbol{D}$ (the atoms, in our context) form a basis for $\mathbb{R}^m$, i.e., the dictionary has full rank; otherwise, it would result that $A = 0$ (there is a $\boldsymbol{y}$ orthogonal on all atoms) and hence $\boldsymbol{D}$ could not be a frame. If $A = B$, then the frame is *tight*; this also means that the rows of $\boldsymbol{D}$ are orthogonal and have the same norm; the singular values of $\boldsymbol{D}$ are all equal. If all the columns of the tight frame $\boldsymbol{D}$ have unit norm, then $\boldsymbol{D}$ is a *unit norm tight frame* (UNTF).

A frame is equiangular if all the absolute scalar products $|\boldsymbol{d}_i^T\boldsymbol{d}_j|$, $i \neq j$, are equal. Equiangular frames may exist only if the dimensions m and n satisfy certain conditions; for example, a necessary (but not sufficient) condition is $n \leq m(m +$

1)/2. Equiangular frames provide a lower bound for the mutual coherence (1.25) of a frame. All frames satisfy the inequality

$$M \geq \sqrt{\frac{n-m}{m(n-1)}}, \tag{4.15}$$

known also as the Welch bound [203], but the equality holds only for equiangular frames. They may be seen as a generalization of orthogonal matrices (and bases, in general). In sparse representations, equiangular frames offer the best recovery guarantees, but they have relatively reduced flexibility in adapting to a certain class of signals. However, they are extremely useful, e.g., in coding theory.

A *Grassmannian* frame has minimal coherence; note that, given dimensions m and n, Grassmannian frames always exist; as already mentioned, they are not necessarily equiangular.

Example 4.7 Consider a dictionary $\boldsymbol{D} \in \mathbb{R}^{m \times n}$, with $n = km$, where $k \geq 1$ can be seen as a factor of overcompleteness. Figure 4.2 shows the values of the Welch bound (4.15) for dictionaries with number of atoms $n \in \{16, 64, 256\}$ and overcompleteness factor k going from 1 (when the bound is trivially zero, equiangular frames being orthogonal matrices) to 10, which is a rather large value for common applications. We see that the Welch bound has quite low values. Since (4.15) is in fact $\sqrt{(k-1)/(km-1)}$, the asymptotic value of the bound is $1/\sqrt{m}$. The purpose stated in the beginning of the chapter, to design dictionaries with low mutual coherence, seems not excessively contradictory with the minimum requirements imposed by the Welch bound. ■

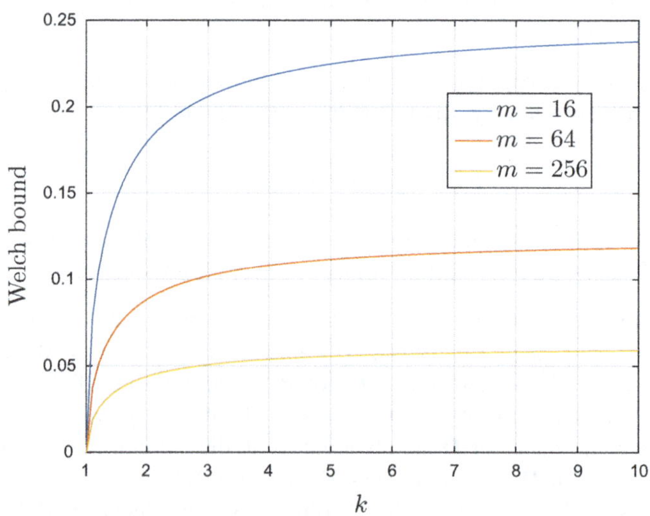

Fig. 4.2 Welch bound values for $m \times km$ frames

Definition 4.8 Let $\boldsymbol{D}$ be a frame and $\boldsymbol{G} = \boldsymbol{D}^T\boldsymbol{D}$ its associated Gram matrix. The Frobenius norm of the Gram matrix, $\|\boldsymbol{G}\|_F$, is called frame *potential.*

For a dictionary with unit norm atoms, the potential is an indicator of the average value of the scalar products between atoms (since these are the elements of $\boldsymbol{G}$). Excluding the diagonal elements, which are all equal to one due to atom normalization, we can define a root mean square (RMS) coherence

$$\tilde{M} = \frac{\|\boldsymbol{D}^T\boldsymbol{D} - \boldsymbol{I}\|_F}{\sqrt{n(n-1)}}. \tag{4.16}$$

While the mutual coherence (1.25) is an indicator of the worst case behavior of a dictionary, the RMS coherence could be a better indicator of average performance, although there are not direct results in this sense besides some practical evidence. Moreover, the RMS coherence seems easier to manipulate. Indeed, the following result shows the class of matrices that have minimum RMS coherence.

Proposition 4.9 ([38, 133, 197]) *A dictionary* $\boldsymbol{D}$ *with normalized atoms for which the RMS coherence (4.16) is minimum is a unit norm tight frame.*

Proof Let $\boldsymbol{D} = \boldsymbol{U}\boldsymbol{\Sigma}\boldsymbol{V}^T$ be the singular value decomposition of the dictionary, where $\boldsymbol{U} \in \mathbb{R}^{m\times m}$ and $\boldsymbol{V} \in \mathbb{R}^{n\times n}$ are orthogonal matrices and $\boldsymbol{\Sigma} \in \mathbb{R}^{m\times n}$ is diagonal. Due to the unit norm atom condition, we have

$$\|\boldsymbol{D}\|_F^2 = \sum_{j=1}^{n} \|\boldsymbol{d}_j\|^2 = n.$$

The singular value decomposition gives

$$\|\boldsymbol{D}\|_F^2 = \sum_{i=1}^{m} \sigma_i^2.$$

So, the sum of the squares of the singular values is constant

$$\sum_{i=1}^{m} \sigma_i^2 = n. \tag{4.17}$$

Using again the SVD, we get

$$\begin{aligned} \left\|\boldsymbol{D}^T\boldsymbol{D} - \boldsymbol{I}\right\|_F^2 &= \left\|\boldsymbol{V}\boldsymbol{\Sigma}^T\boldsymbol{\Sigma}\boldsymbol{V}^T - \boldsymbol{I}\right\|_F^2 = \left\|\boldsymbol{V}(\boldsymbol{\Sigma}^T\boldsymbol{\Sigma} - \boldsymbol{I})\boldsymbol{V}^T\right\|_F^2 \\ &= \sum_{i=1}^{m}(\sigma_i^4 - 2\sigma_i^2) + n = \sum_{i=1}^{m}\sigma_i^4 - n. \end{aligned} \tag{4.18}$$

The minimum is obtained when the singular values are equal:

$$\sigma_1 = \ldots = \sigma_m \triangleq c = \sqrt{\frac{n}{m}} \tag{4.19}$$

Indeed, putting $\sigma_i^2 = c^2 + \theta_i$, it follows from (4.17) that $\sum_{i=1}^m \theta_i = 0$ and so

$$\sum_{i=1}^m \sigma_i^4 = \sum_{i=1}^m (c^4 + 2c^2\theta_i + \theta_i^2) = \frac{n^2}{m} + \sum_{i=1}^m \theta_i^2.$$

The minimum is attained when $\theta_i = 0$, $i = 1 : m$.

It results that an optimal dictionary has the form

$$\boldsymbol{D} = \sqrt{\frac{n}{m}} \boldsymbol{U} \boldsymbol{V}_1^T, \tag{4.20}$$

where $\boldsymbol{V}_1 \in \mathbb{R}^{n \times m}$ contains the first m columns of $\boldsymbol{V}$; hence, the rows of $\boldsymbol{V}_1^T$ are orthogonal. The columns of the optimal dictionary have unit norm by assumption and $\boldsymbol{D}\boldsymbol{D}^T = c^2\boldsymbol{I}$, so $\boldsymbol{D}$ has orthogonal rows and hence is a UNTF. ■

Remark 4.10 The above proposition gives an immediate proof to the Welch bound (4.15). By replacing the optimal value (4.19) of the singular values in (4.18), it results that a UNTF satisfies

$$\|\boldsymbol{D}^T\boldsymbol{D} - \boldsymbol{I}\|_F^2 = \frac{n(n-m)}{m}.$$

Hence, the minimum RMS coherence (4.16) is equal to the Welch bound (4.15). Consequently, the minimum value of the mutual coherence, which may be attained when all (distinct) atom absolute products are equal, cannot be lower than the Welch bound. ■

Remark 4.11 (How to Generate a Random UNTF) A tight frame needs only to have orthogonal rows, so it is easy to generate. A short to describe method is as follows. We take a random $m \times n$ matrix $\boldsymbol{\Delta}$ (or another full rank matrix) and compute the reduced QR decomposition $\boldsymbol{\Delta}^T = \boldsymbol{Q}\boldsymbol{R}$, with $\boldsymbol{Q} \in \mathbb{R}^{n \times m}$ having orthogonal columns (with unit norm). Then $\boldsymbol{Q}^T$ is a tight frame with bounds $A = B = 1$.

If we want a UNTF, the first step is to multiply the above matrix with the constant from (4.19). The dictionary $\boldsymbol{D} = \sqrt{n/m}\,\boldsymbol{Q}^T$ has the correct singular values (and Frobenius norm), but the atoms have not unit norm. To obtain the UNTF, we apply the algorithm from [49], based on the idea that if $\boldsymbol{\Gamma}$ is an orthogonal matrix, then $\tilde{\boldsymbol{D}} = \boldsymbol{D}\boldsymbol{\Gamma}$ has the same singular values and Frobenius norm as $\boldsymbol{D}$, but the atom norms are changed. In particular, if $\boldsymbol{\Gamma}$ is a Givens rotation, only two atoms are affected; the multiplication reduces to

$$[\tilde{\boldsymbol{d}}_i\ \tilde{\boldsymbol{d}}_j] = [\boldsymbol{d}_i\ \boldsymbol{d}_j] \begin{bmatrix} c & s \\ -s & c \end{bmatrix}. \tag{4.21}$$

Imposing the condition that the first atom has unit norm,

$$\left\| \tilde{\boldsymbol{d}}_i \right\| = \left\| c\boldsymbol{d}_i - s\boldsymbol{d}_j \right\| = 1, \tag{4.22}$$

denoting

$$g_{ii} = \|\boldsymbol{d}_i\|, \quad g_{jj} = \left\|\boldsymbol{d}_j\right\|, \quad g_{ij} = \boldsymbol{d}_i^T \boldsymbol{d}_j, \tag{4.23}$$

and taking into account that $c^2 + s^2 = 1$, the rotation is given by

$$c = \frac{1}{\sqrt{1+t^2}}, \quad s = ct, \tag{4.24}$$

where t (the tangent of the angle defining the rotation) is

$$t = \frac{g_{ij} + \text{sign}(g_{ij})\sqrt{g_{ij}^2 - (g_{ii} - 1)(g_{jj} - 1)}}{g_{jj} - 1}. \tag{4.25}$$

A real solution exists if $(g_{ii} - 1)(g_{jj} - 1) < 0$, which means that one of the atoms has norm larger than 1, while the other has norm smaller than 1. This can be always ensured as long as not all atoms have unit norm. Hence, it is always possible to obtain a new atom with unit norm and so only $n-1$ rotations are necessary to obtain the UNTF. The whole procedure is summarized in Algorithm 4.1. The justification of (4.25) can be found in [49].

However, reminding our DL context, we should note that such a construction does not give a dictionary that is adapted to some set of training signals. ■

Example 4.12 To have an image of how the distribution of atom products might look for a UNTF, we generate $\boldsymbol{D}$ as above, starting from a matrix with normally distributed elements. We then compute the absolute values of the elements of $\boldsymbol{G} = \boldsymbol{D}^T\boldsymbol{D}$, keeping only those below the diagonal; the diagonal contains only values of 1 corresponding to atom norms, and the matrix $\boldsymbol{G}$ is symmetric. We sort increasingly the selected values; their number is $n(n-1)/2$. Figure 4.3 shows a plot of the sorted atom products, for $m = 64$, $n = 192$; the overcompleteness factor is thus 3; the horizontal line marks the Welch bound, equal to 0.1023 (this would be the distribution of an equiangular tight frame, if it existed for such dimensions). We note that the mutual coherence is quite large with respect to the Welch bound (about four times larger); however, there are not that many large atom products: only less than 5% are larger than twice the Welch bound. Since also the atom products are

Algorithm 4.1: Generate Random UNTF

Data: sizes $m < n$
Result: UNTF $\boldsymbol{D} \in \mathbb{R}^{m \times n}$

1 Generate random $\boldsymbol{\Delta} \in \mathbb{R}^{m \times n}$ and compute QR factorization $\boldsymbol{\Delta}^T = \boldsymbol{Q}\boldsymbol{R}$, $\boldsymbol{Q} \in \mathbb{R}^{n \times m}$
2 Initialize $\boldsymbol{D} = \sqrt{n/m}\,\boldsymbol{Q}^T$
3 **for** $i = 1$ **to** $n - 1$ **do**
4   **if** $\|\boldsymbol{d}_i\| \neq 1$ **then**
5   Find $j > i$ such that $\|\boldsymbol{d}_i\| < 1 < \|\boldsymbol{d}_j\|$ or $\|\boldsymbol{d}_j\| < 1 < \|\boldsymbol{d}_i\|$
6   Compute c, s like in (4.24), using (4.23) and (4.25)
7   Set $[\boldsymbol{d}_i\ \boldsymbol{d}_j] \leftarrow [\boldsymbol{d}_i\ \boldsymbol{d}_j]\begin{bmatrix} c & s \\ -s & c \end{bmatrix}$

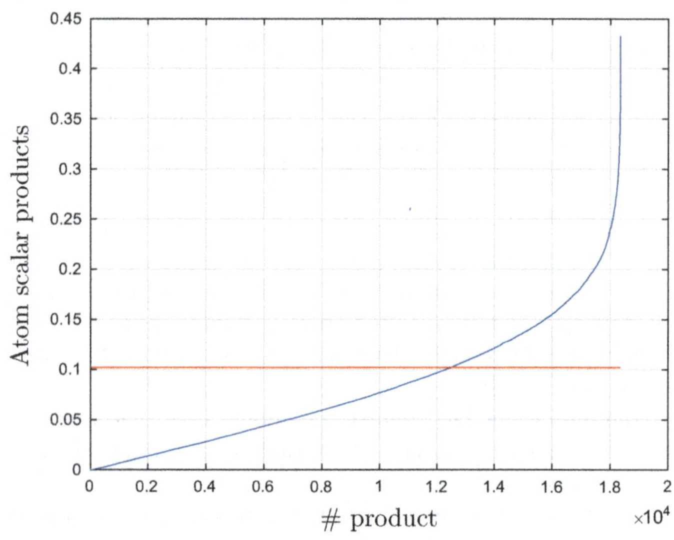

Fig. 4.3 The absolute values of atom products for a random UNTF. The horizontal line corresponds to the Welch bound

not close to 1, we expect good sparse representation recovery properties from such a dictionary. Of course, this is a single example, but its randomness shows rather generic properties; also, the design effort is minimal; better coherence properties of UNTFs can be obtained by optimization, see, e.g., [58, 163, 166, 192, 211]. ■

4.4 Joint Optimization of Error and Coherence

In this section, we describe DL algorithms that jointly optimize the representation error and the coherence of the dictionary. Since the Frobenius norm is easily manipulable in optimization algorithms, a promising idea is to define the penalty function h from (4.1) as the squared RMS coherence (4.16). This has only an indirect effect on the mutual coherence (1.25), but the previous section suggests that reasonably good results can be expected in obtaining incoherent dictionaries. Hence, the objective of the DL problem becomes [176]

$$\min_{\boldsymbol{D},\boldsymbol{X}} \|\boldsymbol{Y} - \boldsymbol{D}\boldsymbol{X}\|_F^2 + \gamma \|\boldsymbol{D}^T\boldsymbol{D} - \boldsymbol{I}\|_F^2 \tag{4.26}$$

where $\gamma > 0$ is a trade-off parameter. Remind that the diagonal elements of the penalty should be zero if normalization is enforced. Even if atoms norms are left free, they should be not too far from 1 due to the higher penalization of such values.

Example 4.13 To illustrate the trade-off offered by the problem (4.26), we present several dictionaries obtained with one of the methods presented later in this section for several values of γ. The relevant size parameters are $m = 64$, $n = 192$, $s = 4$, $N = 1024$; the signals are synthetically generated; these details are not really relevant, since the same behavior occurs for most parameter values.

Figure 4.4 shows the absolute values of atom products (compare with Fig. 4.3) and the error evolution for three values of γ: 0.1, 1, and 10. As expected, a higher γ value leads to more incoherence but to worse error. The final mutual coherences are 0.991, 0.612, 0.567, while the RMS coherences are 0.1234, 0.1077, 0.1027. Since the Welch bound is 0.1023, it means that the dictionary obtained with $\gamma = 10$ is almost a UNTF. This is what we should expect for large values of γ in (4.26). One

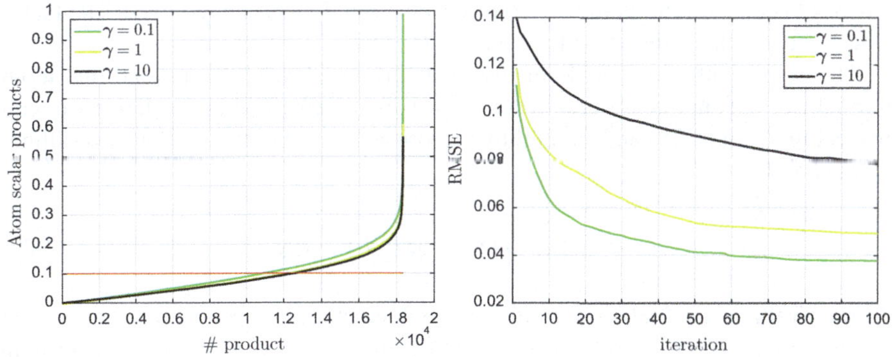

Fig. 4.4 Left: absolute values of atom products for dictionaries optimized with (4.26). Right: evolution of the approximation error for the same dictionaries

may hope that this UNTF is among the most adapted to represent the given signals, but there is no guarantee to it.

This example shows also that we must be careful when choosing the trade-off constant γ. A too small value has almost no effect on coherence, while a too large one may increase the error too much but give only marginal benefits in terms of incoherence. ■

Gradient-Based Optimization Using the trace derivative rules (3.14), the gradient of (4.26) is

$$2(\boldsymbol{D}\boldsymbol{X} - \boldsymbol{Y})\boldsymbol{X}^T + 4\gamma \boldsymbol{D}(\boldsymbol{D}^T\boldsymbol{D} - \boldsymbol{I}). \tag{4.27}$$

Comparing with (3.11), we remark that the gradient needs more operations although the second term (corresponding to the mutual coherence penalty) involves smaller matrices than the first, where a dimension is the number N of signals. So, the road to gradient-based optimization is clearly open; optimizing (4.26) is not much more complicated than optimizing (2.2).

In [176], a few iterations of the limited-memory BFGS algorithm are used in each dictionary update step that improves (4.26) for fixed representations. BFGS is a quasi-Newton method that approximates the inverse Hessian using only gradient information. Atom normalization is performed only at the end of the whole optimization process. The algorithm is called IDL (without mentioning whose initial is "I").

Coordinate Descent With the notation from Proposition 3.3, let us adapt to (4.26) the methods that update atoms successively. For optimizing atom $\boldsymbol{d}$, the relevant optimization objective is

$$\|\boldsymbol{F} - \boldsymbol{d}\boldsymbol{x}^T\|_F^2 + 2\gamma\|\bar{\boldsymbol{D}}^T\boldsymbol{d}\|^2 \tag{4.28}$$

where $\bar{\boldsymbol{D}}$ is the current dictionary from which atom $\boldsymbol{d}$ has been removed. The factor 2 accounts for the symmetry of the matrix $\boldsymbol{D}^T\boldsymbol{D}$.

The equivalent of the SGK method was proposed in [120]; remind that atom normalization is not considered when optimizing (4.28). Taking the derivative with respect to $\boldsymbol{d}_j$ in (4.28) and setting it to zero leads to the optimal (when everything else is fixed) atom

$$\boldsymbol{d} = \left(2\gamma\bar{\boldsymbol{D}}\bar{\boldsymbol{D}}^T + \|\boldsymbol{x}\|^2\boldsymbol{I}\right)^{-1}\boldsymbol{F}\boldsymbol{x}. \tag{4.29}$$

Although this is an analytic formula, its implementation means solving an $m \times m$ linear system, hence its complexity is much higher than that of the SGK formula (3.24).

An extension of this approach is made in [139] by optimizing simultaneously groups of atoms instead of a single atom. In this case, a Sylvester equation results instead of (4.29); it still can be solved efficiently but it is more expensive than solving linear systems. Moreover, generalizing (4.28), which means replacing the vector $\boldsymbol{d}$ with a matrix $\tilde{\boldsymbol{D}}$, accounts only for the scalar products between the atoms in the selected group and other atoms. The correct generalization should include the scalar products between the atoms from the group, that is the term $\gamma \left\| \tilde{\boldsymbol{D}}^T \tilde{\boldsymbol{D}} - \boldsymbol{I} \right\|_F^2$ should be added to the objective. This is solved in [139] by iteratively solving Sylvester equations that build a new $\tilde{\boldsymbol{D}}$ from the previous one.

To derive an equivalent of K-SVD, let us first notice that optimizing (4.28) with respect to the representation only leads to (3.39). Inserting this expression into (4.28), and also imposing unit norm, it results that the atom $\boldsymbol{d}$ must minimize the function

$$\|\boldsymbol{F}\|_F^2 - \boldsymbol{d}^T(\boldsymbol{F}\boldsymbol{F}^T - 2\gamma\bar{\boldsymbol{D}}\bar{\boldsymbol{D}}^T)\boldsymbol{d}. \tag{4.30}$$

This happens when $\boldsymbol{d}$ is the eigenvector corresponding to the maximum eigenvalue of the matrix

$$\boldsymbol{H} = \boldsymbol{F}\boldsymbol{F}^T - 2\gamma\bar{\boldsymbol{D}}\bar{\boldsymbol{D}}^T.$$

So, the "incoherent" K-SVD algorithm would compute this maximum eigenvector and then update the representations through (3.39). Note that when $\gamma = 0$, this is indeed equivalent to the K-SVD relations (3.35).

To adapt AK-SVD, we should employ (4.29), which is a departure from the low complexity principle that governs the algorithm. It is more in the spirit of the method to apply a single iteration of the power method on the matrix $\boldsymbol{H}$, starting with the previous version of the atom. So, we compute the new atom with

$$\boldsymbol{d} \leftarrow \boldsymbol{F}\boldsymbol{x} - 2\gamma\bar{\boldsymbol{D}}\bar{\boldsymbol{D}}^T\boldsymbol{d} \tag{4.31}$$

then normalize the atom. Note that we have replaced $\boldsymbol{F}^T\boldsymbol{d}_j$ with the current representations, as indicated by (3.39). Then the representations are updated via (3.39). This incoherent AK-SVD (abbreviated AK-SVDc) was proposed in [100] and is summarized in Algorithm 4.2. For brevity, we took a single inner iteration ($I = 1$, see Algorithm 3.7). The multiplication in line 4 costs an extra $O(mn^2)$ operations per DL iteration with respect to AK-SVD. Although the number N of signals is not involved, this may be not negligible compared with $O(smN)$, which is the cost of plain AK-SVD.

4.5 Optimizing an Orthogonal Dictionary

Although usually not directly interesting, an orthogonal dictionary will be useful as an auxiliary tool. (Note that in this case $m = n$.) The minimization of the representation error with such a dictionary is called orthogonal Procrustes problem.

Algorithm 4.2: AK-SVDc (Incoherent AK-SVD) Dictionary Update

Data: signals set $\boldsymbol{Y} \in \mathbb{R}^{m \times N}$
current dictionary $\boldsymbol{D} \in \mathbb{R}^{m \times n}$
representation matrix $\boldsymbol{X} \in \mathbb{R}^{n \times N}$
Result: updated dictionary $\boldsymbol{D}$

Compute error $\boldsymbol{E} = \boldsymbol{Y} - \boldsymbol{D}\boldsymbol{X}$
for $j = 1$ **to** n **do**
  Modify error: $\boldsymbol{F} = \boldsymbol{E}_{\mathcal{I}_j} + \boldsymbol{d}_j \boldsymbol{X}_{j,\mathcal{I}_j}$
  Update atom: $\boldsymbol{d}_j = \boldsymbol{F}\boldsymbol{X}_{j,\mathcal{I}_j}^T - 2\gamma \bar{\boldsymbol{D}}\bar{\boldsymbol{D}}^T \boldsymbol{d}_j$
  Normalize atom: $\boldsymbol{d}_j \leftarrow \boldsymbol{d}_j / \|\boldsymbol{d}_j\|$
  Update representation: $\boldsymbol{X}_{j,\mathcal{I}_j}^T = \boldsymbol{F}^T \boldsymbol{d}_j$
  Recompute error: $\boldsymbol{E}_{\mathcal{I}_j} = \boldsymbol{F} - \boldsymbol{d}_j \boldsymbol{X}_{j,\mathcal{I}_j}$

Proposition 4.14 *Given the matrices* $\boldsymbol{Y}, \boldsymbol{X} \in \mathbb{R}^{m \times N}$*, if* $\boldsymbol{Q} \in \mathbb{R}^{m \times m}$ *is orthogonal, then the approximation error* $\|\boldsymbol{Y} - \boldsymbol{Q}\boldsymbol{X}\|_F$ *is minimized by*

$$\boldsymbol{Q} = \boldsymbol{V}\boldsymbol{U}^T, \tag{4.32}$$

where the matrices $\boldsymbol{U}$, $\boldsymbol{V}$ *are obtained from the singular value decomposition*

$$\boldsymbol{X}\boldsymbol{Y}^T = \boldsymbol{U}\boldsymbol{\Sigma}\boldsymbol{V}^T. \tag{4.33}$$

Here $\boldsymbol{\Sigma}$ *is diagonal,* $\boldsymbol{U}$ *and* $\boldsymbol{V}$ *are orthogonal and all are* $m \times m$.

Proof The relation $\boldsymbol{Q}^T\boldsymbol{Q} = \boldsymbol{I}$ simplifies the representation error to

$$\|\boldsymbol{Y} - \boldsymbol{Q}\boldsymbol{X}\|_F^2 = \|\boldsymbol{Y}\|_F^2 + \|\boldsymbol{X}\|_F^2 - 2\mathrm{tr}(\boldsymbol{Y}^T\boldsymbol{Q}\boldsymbol{X}) = \|\boldsymbol{Y}\|_F^2 + \|\boldsymbol{X}\|_F^2 - 2\mathrm{tr}(\boldsymbol{Q}\boldsymbol{X}\boldsymbol{Y}^T).$$

Using (4.33) and denoting $\boldsymbol{W} = \boldsymbol{V}^T\boldsymbol{Q}\boldsymbol{U}$, the error is minimum if

$$\mathrm{tr}(\boldsymbol{Q}\boldsymbol{X}\boldsymbol{Y}^T) = \mathrm{tr}(\boldsymbol{W}\boldsymbol{\Sigma}) = \sum_{i=1}^{m} w_{ii}\sigma_i$$

is maximum. Since $\boldsymbol{W}$ is orthogonal, the maximum is attained when $w_{ii} = 1$, which means that $\boldsymbol{W}$ is the unit matrix. This gives immediately the expression (4.32). ■

Remark 4.15 (Dictionary Rotation) Returning to our usual problem (2.2), it is always possible to replace the dictionary $\boldsymbol{D}$ with a new dictionary $\boldsymbol{Q}\boldsymbol{D}$, where $\boldsymbol{Q}$ is orthogonal. This is called a rotation of the dictionary; multiplication with an orthogonal matrix does not change the norm of an atom, but only its direction. In view of Proposition 4.14, the new dictionary is not worse than the previous one. Indeed, one can take $\boldsymbol{Q}$ like in (4.32), but using now the singular value decomposition

$$\boldsymbol{D}\boldsymbol{X}\boldsymbol{Y}^T = \boldsymbol{U}\boldsymbol{\Sigma}\boldsymbol{V}^T \tag{4.34}$$

instead of (4.33). It results that, for this optimal $\boldsymbol{Q}$, the representation errors satisfy

$$\|\boldsymbol{Y} - \boldsymbol{Q}\boldsymbol{D}\boldsymbol{X}\|_F \leq \|\boldsymbol{Y} - \boldsymbol{D}\boldsymbol{X}\|_F.$$

Moreover, due to the orthogonality of $\boldsymbol{Q}$, the scalar products of the atoms are the same, since $(\boldsymbol{Q}\boldsymbol{D})^T(\boldsymbol{Q}\boldsymbol{D}) = \boldsymbol{D}^T\boldsymbol{D}$. So, the mutual coherence of the dictionary is not changed. ■

4.6 Imposing Explicit Coherence Bounds

Besides dealing with the RMS coherence, the approach based on (4.26) optimizes it together with the representation error, and hence cannot control directly its values. A different viewpoint was proposed in [10]: to work directly with the mutual coherence (1.25) and to impose a strict bound on its values. The optimization problem is formulated as

$$\begin{aligned} \min_{\boldsymbol{D},\boldsymbol{X}} \quad & \|\boldsymbol{Y} - \boldsymbol{D}\boldsymbol{X}\|_F^2 \\ \text{s.t.} \quad & |\boldsymbol{d}_i^T\boldsymbol{d}_j| \leq M_0, \ \forall i \neq j \end{aligned} \tag{4.35}$$

where M_0 is a given coherence bound. Of course, the bound must be larger than the Welch bound (4.15) in order to have a feasible problem.

The method proposed in [10] can be used with any DL algorithm that has the two-step structure of Algorithm 3.1. The idea is to add a third step after sparse coding and dictionary update, with the purpose of reducing the coherence below the imposed bound without losing too much in representation error. The third step affects only the dictionary and is composed of two operations: atom decorrelation and dictionary rotation. The latter operation was already presented in Remark 4.15; the former is explained below.

Assume that we have a dictionary $\boldsymbol{D}$ and we want to make it satisfy the constraint of (4.35) disregarding the increase in representation error. However, we would want to find an incoherent dictionary that is as near as possible from the original one, hoping that so the representation error will not increase too much. The decorrelation algorithm from [191] performs successive projections on two sets. For a dictionary, the Gram matrix

$$\boldsymbol{G} = \boldsymbol{D}^T\boldsymbol{D} \tag{4.36}$$

is symmetric, positive semidefinite, has rank m, its diagonal elements are equal to 1 and the off-diagonal elements are inner products of the atoms; if the dictionary satisfies the mutual coherence bound, then the off-diagonal elements of $\boldsymbol{G}$ are less than M_0 in absolute value. The two sets share some of the above properties and together they have all the properties.

The first set is

$$\mathcal{K} = \{\boldsymbol{G} \in \mathbb{R}^{n \times n} \mid \boldsymbol{G} = \boldsymbol{G}^T,\ g_{ii} = 1,\ |g_{ij}| \le M_0,\ \forall i, j\} \tag{4.37}$$

and imposes symmetry and the element value conditions. The second set is

$$\mathcal{P} = \{\boldsymbol{G} \in \mathbb{R}^{n \times n} \mid \boldsymbol{G} = \boldsymbol{G}^T,\ \boldsymbol{G} \succ 0,\ \mathrm{rank}(\boldsymbol{G}) \le m\} \tag{4.38}$$

and imposes symmetry and the spectral conditions. Note that the first set is convex; however, the second set is not convex because of the rank condition.

We now define projection operations onto the two sets. A matrix $\boldsymbol{G}$ is projected onto $\mathcal{K}$, operation denoted $\boldsymbol{K} = \mathrm{proj}_{\mathcal{K}}(\boldsymbol{G})$, by

$$\begin{aligned} k_{ii} &= 1 \\ k_{ij} &= \begin{cases} g_{ij}, & \text{if } |g_{ij}| \le M_0 \\ M_0, & \text{if } g_{ij} \ge M_0, \\ -M_0, & \text{if } g_{ij} \le -M_0 \end{cases} \qquad i \ne j \end{aligned} \tag{4.39}$$

So, basically, this is hard thresholding to the mutual coherence bound. The result is a symmetric matrix, but not necessarily positive semidefinite or with rank m.

The projection onto the spectral constraints set $\mathcal{P}$, denoted $\mathrm{proj}_{\mathcal{P}}(\boldsymbol{G})$, is defined by the following operations. First, the eigenvalue decomposition

$$\boldsymbol{G} = \boldsymbol{V}\boldsymbol{\Lambda}\boldsymbol{V}^T \tag{4.40}$$

is computed, where Λ is diagonal, with diagonal elements sorted decreasingly and $\boldsymbol{V}$ is orthogonal. Then, the spectrum is truncated to the first (and largest) m eigenvalues (provided they are positive) through

$$[\mathrm{Thresh}(\boldsymbol{\Lambda}, m)]_{ii} = \begin{cases} \lambda_{ii}, & \text{if } \lambda_{ii} > 0 \text{ and } i \le m \\ 0, & \text{if } \lambda_{ii} \le 0 \text{ or } i > m \end{cases} \tag{4.41}$$

Finally, the projection is

$$\mathrm{proj}_{\mathcal{P}}(\boldsymbol{G}) = \boldsymbol{V} \cdot \mathrm{Thresh}(\boldsymbol{\Lambda}, m) \cdot \boldsymbol{V}^T. \tag{4.42}$$

This is a rank-m positive semidefinite matrix, but its elements do not necessarily satisfy the coherence constraints.

Successive projections onto two sets are guaranteed to find an element in their intersection if both sets are convex; in general, the intersection element that is found is not the nearest from the starting point; however, it is usually a good approximation of the nearest point. Since $\mathcal{P}$ is not convex, the above property is not guaranteed by successive projections on $\mathcal{K}$ and $\mathcal{P}$, but it was observed that in most cases the result is in or very near the intersection of these two sets. In such case, due to (4.36) and (4.40), the new dictionary is immediately computed as

$$\boldsymbol{D} = [\text{Thresh}(\boldsymbol{\Lambda}, m)]^{1/2} \cdot \boldsymbol{V}^T \tag{4.43}$$

and (almost) satisfies the constraints of (4.35). Since M_0 is anyway a heuristically-chosen bound on the mutual coherence, a slight violation of the constraints of (4.35) is non-consequential. (A similar projection approach was adopted in [192] for the design of incoherent UNTF.)

Algorithm 4.3 summarizes a DL iteration based on the above ideas. The algorithm is named IPR (Iterative Projections and Rotations) after its decorrelation step. Sparse coding and dictionary update can be performed with the preferred algorithms and in principle do not interfere with decorrelation. The two projections and the rotation are performed a prescribed number of times or until some stopping criterion is met, although a few iterations seem to suffice (five iterations were used in [10]). Although the optimal rotation could be computed only once, after all the projections (since it does not change the inner products of the atoms), it was observed that better representation errors are obtained when the rotation is performed in the loop.

A smoother convergence may be ensured if the projection in line 5 is replaced by a weighted average with the current Gram matrix:

$$\boldsymbol{G} \leftarrow \alpha \boldsymbol{G} + (1-\alpha)\text{proj}_{\mathcal{K}}(\boldsymbol{G}),$$

with $\alpha \in (0, 1)$. With such an approach, the attempt to reach the target mutual coherence M_0 is made gradually; it is clear that several iterations are required to get near it; however, the decrease of the actual mutual coherence is more likely to be monotonous and the overall result is usually better than the brutal approach of imposing directly M_0. The value of α recommended by most practitioners is around 0.5.

Another version of the algorithm, proposed in [192], is to average the projections of the current dictionary on $\mathcal{K}$ and $\mathcal{P}$. In fact, in [192], there are four projections; $\mathcal{K}$ is seen as the intersections of two sets: in one $\boldsymbol{G}$ has unit diagonal, in the other $\boldsymbol{G}$ has bounded off-diagonal elements; $\mathcal{P}$ is replaced with the intersection of the set of rank-m matrices with the set of UNTF. However, from the limited numerical experience it appears that the averaged projections approach is slower than alternating projections.

The cost of IPR is much higher than that of a usual DL algorithm. Only the projection on $\mathcal{K}$ in line 5 and the thresholding in line 7 are negligible with respect to the dictionary update costs. The matrix multiplications in lines 4, 8 (there are three

Algorithm 4.3: IPR Iteration

Data: signals set $\boldsymbol{Y} \in \mathbb{R}^{m \times N}$
current dictionary $\boldsymbol{D} \in \mathbb{R}^{m \times n}$
desired mutual coherence M_0
Result: updated dictionary $\boldsymbol{D}$

1 Sparse coding: compute sparse representations $\boldsymbol{X}$ using OMP
2 Dictionary update: compute new dictionary $\boldsymbol{D}$ and possibly update the nonzeros of $\boldsymbol{X}$ with an algorithm of choice
3 **for** $it = 1$ **to** Nit **do**
4   Compute Gram matrix: $\boldsymbol{G} = \boldsymbol{D}^T \boldsymbol{D}$
5   Project to bound coherence: $\boldsymbol{G} \leftarrow \text{proj}_{\mathcal{K}}(\boldsymbol{G})$, using (4.39)
6   Compute eigenvalue decomposition: $\boldsymbol{G} = \boldsymbol{V} \boldsymbol{\Lambda} \boldsymbol{V}^T$
7   Threshold spectrum for new dictionary: $\boldsymbol{D} = [\text{Thresh}(\boldsymbol{\Lambda}, m)]^{1/2} \cdot \boldsymbol{V}^T$
8   Compute optimal rotation: $\boldsymbol{Q} = \boldsymbol{V}\boldsymbol{U}^T$ using the SVD $\boldsymbol{D}\boldsymbol{X}\boldsymbol{Y}^T = \boldsymbol{U}\boldsymbol{\Sigma}\boldsymbol{V}^T$
9   Set dictionary: $\boldsymbol{D} \leftarrow \boldsymbol{Q}\boldsymbol{D}$

multiplications here) and 9 have complexities of $O(mn^2)$, $O(m^3 + smN + mn^2)$, $O(m^2 n)$, respectively. The eigenvalue decomposition in line 6 is $O(n^3)$ and the SVD in line 8 is $O(m^3)$, but with large coefficients hidden by the $O(\cdot)$ notation. So, we should expect an execution time that is an order of magnitude higher than that of the algorithms from Chap. 3.

Example 4.16 We illustrate the behavior of Algorithm 4.3 in Fig. 4.5, using the same size parameters as in Example 4.13. Blue curves show IPR with AK-SVD as underlying dictionary update method. The desired mutual coherence is $M_0 = 0.2$, which is a relatively low value. The fact that many scalar products between atoms are around this value shows that IPR is effective in achieving its goal. Although imperfect, the shape of the curve resembles typical results for min-max (or infinity norm) optimization, where the variable distribution is flattened by the imposed bound.

Since only few projections iterations are used (we took $Nit = 5$), the actual final mutual coherence is slightly larger than the bound, precisely 0.2138. However, this is a quite satisfactory approximation. The RMS coherence is 0.1158, which is relatively far from the Welch bound 0.1023, meaning that the designed dictionary is not near from a UNTF.

The error evolution is relatively smoothly decreasing. Comparing with the case $\gamma = 1$ from Fig. 4.4, we see that the final error level is similar. The difference resides in the distribution of the atom products and is natural: each method is better at the measure it optimizes. The methods from Sect. 4.4 give lower RMS coherence, while IPR gives lower mutual coherence.

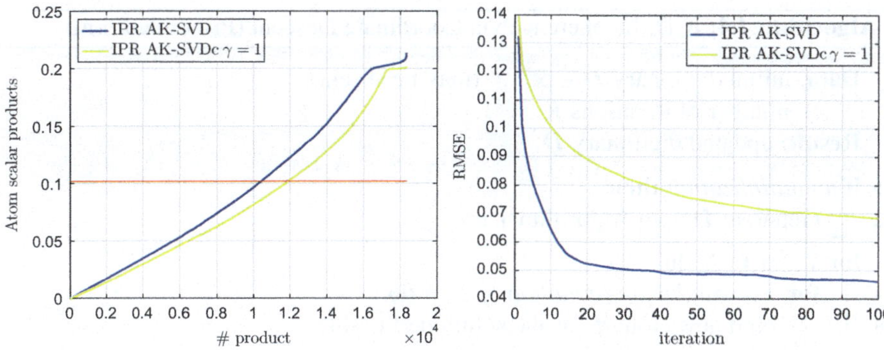

Fig. 4.5 Left: absolute values of atom products for a dictionary optimized with the IPR Algorithm 4.3, with AK-SVD (blue) and AK-SVDc (yellow) for dictionary update. Right: evolution of the approximation error

One can seek the best of both worlds by running IPR with the incoherent AK-SVD. The results are shown in yellow in Fig. 4.5, for $\gamma = 1$. The atom products are obviously smaller (the yellow curve is below the blue one). The RMS coherence is 0.1024, which means that the designed dictionary is practically a UNTF. The mutual coherence is 0.2023, i.e., only slightly larger than the bound M_0. The error is obviously larger when AK-SVDc is used; however, comparing with the near-UNTF dictionary designed in Example 4.13 (black curve in Fig. 4.4), the error is clearly better. Hence, overall, IPR with AK-SVDc achieves a better trade-off. ■

4.7 Atom-by-Atom Decorrelation

By choosing a right value for the bound M_0 and repeating the decorrelation operations of IPR, namely lines 4–7 of Algorithm 4.3, one can attempt finding Grassmannian frames. This is also the strategy from [192] for computing UNTFs. Conversely, almost any iterative method for computing (approximate) Grassmannian frames can be used in the DL context. We present here the principle of another way to perform decorrelation, generalized from [163].

Unlike the IPR approach, which changes all the atoms simultaneously, we adopt the atom-by-atom optimization technique already familiar from many of the DL algorithms presented in the previous chapter, which is in fact a coordinate descent method. We borrow the notation from Sect. 4.4 and particularly (4.28): $\boldsymbol{d}$ is the current atom, whose optimization we seek, and $\bar{\boldsymbol{D}}$ is the current dictionary (or frame) with atom $\boldsymbol{d}$ removed. As we want to optimize the mutual coherence, with only $\boldsymbol{d}$ as variable and all other atoms fixed, the optimal atom is the solution of

Algorithm 4.4: ICD: Incoherence via Coordinate Descent (Principle Form)

Data: initial dictionary $\boldsymbol{D} \in \mathbb{R}^{m \times n}$ (may be absent)
number of iterations K
Result: updated dictionary $\boldsymbol{D}$

if *no initialization* **then**
  Generate $\boldsymbol{D}$ with Algorithm 4.1
for $k = 1$ **to** K **do**
  for $j =$ *random permutation of* $2 : n$ **do**
    Find new atom $\boldsymbol{d}_j$ as the solution of (4.44)

$$
\begin{aligned}
\min_{\boldsymbol{d}} \quad & \|\bar{\boldsymbol{D}}^T \boldsymbol{d}\|_\infty \\
\text{s.t.} \quad & \|\boldsymbol{d}\| = 1
\end{aligned}
\tag{4.44}
$$

The principle form of an algorithm based on (4.44) is presented in Algorithm 4.4, named ICD (incoherence via coordinate descent). In the absence of a better initialization, we can always start from a random UNTF generated by Algorithm 4.1. The number K of iterations can be large if we aim finding an (approximate) Grassmannian frame or small if the algorithm is run for decorrelation in a DL context. In the atom update loop we take into account that one atom can always be fixed, since a rotation can be applied to give an atom any direction we wish (angles between atoms are invariant to rotation); without loss of generality, $\boldsymbol{d}_1$ is fixed. The atoms are swept in random order, since this usually improves convergence speed and can achieve better results.

Unfortunately, (4.44) is not a convex problem. Also, we wish to find a quick solution, since the problem is solved many times. In [163], the problem was relaxed to

$$
\begin{aligned}
\min_{\tilde{\boldsymbol{d}}} \quad & \|\bar{\boldsymbol{D}}^T \tilde{\boldsymbol{d}}\|_\infty \\
\text{s.t.} \quad & \left\|\tilde{\boldsymbol{d}} - \boldsymbol{d}\right\| \le \rho
\end{aligned}
\tag{4.45}
$$

So, we relax the unit norm condition and let the atom float in a ball around its previous value; the radius ρ of the ball is computed such that the new atom stays near the old one, but has freedom enough to reduce its mutual coherence with the other atoms. After solving (4.45), we put $\boldsymbol{d} \leftarrow \tilde{\boldsymbol{d}} / \left\|\tilde{\boldsymbol{d}}\right\|$. The ICD algorithm with (4.45) instead of (4.44) has excellent results in the design of incoherent frames, in the sense that it obtained the best mutual coherence values for many given frame sizes. Of course, the algorithm is not optimal neither locally (in the sense that each atom update is not optimal) nor globally.

Recent advances along the same basic idea (4.44) can be found in [58, 166]. The topic is still open.

Problems

P 4.1 Consider a dictionary $\boldsymbol{D}$ of size $m \times n$, where $n = km$, $k \in \mathbb{N}$. The dictionary is made of k concatenated orthogonal $m \times m$ blocks. Show that $\boldsymbol{D}$ is a tight frame. What is the value of the frame bounds A and B from (4.14)?

P 4.2 We consider the standard DL problem (2.2) where we seek normalized dictionary $\boldsymbol{D}$ and sparse representations $\boldsymbol{X}$ such that $\|\boldsymbol{Y} - \boldsymbol{D}\boldsymbol{X}\|_F^2$ is minimum, but we want the dictionary to be a UNTF. We adapt the standard alternate optimization approach from Algorithm 3.1 as follows:

1. The initialization is a random UNTF $\boldsymbol{D}$ generated with Algorithm 4.1.
2. Sparse coding is done as usual, producing new representations $\boldsymbol{X}$.
3. Dictionary update consists of designing an orthogonal matrix $\boldsymbol{Q}$ that minimizes $\|\boldsymbol{Y} - \boldsymbol{Q}\boldsymbol{D}\boldsymbol{X}\|_F^2$ as shown in Remark 4.15, and putting $\boldsymbol{D} \leftarrow \boldsymbol{Q}\boldsymbol{D}$.

Discuss the properties of the algorithm. Is it optimal or not?

P 4.3 Although regularization and incoherence constraints serve to partly similar goals, they also have disjoint implications. Hence, one may consider joining them, by adding the regularization term $\mu \|\boldsymbol{X}\|_F^2$ to the objectives of the optimization problems (4.26) and (4.35). Describe the modifications implied by this change to the IDL, AK-SVDc, and IPR algorithms.

Argue that, no matter the algorithm, the most sensible use of this combination is to set μ to zero after a few iterations, thus eliminating regularization.

P 4.4 Give the "right" version of Proposition 4.14: which orthogonal matrix $\boldsymbol{Q}$ minimizes the error $\|\boldsymbol{Y} - \boldsymbol{X}\boldsymbol{Q}\|$, where $\boldsymbol{Y}, \boldsymbol{X}$ are given?

P 4.5 Propose an algorithm that combines the objective of (4.26) and the constraints of (4.35). So, RMS coherence is minimized via the objective and mutual coherence is explicitly bounded. How would you choose the constants γ and M_0?

Chapter 5
Other Views on the DL Problem

Abstract The dictionary learning problem can be posed in different ways, as we have already seen. In this chapter we first take a look at the DL problem where the sparsity level is not bounded for each training signal; instead, we bound the average sparsity level. This allows better overall representation power, due to the ability to place the nonzeros where they are most needed. The simplest way to pose the problem is to combine the error objective with an ℓ_1 penalty that encourages sparsity in the whole representation matrix $\boldsymbol{X}$. Several algorithms can solve this problem; we present those based on coordinate descent in AK-SVD style, on majorization and on proximal gradient. The latter approach can also be used with a 0-norm penalization. Other modifications of the objective include the addition of a regularization term (elastic net) or of a coherence penalty. Another view is given by task-driven DL, where the optimization objective is taken directly from the application and the sparse representation is only an intermediary tool. Returning to the standard DL problem, we present two new types of algorithms. One is based on selection: the atoms are chosen from a pool of candidates and so are no longer free variables. The other is online DL, where the training signals are assumed to be available in small bunches and the dictionary is updated for each bunch; online DL can thus adapt the dictionary to a time-varying set of signals, following the behavior of the generating source. Two online algorithms are presented, one based on coordinate descent, the other inspired by the classic recursive least squares (RLS). Finally, we tackle the DL problem with incomplete data, where some of the signals elements are missing, and present a version of AK-SVD suited to this situation.

5.1 Representations with Variable Sparsity Levels

The basic DL problem (2.2) imposes the same number s of nonzeros in each column of the matrix $\boldsymbol{X}$; each signal is represented with at most s atoms, typically with exactly s atoms. This may be an advantage, for example, in compression, where only the atoms indices and the representation coefficients values need to be stored, but not their number, which is always s. However, in other applications like denoising (see Sect. 2.4.1), one may obtain better results by allowing different number of atoms

B. Dumitrescu, P. Irofti, *Dictionary Learning Algorithms and Applications*,
https://doi.org/10.1007/978-3-319-78674-2_5

in the representation of each signal. For the same total number of nonzeros in $\boldsymbol{X}$, e.g. Ns in (2.2), it is clear that allowing different sparsity levels $s_\ell = \|\boldsymbol{x}_\ell\|_0$ for each column would generally give a lower error $\|\boldsymbol{Y} - \boldsymbol{D}\boldsymbol{X}\|_F$. This should be true especially for low values of s.

We define the average sparsity level (ASL) by

$$\bar{s} = \frac{1}{N}\|\boldsymbol{X}\|_0 = \frac{1}{N}\sum_{\ell=1}^{N}\|\boldsymbol{x}_\ell\|_0 . \tag{5.1}$$

(We naturally extend the 0-norm to matrices, hence $\|\boldsymbol{X}\|_0$ is the number of nonzeros elements of the matrix $\boldsymbol{X}$.) If the number of nonzeros is the same for each signal representation, like in (2.2), then $\bar{s} = s$.

With variable sparsity levels, a first way of posing the problem is by imposing a fixed error level per signal. Denoting ς the desired RMSE (2.10), the optimization problem can be written as

$$\begin{aligned} \min_{\boldsymbol{D},\boldsymbol{X}} \quad & \|\boldsymbol{Y} - \boldsymbol{D}\boldsymbol{X}\|_F^2 \\ \text{s.t.} \quad & \|\boldsymbol{y}_\ell - \boldsymbol{D}\boldsymbol{x}_\ell\| \le \sqrt{m}\varsigma, \ 1 \le \ell \le N \end{aligned} \tag{5.2}$$

An implicit objective here is to obtain low values of the ASL or of $\|\boldsymbol{X}\|_0$, but it is hard to quantify this as an explicit constraint. With this sparsity constraint in mind, the only modification necessary to solve (5.2) with most of the algorithms presented in the previous chapters is to change the stopping criterion of OMP from number of nonzeros to error level; sparse coding for signal $\boldsymbol{y}_\ell$ is implemented by the call OMP($\boldsymbol{D}$, $\boldsymbol{y}_\ell$, m, $\sqrt{m}\varsigma$) (see Sect. 1.3 for the syntax of the OMP function). Although it looks redundant to minimize the error and constrain the same error, the DL process is likely to train atoms such that the error constraint is met with fewer atoms and hence ASL decreases as DL progresses. However, it is clearly difficult to find the correlation between the error level and the ASL other than by repeatedly solving the problem until a convenient compromise is obtained. Moreover, an important drawback is that imposing exactly the same error level to all representations may be too restrictive and so counterproductive in many applications.

In any case, a clear improvement of (5.2) is to impose simultaneously error and sparsity level restrictions, and thus transform it into

$$\begin{aligned} \min_{\boldsymbol{D},\boldsymbol{X}} \quad & \|\boldsymbol{Y} - \boldsymbol{D}\boldsymbol{X}\|_F^2 \\ \text{s.t.} \quad & \|\boldsymbol{y}_\ell - \boldsymbol{D}\boldsymbol{x}_\ell\| \le \sqrt{m}\varsigma, \ 1 \le \ell \le N \\ & \|\boldsymbol{x}_\ell\|_0 \le s \end{aligned} \tag{5.3}$$

Again, only sparse coding has to be changed into the call OMP($\boldsymbol{D}$, $\boldsymbol{y}_\ell$, s, $\sqrt{m}\varsigma$). In order to achieve a good compromise, the maximum sparsity level s has to be larger than the desired ASL; also, the maximum error should be larger than the desired average error. It is certainly difficult to choose good values from the beginning, hence the problem (5.3) still needs to be solved several times.

A more flexible optimization problem is obtained [116, 129, 148, 168, 210] by adding an ℓ_1 penalty to the representation error:

$$\min_{\boldsymbol{D},\boldsymbol{X}} \quad \|\boldsymbol{Y} - \boldsymbol{D}\boldsymbol{X}\|_F^2 + \lambda \|\boldsymbol{X}\|_{1,1} \tag{5.4}$$

where $\lambda > 0$ is a trade-off parameter between error and sparsity and the sparsifying term is defined with the norm

$$\|\boldsymbol{X}\|_{1,1} = \sum_{\ell=1}^{N} \|\boldsymbol{x}_\ell\|_1 = \sum_{\ell=1}^{N}\sum_{i=1}^{m} |x_{i\ell}|. \tag{5.5}$$

(Note that $\|\boldsymbol{X}\|_{1,1}$ is different from the matrix 1-norm $\|\boldsymbol{X}\|_1 = \max_\ell \sum_{i=1}^{m} |x_{i\ell}|$.) Tuning λ may be not immediate, and so the problem (5.4) must be solved several times, but the compromise between error and sparsity is made automatically. When solving the problem with different values of λ, assuming that the values are taken in increasing or decreasing order, a good initialization for the current λ is the solution $(\boldsymbol{D}, \boldsymbol{X})$ obtained for the previous λ. Obviously, a larger λ produces a sparser solution.

A possible drawback of (5.4) is that its naturally associated sparse coding algorithm is based on solving (1.9), hence has higher complexity than OMP. Since the dictionary is optimized for this type of coding, if faster solutions are needed in an application, OMP will probably give worse results than with a dictionary learned with OMP as sparse coder.

Remark 5.1 A legitimate question is whether the solution of (5.4) coincides with the solution of the basic DL problem (2.2). A partial answer is given in [85] in the *noiseless* case for the related problem

$$\begin{aligned} \min_{\boldsymbol{D},\boldsymbol{X}} \quad & \|\boldsymbol{X}\|_{1,1} \\ \text{s.t.} \quad & \boldsymbol{Y} = \boldsymbol{D}\boldsymbol{X} \\ & \|\boldsymbol{d}_j\| = 1,\ j = 1 : n \end{aligned} \tag{5.6}$$

Hugely simplifying the discussion, this problem has a local minimum in the true dictionary if the true representation is sparse enough, the dictionary is sufficiently incoherent and the number N of training signals is large enough. So, the result is similar to that from Remark 2.6, but the uniqueness issue is not solved. (Like there, atoms permutations and sign flips are not considered a different solution.)

In the noisy case, it was shown in [104] under a probabilistic model that (5.4) has a local minimum in a neighborhood of the true dictionary (which is the most we could expect). The hypotheses are similar in nature with those already mentioned above, but quite different in form. ■

5.2 A Simple Algorithm for DL with ℓ_1 Penalty

The straightest idea for solving (5.4) is the basic alternating optimization approach from Algorithm 3.1. The sparse coding step is performed with an algorithm specialized in optimizing (5.4) with fixed $\boldsymbol{D}$; if we look at each signal, the problem is basis pursuit (1.9). Then, the dictionary update can be made in several ways.

For example, if the SGK strategy from Sect. 3.3.2 is adopted, in which one atom is updated while all the other variables are fixed, the ℓ_1 penalty has no impact, since $\boldsymbol{X}$ is fixed; so the dictionary update of SGK is unchanged and the only modification in Algorithm 3.5 is in the sparse coding step. Another algorithm that works similarly was proposed in [129], but the dictionary update is made in online fashion and will be presented later independently. The same algorithm structure can be found in [116], where the dictionary is updated by minimizing the error norm, with the constraints $\|\boldsymbol{d}_j\| \leq 1$, by the Newton or conjugate gradient on the Lagrange dual function. (Note that this is not MOD, which does not impose any atom norm constraints; obviously, MOD could be used as well in the current context.)

We detail here the algorithm from [168], which works in AK-SVD style and hence its dictionary update stage optimizes each atom and the corresponding representation separately, thus belonging to the block coordinate descent family. The main difference from AK-SVD is that now the nonzero positions are no longer fixed in the dictionary update, but can be optimized as well; with the notation from Sect. 3.3.2, we take now $\mathcal{I}_j = 1 : N$. So, when optimizing atom $\boldsymbol{d}_j$, the representation error (3.19) without this atom is simply

$$\boldsymbol{F} = \boldsymbol{Y} - \sum_{i \neq j} \boldsymbol{d}_i \boldsymbol{x}_i^T. \tag{5.7}$$

(Remind that $\boldsymbol{x}_i^T$ is the ith row of $\boldsymbol{X}$.) When all the other atoms and representations are fixed, the problem (5.4) becomes

$$\min_{\boldsymbol{d}_j, \boldsymbol{x}_j^T} \quad \|\boldsymbol{F} - \boldsymbol{d}_j \boldsymbol{x}_j^T\|_F^2 + \lambda \|\boldsymbol{x}_j^T\|_1 \tag{5.8}$$

If the representation $\boldsymbol{x}_j^T$ is fixed, then the optimal normalized atom is the same as in (3.30), see Remark 3.5, namely

$$\boldsymbol{d}_j = \boldsymbol{F}\boldsymbol{x}_j^T / \|\boldsymbol{F}\boldsymbol{x}_j^T\|. \tag{5.9}$$

When the atom is fixed, then the problem (5.8) becomes

$$\min_{\boldsymbol{x}_j^T} \sum_{\ell=1}^{N} \left(\|\boldsymbol{f}_\ell - \boldsymbol{d}_j x_{j\ell}\|^2 + \lambda |x_{j\ell}| \right) \tag{5.10}$$

and so it is decoupled on the elements of $\boldsymbol{x}_j^T$. Optimizing

$$\|\boldsymbol{f}_\ell - \boldsymbol{d}_j \boldsymbol{x}_{j\ell}\|^2 + \lambda |x_{j\ell}| = \|\boldsymbol{f}_\ell\|^2 + \boldsymbol{x}_{j\ell}^2 - 2\boldsymbol{f}_\ell^T \boldsymbol{d}_j \boldsymbol{x}_{j\ell} + \lambda |x_{j\ell}|$$

leads to the solution (see proof of Proposition 1.8)

$$\boldsymbol{x}_{j\ell} = \text{Soft\_th}\left(\boldsymbol{f}_\ell^T \boldsymbol{d}_j, \frac{\lambda}{2} \right), \tag{5.11}$$

where the soft thresholding operator is defined in (1.22).

Putting together the above operations we obtain Algorithm 5.1, named Atom Profile Updating (APrU) DL. In principle, the sparse coding stage can be implemented with any algorithm. However, it was observed in [168] that using FISTA, described in Sect. 1.4, which is specialized in solving (5.4) (with fixed dictionary), gives much better results than when using OMP. This is natural, since FISTA and the dictionary update of APrU cooperate in optimizing the same objective, while OMP has no direct relation with this objective. Note that each step of APrU decreases the objective (5.4), since it represents an optimal descent on the atom or the representations (for the latter, in both lines 1 and 7; note that sparse coding works on the columns of $\boldsymbol{X}$, while the update in line 7 works on the rows). Remind, however, that the error itself is not necessarily optimal.

A variation of the algorithm, named One Stage (OS) DL in [168] is to completely eliminate the sparse coding stage. Indeed, since the dictionary update stage is able to change the nonzero positions in $\boldsymbol{X}$, then sparse coding is no longer necessary; however, an initialization of the representation matrix $\boldsymbol{X}$ is needed; it can be obtained with either FISTA or OMP, the latter called with the desired average sparsity level. The results given by OS appear to be only slightly worse than those of APrU, but OS is significantly faster.

An improvement is to end the algorithm with a few AK-SVD dictionary update steps that assume that the nonzeros structure of $\boldsymbol{X}$ is fixed, as given by the last iteration of APrU or OS; this ensures that only the error is optimized in (5.4), without the ℓ_1 penalty term. Penalization is useful for inducing sparsity, but its role stops when sparsity was indeed obtained and hence penalization actually biases the solution.

The complexity of APrU is not much higher than that of some algorithms presented in Chap. 3; for example, compared to AK-SVD, the dictionary update has the same operations, to which only the soft thresholding is added. In the sparse coding stage, APrU is more complex when FISTA is used instead of OMP. In the OS variant, APrU avoids the sparse coding stage and thus can be even faster than AK-SVD. Of course, this conclusion is valid for a single run; however, APrU needs typically several runs with different values of λ before obtaining the desired ASL.

Algorithm 5.1: One Iteration of APrU

Data: current dictionary $\boldsymbol{D} \in \mathbb{R}^{m \times n}$
signals set $\boldsymbol{Y} \in \mathbb{R}^{m \times N}$
Result: updated dictionary $\boldsymbol{D}$

Sparse coding: compute sparse representations $\boldsymbol{X}$
Compute error $\boldsymbol{E} = \boldsymbol{Y} - \boldsymbol{D}\boldsymbol{X}$
for $j = 1$ **to** n **do**
  Modify error: $\boldsymbol{F} = \boldsymbol{E} + \boldsymbol{d}_j \boldsymbol{x}_j^T$
  Update atom: $\boldsymbol{d}_j = \boldsymbol{F}\boldsymbol{x}_j^T / \|\boldsymbol{F}\boldsymbol{x}_j^T\|$
  for $\ell = 1$ **to** N **do**
    Update representation coefficient: $x_{j\ell} = \text{Soft\_th}\left(\boldsymbol{f}_\ell^T \boldsymbol{d}_j, \frac{\lambda}{2}\right)$
  Recompute error: $\boldsymbol{E} = \boldsymbol{F} - \boldsymbol{d}_j \boldsymbol{x}_j^T$

Example 5.2 We illustrate the advantages of a nonuniform distribution of nonzeros by comparing AK-SVD and the OS version of the APrU Algorithm 5.1. For the same ASL, APrU should produce a smaller overall error than AK-SVD because it has the ability to better distribute the nonzeros in order to reduce the error. Let us see if this really happens.

We run the DL algorithm on a matrix $\boldsymbol{Y}$ containing patches from images, with $m = 64$, $N = 8192$. The dictionary has $n = 128$ atoms and the initial dictionary is a random collection of training patches. For all the runs described below we use the same $\boldsymbol{Y}$ and the same initial dictionary. We run AK-SVD with several values of the sparsity level s. For APrU, we tune the trade-off parameter λ such that an ASL approximately equal to s is finally obtained; the matrix $\boldsymbol{X}$ is initialized via OMP($\boldsymbol{D}, \boldsymbol{y}_\ell, s, 0$), like in the first sparse coding stage of AK-SVD. Both AK-SVD and APrU are run for $K = 50$ iterations, which is enough for near-convergence. Subsequently, the APrU result is refined with 5 more AK-SVD dictionary update iterations with fixed nonzero structure.

The results are shown in Table 5.1. We see that for each value of s, APrU can produce a lower RMSE than AK-SVD, with an ASL that is slightly smaller than s. The improvement is somewhat larger for smaller s, but not dramatically. The values of λ have been found with only few tries (of course, a bit of experience helps in reducing the number of tries). We note that the final refinement of APrU is very important in reducing the error, which is otherwise larger than that of AK-SVD, due to the optimization of the penalized error.

The reason why APrU produces a smaller error is almost entirely due to the way of posing the problem, not to algorithmic superiority. The implicit flexibility in allocating the nonzeros is the key. For example, when $s = 10$, the number of nonzeros used by a representation (i.e., the number of nonzeros on a column of $\boldsymbol{X}$) varies between 0 and 37, while for AK-SVD and the standard DL problem it is a flat 10. ■

Table 5.1 Comparison between AK-SVD and APrU for image representation

	s	2	3	4	6	8	10	12
AK-SVD	RMSE/10^{-5}	9.16	7.76	6.87	5.66	4.96	4.47	4.12
APrU	RMSE/10^{-5}	7.89	6.53	5.76	4.85	4.31	3.91	3.60
	ASL	1.80	2.87	3.87	5.90	7.81	9.80	11.74
	$\lambda/10^{-4}$	12	8	6	4	3	2.4	2

5.3 A Majorization Algorithm

We present here another algorithm for DL with ℓ_1 penalty, based on majorization and developed in [210]. The main idea behind majorization techniques is to replace the objective with a function that is always larger, but has better optimization properties. We denote $f_\lambda(\boldsymbol{D}, \boldsymbol{X})$ the objective of (5.4) and $(\boldsymbol{D}_0, \boldsymbol{X}_0)$ the current values of the dictionary and representations. The new pair $(\boldsymbol{D}, \boldsymbol{X})$ is found by alternate minimization.

For fixed $\boldsymbol{D} = \boldsymbol{D}_0$, the next representation matrix $\boldsymbol{X}$ is computed by minimizing

$$f_\lambda(\boldsymbol{D}_0, \boldsymbol{X}) + c_D \|\boldsymbol{X} - \boldsymbol{X}_0\|_F^2 - \|\boldsymbol{D}_0(\boldsymbol{X} - \boldsymbol{X}_0)\|_F^2 . \tag{5.12}$$

If the constant c_D satisfies the condition $c_D \geq \|\boldsymbol{D}_0\|_2^2$, then the function (5.12) majorizes $f_\lambda(\boldsymbol{D}_0, \boldsymbol{X})$. Also, the function is equal to f_λ when $\boldsymbol{X} = \boldsymbol{X}_0$. The form of (5.12) also ensures that the optimal $\boldsymbol{X}$ is not far from $\boldsymbol{X}_0$, thus smoothing the optimization process.

The minimum of (5.12) is attained for

$$\boldsymbol{X} = \text{Soft\_th}\left(\frac{1}{c_D}\left(\boldsymbol{D}_0^T\boldsymbol{Y} + (c_D\boldsymbol{I} - \boldsymbol{D}_0^T\boldsymbol{D}_0)\boldsymbol{X}_0\right), \frac{\lambda}{2}\right), \tag{5.13}$$

where the soft thresholding operator (1.22) works at element level if the argument is a matrix. The computation effort is not much larger than for APrU, with the remark that the whole representation matrix is computed here. Finding a good value c_D needs some extra operations, but the approximation of the largest singular eigenvalue of a matrix can be done efficiently.

The dictionary update is done similarly, but with some complications if normalization is sought. It is proposed in [210] to replace atom normalization with a boundedness condition $\boldsymbol{d}_j^T\boldsymbol{d}_j \leq 1$; this is in fact enough to keep the dictionary well behaved, as long as the representations are also bounded, which is implicitly ensured by the presence of the term $\|\boldsymbol{X}\|_{1,1}$ in the objective (note that the term is active only in the sparse representations update). With fixed $\boldsymbol{X} = \boldsymbol{X}_0$ and with Lagrange multipliers for the inequality constraints, the function to be minimized is

$$\|Y - DX_0\|_F^2 + c_X \|D - D_0\|_F^2 - \|(D - D_0)X_0\|_F^2 + \sum_{j=1}^{n} \theta_j (d_j^T d_j - 1). \quad (5.14)$$

This is the Lagrangian of the function formed by the first three terms, which is convex in D, and $\theta_j \geq 0$ are the multipliers. Due to convexity, the optimal multipliers θ_j and atoms d_j satisfy the KKT condition

$$\theta_j (d_j^T d_j - 1) = 0. \quad (5.15)$$

So, either the atom has norm equal to one or the multiplier is zero. Setting to zero the gradient of (5.14), the optimal dictionary is

$$D = \frac{1}{c_X} \left(Y X_0^T + D_0 (c_X I - X_0 X_0^T) \right) \left(\frac{1}{c_X} \Theta + I \right)^{-1} \triangleq B \left(\frac{1}{c_X} \Theta + I \right)^{-1}, \quad (5.16)$$

where $\Theta = \operatorname{diag}(\theta_j)$. Since $1/c_X \Theta + I \succeq I$ is diagonal, each of its diagonal elements acts as a scaling factor for the columns of the matrix B. In view of (5.15), either $\|b_j\| < 1$ and then $\theta_j = 0$ and $d_j = b_j$, or $\|b_j\| \geq 1$ and $d_j = b_j / \|b_j\|$ (and θ_j assumes the positive value that makes possible the latter equality).

A similar approach can be used if atom normalization is strictly imposed. The multipliers θ_j can also be negative, but essentially it is enough to normalize the columns of B to obtain the current dictionary.

The DL algorithm from [210] simply iterates (5.16) (with the scaling described above) and (5.13), possibly repeating the use of each relation several times before going to the other. If the constants c_X and c_D are correctly chosen, the process converges to a local minimum.

5.4 Proximal Methods

Proximal algorithms have seen a great resurgence in the latest years, due to their ability to solve large scale optimization problems while offering convergence guarantees for many types of problems. We will discuss them only as much as necessary for solving DL problems, starting with (5.4); see [143] for an introduction to proximal algorithms and also [7] for a broader view on optimization with sparsity inducing penalties; the bibliographies of the papers cited below also contain valuable references. We have already touched the issue of proximal algorithms in Sect. 1.4, but here the presentation will be more general.

Split Objective Among others, proximal algorithms can be applied to problems that have the form

$$\min_{\xi} \quad f(\xi) + g(\xi) \quad (5.17)$$

where f is differentiable and g is convex but not necessarily smooth or even continuous; if f is also convex, then the problem becomes much easier, but this is not our case. The variable $\boldsymbol{\xi}$ lies on a compact set. In our case, the variable comprises the dictionary and the representations, $\boldsymbol{\xi} = (\boldsymbol{D}, \boldsymbol{X})$; for the dictionary, the compact set is that defined by $\|\boldsymbol{d}_j\| \leq 1$, i.e., normalization is performed only after the DL algorithm, like in the previous method from this section. The representations are artificially bounded by

$$|x_{ij}| \leq T, \tag{5.18}$$

where T is a large constant that is usually irrelevant. We trivially define projections on these compact (and convex) sets; for an atom $\boldsymbol{d}_j$, the projection is

$$\mathcal{P}_D(\boldsymbol{d}_j) = \begin{cases} \boldsymbol{d}_j, & \text{if } \|\boldsymbol{d}_j\| \leq 1 \\ \boldsymbol{d}_j / \|\boldsymbol{d}_j\|, & \text{otherwise} \end{cases}$$

For a representation coefficient, the projection is

$$\mathcal{P}_X(x_{ij}) = \begin{cases} x_{ij}, & \text{if } |x_{ij}| \leq T \\ \text{sign}(x_{ij})T, & \text{otherwise} \end{cases}$$

The extension of the projection operators to the full matrices $\boldsymbol{D}$ and $\boldsymbol{X}$ is immediate.

Transforming (5.4) into (5.17) leads to the functions

$$f(\boldsymbol{D}, \boldsymbol{X}) = \|\boldsymbol{Y} - \boldsymbol{D}\boldsymbol{X}\|_F^2, \quad g(\boldsymbol{D}, \boldsymbol{X}) = \lambda \|\boldsymbol{X}\|_{1,1}. \tag{5.19}$$

We give now a few definitions and results regarding the proximal operator and algorithms that use it.

Definition 5.3 The proximal operator associated with the function g is

$$\text{prox}_{tg}(\boldsymbol{v}) = \arg\min_{\boldsymbol{\xi}} \left(g(\boldsymbol{\xi}) + \frac{1}{2t} \|\boldsymbol{\xi} - \boldsymbol{v}\|^2 \right), \tag{5.20}$$

where $t > 0$ is a constant.

The constant t will soon get the significance of step size, in the context of proximal algorithms. It is clear that if $g(\boldsymbol{v})$ is minimum, then $\text{prox}_{tg}(\boldsymbol{v}) = \boldsymbol{v}$, hence $\boldsymbol{v}$ is a fixed point of the proximal operator. For other $\boldsymbol{v}$, the proximal operator compromises between the minimum of g and the current position $\boldsymbol{v}$. We can see the proximal operator as a regularization of the function g depending on the current point.

Remark 5.4 Let us compute the proximal operators for a few simple penalty functions, where the definition (5.20) gives immediately the result. The reader is invited to fill in the details.

1. The usual 2-norm regularization (just for practice):

$$g(\boldsymbol{\xi}) = \frac{1}{2}\|\boldsymbol{\xi}\|^2 \Rightarrow \text{prox}_{tg}(\boldsymbol{v}) = \frac{1}{1+t}\boldsymbol{v}$$

2. The ℓ_1-norm (useful in our current DL setup):

$$g(\boldsymbol{\xi}) = \|\boldsymbol{\xi}\|_1 \Rightarrow \text{prox}_{tg}(\boldsymbol{v}) = \text{Soft\_th}(\boldsymbol{v}, t) \tag{5.21}$$

3. The 0-norm (useful later):

$$g(\boldsymbol{\xi}) = \|\boldsymbol{\xi}\|_0 \Rightarrow \text{prox}_{tg}(\boldsymbol{v}) = \text{Hard\_th}(\boldsymbol{v}, \sqrt{2t}), \tag{5.22}$$

   where the function

$$\text{Hard\_th}(\xi, t) = \begin{cases} \xi, \text{ if } |\xi| > t \\ 0, \text{ otherwise} \end{cases} \tag{5.23}$$

   represents the *hard thresholding* operator. The operator is defined for scalars in (5.23); extension to vectors or matrices is done elementwise.

We prove only (5.22), for the scalar case; generalization to vectors is immediate. If $\xi \neq 0$, then the function to be minimized in (5.20) is $1 + (\xi - v)^2/2t$, whose minimum is $\xi = v$; the optimal value of the function is 1. If $\xi = 0$, the function is $v^2/2t$. So, if $v^2/2t < 1$, then $\text{prox}_{tg}(v) = 0$, otherwise $\text{prox}_{tg}(v) = v$; this is in fact (5.22). ■

The proximal gradient method for solving (5.17) is defined by the iteration

$$\boldsymbol{\xi} \leftarrow \text{prox}_{tg}\left(\boldsymbol{\xi} - t\frac{\partial f(\boldsymbol{\xi})}{\partial \boldsymbol{\xi}}\right). \tag{5.24}$$

So, this is the combination of a gradient descent on the function f with a regularized minimization of the function g. The step size t may be variable and has to be small enough; alternatively, one can use line search for a nearly optimal t. Under mild conditions, the iteration defined by (5.24) converges; if f is convex, then it converges to the (global) minimum.

The iteration (5.24) was used in [148] to solve (5.4). For this DL problem, with functions defined in (5.19), the iteration can be split on the variables $\boldsymbol{D}$ and $\boldsymbol{X}$. Reminding the formulas (3.11) and (3.15) for the gradient of the error function f, the proximal gradient method (5.24) gives the following update rules.

For the dictionary, since the function g depends only on $\boldsymbol{X}$, we have a simple gradient step:

$$D \leftarrow \mathcal{P}_D\left(D + t(Y - DX)X^T\right). \tag{5.25}$$

The only difference from the Sparsenet Algorithm 3.4 is that normalization is seen differently here by allowing atom norms to be also subunitary.

The representations have a slightly more complicated update formula, involving the soft thresholding operator (1.22):

$$X \leftarrow \mathcal{P}_X\left(\text{Soft\_th}\left(X + tD^T(Y - DX), t\lambda\right)\right). \tag{5.26}$$

Iterating (5.25) and (5.26) gives the DL algorithm for solving (5.4). Indications on how to adaptively compute the step size and other details are given in [148]. Besides the step size computation, the complexity of this method is similar to that of APrU, which still holds the advantage of not needing any extra parameter.

ℓ_0 Penalty [9] Using directly the sparsity level in the objective appears to complicate the problem a lot from the optimization viewpoint, although it is more meaningful than the ℓ_1-norm. However, the proximal operator can tame the 0-norm and so we consider the problem

$$\min_{D,X} \quad \|Y - DX\|_F^2 + \lambda\|X\|_0 \tag{5.27}$$

The proximal gradient method applied to this problem gives the same dictionary update (5.25) as for the ℓ_1 penalty (since, again, the penalty function does not depend on D). Using (5.22), the representation update is (compare with (5.26))

$$X \leftarrow \mathcal{P}_X\left(\text{Hard\_th}\left(X + tD^T(Y - DX), \sqrt{2t\lambda}\right)\right). \tag{5.28}$$

Note that we still impose the representation bounds. The step size t needs to be small enough and can be varied.

Remark 5.5 (Atom-by-Atom Update) The update (5.25) changes the whole dictionary in a single operation, hence it can be viewed as a parallel update of all atoms. However, we can update one atom at a time by simply applying (5.25) one column at a time. Since the new atoms are immediately used in the update of the next atoms, this Gauss–Seidel-like update gives potentially better result, with the extra effort of recomputing the error matrix (which can be done efficiently, see the algorithms in Chap. 3). This is the scheme actually proposed in [9], where the convergence of the proximal gradient method is proved in this case, under mild technical conditions. ■

5.5 A Gallery of Objectives

Besides the problems presented in the previous sections, the DL literature contains other ways of setting the objective of the DL problem. Some of them are obtained by simply combining two or more of the penalty functions already discussed; others have a different character. We present here a short list of objectives and hints to the corresponding algorithms.

Elastic Net Regularization The ℓ_1-penalty (5.4) can be combined with regularization, as proposed in [174], giving the DL problem

$$\min_{\boldsymbol{D},\boldsymbol{X}} \quad \|\boldsymbol{Y}-\boldsymbol{D}\boldsymbol{X}\|_F^2 + \lambda\|\boldsymbol{X}\|_{1,1} + \mu\|\boldsymbol{X}\|_F^2 \tag{5.29}$$

The problem stays convex in each of the arguments, $\boldsymbol{D}$ and $\boldsymbol{X}$, when the other is constant. Several algorithms can be adapted for solving (5.29). Let us derive an algorithm in the style of APrU Algorithm 5.1. Using the same notation as in Sect. 5.2, the optimization of atom $\boldsymbol{d}_j$ and of the corresponding representation coefficients $\boldsymbol{x}_j^T$, when everything else is fixed, is

$$\min_{\boldsymbol{d}_j,\boldsymbol{x}_j^T} \quad \|\boldsymbol{F}-\boldsymbol{d}_j\boldsymbol{x}_j^T\|_F^2 + \lambda\|\boldsymbol{x}_j^T\|_1 + \mu\|\boldsymbol{x}_j^T\|^2 \tag{5.30}$$

The update (5.9) of the atom $\boldsymbol{d}_j$ is unchanged. When the atom is fixed, then (5.30) amounts to minimizing

$$\sum_{\ell=1}^{N}\left((1+\mu)\boldsymbol{x}_{j\ell}^2 - 2\boldsymbol{f}_\ell^T\boldsymbol{d}_j\boldsymbol{x}_{j\ell} + \lambda|x_{j\ell}|\right)$$

and the solution is

$$\boldsymbol{x}_{j\ell} = \frac{1}{1+\mu}\text{Soft\_th}\left(\boldsymbol{f}_\ell^T\boldsymbol{d}_j, \frac{\lambda}{2}\right). \tag{5.31}$$

So, comparing with (5.11), we note that regularization leads to shrinking the coefficients with $1+\mu$, as already noticed in Sect. 4.2.

Coherence Penalty As seen in Chap. 4, there are several ways of keeping coherence under control. Some of them can be easily combined with the ℓ_0 or ℓ_1 penalties. For example, it was proposed in [9] to solve

$$\min_{\boldsymbol{D},\boldsymbol{X}} \quad \|\boldsymbol{Y}-\boldsymbol{D}\boldsymbol{X}\|_F^2 + \lambda\|\boldsymbol{X}\|_0 + \gamma\|\boldsymbol{D}^T\boldsymbol{D}-\boldsymbol{I}\|_F^2 \tag{5.32}$$

Like for (5.27), a proximal gradient method was chosen to solve the above problem. The dictionary update step resembles the IDL method from [176], see Sect. 4.4, in the sense that the gradient is used. However, in [9], the update is performed atom by atom; a simple gradient step is applied, with a carefully estimated step size.

Impulsive Noise and ℓ_1 Error [107] The Frobenius norm of the error matrix is the right minimization objective when the noise is Gaussian. However, when the noise has also an impulsive component or has a longer tail, the ℓ_1 norm is more appropriate. For example, in [107], sparse representations are used for text documents classification, in particular for the detection of new topics. With an ℓ_1 error norm, the DL problem can be expressed as

$$\min_{\boldsymbol{D},\boldsymbol{X}} \quad \|\boldsymbol{Y} - \boldsymbol{D}\boldsymbol{X}\|_{1,1} + \lambda\|\boldsymbol{X}\|_{1,1} \tag{5.33}$$

The objective is convex in one of the variables, $\boldsymbol{D}$ or $\boldsymbol{X}$, if the other is fixed. Hence, the alternate optimization idea still works. In each variable, the problem belongs to the linear programming class, hence it can be solved exactly with one of the many existing algorithms. An interesting online algorithm is given in [107], based on the alternating direction method of multipliers (ADMM).

Minimizing the Error with Fixed Total Number of Nonzeros A variation on the theme discussed in Sect. 5.1 can be made by imposing a total budget of nonzero coefficients for the whole representation matrix $\boldsymbol{X}$ or, equivalently, on the average sparsity level (5.1). Denoting S the bound on the total number of nonzeros, the optimization problem is

$$\begin{aligned} \min_{\boldsymbol{D},\boldsymbol{X}} \quad & \|\boldsymbol{Y} - \boldsymbol{D}\boldsymbol{X}\|_F^2 \\ \text{s.t.} \quad & \|\boldsymbol{X}\|_0 \le S \end{aligned} \tag{5.34}$$

The algorithm proposed in [200] has two stages. The first consists of running a standard DL algorithm (MOD is preferred in the cited paper, but any dictionary update method can be accommodated) with a group-level OMP. Instead of coding each signal independently, this OMP acts on all signals by choosing the atom and the signal that have the largest atom-residual scalar product; the support for that signal is increased with the current atom index and the new coefficients and residual are computed like in standard OMP; this operation is repeated S times. Such an OMP could be obtained with the call OMP($\boldsymbol{I}_N \otimes \boldsymbol{D}$, vec($\boldsymbol{Y}$), S, 0), but this would be impractical due to the huge size of the matrix $\boldsymbol{I}_N \otimes \boldsymbol{D}$. An efficient implementation can be obtained if the algorithm is seen as N concurrent versions of the standard OMP that advance one step when their atom-residual product is the best. The memory requirements are moderately large, due to the need to store all residuals, which need an extra $m \times N$ matrix. The computation complexity is also larger than for running N independent OMP algorithms; the extra operations are only for computing the best atom-residual product among the N candidates produced by each individual OMP in the usual way; so, the number of extra operations is $O(SN) = O(\bar{s}N^2)$, where $\bar{s}$ is the ASL (5.1). This extra load may become significant if N is large.

The second stage of the algorithm from [200] is to redistribute the nonzeros such that the total number of nonzeros (and the ASL) stays constant. This is done on the rows of $\boldsymbol{X}$, either on a single row or between two rows; these operations are intertwined with regular dictionary updates. All operations are guaranteed to reduce the representation error without changing the total number of nonzeros, hence the algorithm is guaranteed to converge. We present here only the redistribution of zeros on a single row, whose index is j. Like in Sect. 3.3.2, we optimize atom j (denoted $\boldsymbol{d}$) and representation row j (denoted $\boldsymbol{x}^T$), with the difference that the nonzeros positions are not preset; however, their number s_j is fixed. Like in (5.7), we define by

$$\boldsymbol{F} = \boldsymbol{Y} - \sum_{i \neq j} \boldsymbol{d}_i \boldsymbol{x}_i^T$$

the error matrix when computed with all the other atoms and representations. Unlike in (3.19), now the full rows are considered, since we do not know the positions of the zeros. Assuming $\boldsymbol{d}$ is fixed (and normalized), we want to find the support of the representation $\boldsymbol{x}^T$, i.e., we want to solve the problem

$$\begin{aligned} \min_{\boldsymbol{x}} \quad & \|\boldsymbol{F} - \boldsymbol{d}\boldsymbol{x}^T\|_F^2 \\ \text{s.t.} \quad & \|\boldsymbol{x}\|_0 = s_j \end{aligned} \tag{5.35}$$

If we disregard the sparsity constraint, Proposition 3.8 gives the solution $\boldsymbol{x} = \boldsymbol{F}^T\boldsymbol{d}$. Inserting this expression into (5.35) gives

$$\left\|\boldsymbol{F} - \boldsymbol{d}\boldsymbol{x}^T\right\|_F^2 = \|\boldsymbol{F}\|_F^2 - \left\|\boldsymbol{F}^T\boldsymbol{d}\right\|^2.$$

If $\boldsymbol{x}$ has only s_j nonzeros, then they are in the positions of the largest (in absolute value) s_j elements of $\boldsymbol{F}^T\boldsymbol{d}$. After finding the support, the optimal atom and representations can be computed like in K-SVD, see Proposition 3.7.

The procedure for redistributing the nonzeros between two rows is somewhat more intricate and can be found in [200] together with numerical results that show better errors than K-SVD or error based K-SVD (the latter solving (5.2)) for the same total number of nonzeros. This is expected, as explained in the beginning of this chapter, due to the extra degree of freedom in placing the nonzeros; however, there is no comparison with a method with a similar goal, e.g., one based on the ℓ_1 penalty.

The nonzeros redistributions are alternated with dictionary updates that can also modify the representations, in the style of standard DL methods. In [200], a MOD-like procedure is preferred, but other methods like AK-SVD could be employed as well. Both nonzero redistribution algorithms and dictionary update are guaranteed to reduce the representation error, hence the whole algorithm is guaranteed to converge (to a local minimum).

Classification Oriented Objectives Besides the objectives presented in this chapter and those omitted by the current authors (mostly unwillingly), some DL methods are designed especially for classification and have objectives tuned to this purpose. They will be discussed later in Chap. 8.

5.6 Task-Driven DL

Until now, our unique concern was to find a dictionary such that the training signals are modeled as sparse representations of the atoms. Such an approach is driven by the available data, with no direct connection with a possible application. The idea to consider a *task-driven* DL was proposed in [130] and is an attempt to optimize the dictionary with an objective that models the goal of the application.

Let $f(\boldsymbol{D}, \boldsymbol{X})$ be a function quantifying the sparse representation error, which can be the simple quadratic error (3.10), the ℓ_1 penalized error (5.4), or the elastic net objective (5.29); the first function is simpler, but the other two do not need a special constraint for sparsity; the third function is adopted in [130], due to the better numerical behavior. For a given dictionary, let $\boldsymbol{X}^*$ be the optimal representation that minimizes the error function,

$$\boldsymbol{X}^* = \arg\min_{\boldsymbol{X}} f(\boldsymbol{D}, \boldsymbol{X}) \tag{5.36}$$

We have already seen in Sect. 3.7 a method, SimCO, working with the optimal representation, but now the approach is quite different.

We assume that each signal $\boldsymbol{y}$ is associated with a vector $\boldsymbol{z}$, specific to the application at hand (see below Example 5.6). We want to find a model for $\boldsymbol{z}$, described by a parameter matrix $\boldsymbol{W}$ and the optimal sparse representation $\boldsymbol{x}^*$, which is thus the connection between $\boldsymbol{y}$ and $\boldsymbol{z}$. The simplest example of such model is a regression where ideally $\boldsymbol{z} = \boldsymbol{W}\boldsymbol{x}^*$. In general, we can measure the appropriateness of the model with a convex function $h(\boldsymbol{W})$. For example, in the regression case, we could take

$$h(\boldsymbol{W}) = \left\| \boldsymbol{Z} - \boldsymbol{W}\boldsymbol{X}^* \right\|_F^2 + \gamma \left\| \boldsymbol{W} \right\|_F^2, \tag{5.37}$$

where $\boldsymbol{Z}$ is the matrix containing the N vectors associated with the training signals $\boldsymbol{Y}$; regularization is again used for improving the numerical properties of the optimization process.

Example 5.6 Halftoning a grayscale image means building an image with only black and white pixels that looks similar to the original. Efficient algorithms for this purpose have been proposed already in the 1970s. The inverse operation, to obtain a grayscale image from the halftone version, is more difficult. It can be formulated as follows in the context of task-driven DL. The training set consists of halftone image patches $\boldsymbol{Y}$ and the corresponding grayscale patches $\boldsymbol{Z}$. Assume that, using these training data, we learn (see some algorithmic details below) a dictionary $\boldsymbol{D}$ and a parameter matrix $\boldsymbol{W}$ such that the function (5.37) is minimized. Then, given a test set of halftone patches $\tilde{\boldsymbol{Y}}$, we compute sparse (near-)optimal representations $\tilde{\boldsymbol{X}}^*$ and build the grayscale patches $\tilde{\boldsymbol{Z}} = \boldsymbol{W}\tilde{\boldsymbol{X}}^*$.

Of course, better results are obtained when the patches are overlapped and averages are computed for each pixel, as described in Sect. 2.4.1. The beauty of the approach is the lack of any assumptions and the simplicity of the linear regression modeling the actually nonlinear relation between the halftone and grayscale images. ■

Example 5.7 Another application is super-resolution, where a high resolution image is built from low resolution versions. The training sets $\boldsymbol{Y}$ and $\boldsymbol{Z}$ are made of patches from low resolution and high resolution images, respectively; note that the patches have different sizes (in pixels) to cover the same physical area of an image. After training, the high resolution patches are restored from the representations corresponding to the low resolution patches, similarly to the way the grayscale patches are obtained from halftone patches in the previous example. Super-resolution techniques along these lines are described in [212] and [215] where, due to the fact that the dictionaries $\boldsymbol{D}$ and $\boldsymbol{W}$ are linked by the same representations, the learning problem is named *coupled* DL.

A somewhat similar problem is deblurring, where patches have equal size, but the interpretation of low quality (for the blurred image) and high quality (for the restored image) is similar to that from super-resolution. ■

A simple way of finding the parameter matrix $\boldsymbol{W}$ is to first design a dictionary $\boldsymbol{D}$ with one of the known DL algorithms, then, with the optimal $\boldsymbol{X}^*$ now available, optimize the function $h(\boldsymbol{W})$ with respect to $\boldsymbol{W}$. In particular, if the function has the form (5.37), then the optimization can be made with a regularized DL algorithm, see Sect. 4.2. However, the fact that $\boldsymbol{D}$ is designed regardless of the function h, and hence of the application, may lead to inferior results.

A way to simultaneously consider both objectives is to optimize their combination, like in the problem

$$\min_{\boldsymbol{D},\boldsymbol{W},\boldsymbol{X}} f(\boldsymbol{D}, \boldsymbol{X}) + \alpha h(\boldsymbol{W}, \boldsymbol{X}), \tag{5.38}$$

where $\alpha > 0$ is a trade-off constant and now the function h depends on a free variable $\boldsymbol{X}$, which is optimized; hence, at the end, the value is only an approximation of $\boldsymbol{X}^*$. Another drawback is that optimizing (5.38) only approximates the optimization of h. We will see later in Sect. 8.5 an example of application in this flavor, in the context of classification, where the use of (5.38) leads to a problem called discriminative DL, ingeniously solved with K-SVD. A similar approach was proposed in [212].

Task-driven DL avoids the above pitfalls and poses the optimization problem as

$$\min_{D,W} h(W, D) \tag{5.39}$$

where the dependence on D comes through the optimal X^*. Hence, the dictionary D is implicitly chosen such that h is minimized, with no direct account of the sparse representation error.

The algorithm proposed in [130] (a related but less general one can be found in [215]) appeals to the adequate tool for sparse coding (depending on the error function) and to gradient descent for dictionary and parameter matrix update. It has an online form, like the algorithms discussed later in Sect. 5.8, but the principle can apply as well to the usual form where all signals are available and used from the beginning. The gradient of $h(W, D)$ with respect to W is easy to compute (exercise: retrieve from Chap. 3 the appropriate formulas). The gradient with respect to D is more complicated, but still computable with a formula; the user is directed to the original paper for the details. Using gradient descent, the algorithm has a slower convergence and several parameters whose values have to be chosen, but the results in applications are good.

Remark 5.8 A variation on the theme of (5.38) was proposed in [199], with the argument that forcing the same representation to both the signal and the application-specific vectors may be too restrictive. A looser connection may be obtained with the objective

$$\|Y - DX_1\|_F^2 + \|Z - WX_2\|_F^2 + \beta \|X_2 - VX_1\|_F^2, \tag{5.40}$$

where regularization terms are ignored. The resulting problem is named *semi-coupled* DL. So, the sparse representation matrices are different: X_1 for the signals and X_2 for the associated vectors. However, the representations are connected through a linear transformation V, which is jointly optimized with the dictionaries. The optimization procedure is based on alternating optimization. Sparse coding is implemented through ℓ_1 relaxation. Dictionary update is made separately on D, W, and V; the optimization of the latter is a simple (regularized) least squares problem. The applications are similar to those already discussed in this section. ■

5.7 Dictionary Selection

Although minimizing the representation error like the standard methods, the DL algorithm proposed in [29] is based on a different principle, that of dictionary selection. Unlike in DL, the atoms are not viewed as variables in a continuum like $\mathbb{R}^m$; they are selected from a finite given set that can be described as a matrix $A \in \mathbb{R}^{m \times n_a}$ whose columns are the candidate atoms; their number is n_a. The design of the dictionary becomes a combinatorial optimization problem and it can be viewed as the selection of n atoms from the n_a candidates. Of course, this makes the problem no easier, since it now has a purely discrete character and the optimization methods that are typically used in DL cannot be applied.

A first but simpler issue is to build the set $\boldsymbol{A}$ and to choose its size. A possible option is to favor incoherence and thus build an incoherent frame with n_a atoms; it is hard to recommend a systematic method for choosing the number n_a; it is clear that a larger value offers more options, but can only increase the complexity of the selection, no matter what algorithm we use for selection. The quick method proposed in [29] is to start with the union of some standard transforms and then prune the atoms that have high correlations with other atoms until a preset threshold is met. Alternatively, random atoms are also a good initialization.

The selection is solved with a greedy algorithm. The dictionary is built by selecting the most promising atom from $\boldsymbol{A}$ that was not yet selected; once selected, an atom stays in the dictionary. Such an approach has the attraction of low complexity. It remains to give a precise meaning to "most promising," i.e., to give a criterion for selecting an atom.

The most direct way is to use OMP, since this is the algorithm that will most likely be used as a sparse coder in conjunction with the selected dictionary. However, any other sparse coding algorithm may be used as well. The dictionary selection proposed in [29] is presented in Algorithm 5.2, which is almost self-explaining. Its idea is extremely natural: add to the current dictionary the atom that, together with the already selected atoms, decreases the most the representation error given by the coefficients computed with OMP, run with the desired sparsity level s. Obviously, there is no optimality guarantee in this selection method, but, relying on the qualities and resilience of OMP, the overall performance should be good. The complexity is at least comparable with that of standard DL algorithms; there are $O(nn_a)$ sparse coding steps (indeed, with dictionaries of size smaller than the final one), compared with K for a standard algorithm, where K is the number of iterations.

Another solution, faster than OMP and with better theoretical guarantees but poorer practical behavior, is to replace the representation error with simple projections. If $\tilde{\boldsymbol{D}}$ is a tentative dictionary and $\boldsymbol{y}_\ell$ a signal, we compute $\boldsymbol{z}_\ell = \tilde{\boldsymbol{D}}^T \boldsymbol{y}_\ell$ and select the largest (in absolute value) s elements of $\boldsymbol{z}_\ell$ in a vector $\tilde{\boldsymbol{z}}_\ell$. For the whole training set, the quality measure, to be used in line 7 of Algorithm 5.2, is

$$\rho = \sum_{\ell=1}^{N} \|\tilde{\boldsymbol{z}}_\ell\|^2 .$$

If the candidate atoms were orthogonal, the projections would be the sparse representation coefficients (as later discussed in Sect. 7.3.1), whose magnitude is a good selection measure. Initializing with incoherent frames has thus a good support in this context, as orthogonality is better approximated.

The above selection criterion has a very fast implementation, since it is enough to compute $\boldsymbol{A}^T \boldsymbol{Y}$ to have all the necessary scalar products from the beginning. Then,

Algorithm 5.2: Dictionary Selection

Data: signals set $\boldsymbol{Y} \in \mathbb{R}^{m \times N}$
desired number of atoms n
desired sparsity level s
number of atoms for initialization n_a
Result: selected dictionary $\boldsymbol{D} \in \mathbb{R}^{m \times n}$

Initialization: build incoherent matrix $\boldsymbol{A} \in \mathbb{R}^{m \times n_a}$
Selected indices $\mathcal{D} = \emptyset$, initial dictionary $\boldsymbol{D} = \emptyset$
for $k = 1 : n$ **do**
  for $i = 1 : n_a,\ i \notin \mathcal{D}$ **do**
    Build tentative dictionary $\tilde{\boldsymbol{D}} = [\boldsymbol{D}\ \boldsymbol{a}_i]$
    Compute sparse representations: $\boldsymbol{X} = \text{OMP}(\tilde{\boldsymbol{D}}, \boldsymbol{Y}, \min(k, s), 0)$
    Compute representation error: $\rho_i = \left\| \boldsymbol{Y} - \tilde{\boldsymbol{D}}\boldsymbol{X} \right\|_F$
  Select best candidate atom: $j = \arg\min_i \rho_i$
  Update dictionary: $\boldsymbol{D} \leftarrow [\boldsymbol{D}\ \boldsymbol{a}_j]$
  Update selected indices: $\mathcal{D} \leftarrow \mathcal{D} \cup \{j\}$

the only required operation is the partial sort that finds the largest s elements of a vector.

Example 5.9 We run the dictionary selection Algorithm 5.2 for a simple example with $N = 1000$ signals of size $m = 16$, obtaining dictionaries with $n = 40$ atoms, starting with $n_a = 200$ candidate atoms; the sparsity level is $s = 4$. For comparison, we also run AK-SVD for 100 iterations. The signals are either synthetically generated or taken from images. We generate the candidate atoms $\boldsymbol{A}$ either randomly or by making an arbitrary selection of n_a signals. In all examples and combinations, AK-SVD is faster and produces a lower RMSE. The difference between AK-SVD and the selection method is lower for images than for random signals and lower for signal candidates than for random candidates. The RMSE is at least 30% higher for the selection method, even more than twice higher in the worst case. The mutual coherence of the dictionary obtained by selection is lower than for the dictionary designed by AK-SVD. However, the methods presented in Sect. 4.6 (see Example 4.16) are able to obtain both lower RMSE and mutual coherence than the selection method. ■

This example confirms the intuition that DL can obtain better results than dictionary selection due to its intrinsic freedom in shaping the atoms. However, in applications where the freedom is inherently limited and the universe of possible atoms is finite or there is a priori information that strongly suggests possible atoms, dictionary selection can be useful.

Example 5.10 An interesting particular case of dictionary selection is that where $\boldsymbol{Y} = \boldsymbol{A}$: the dictionary is selected from the available training signals. Algorithm 5.2 still works, with only the removal of the initialization step, no longer necessary. Applications rely on the fact that the selected signals are somehow representative for the whole set, since they are part of the sparse representations of many signals. For example, *video summarization* consists of selecting a relatively small number of frames that are representative for a video; these images can be previewed as a quick illustration of the content. Manually picking the frames is laborious; random selection can do a good job, but occasionally introduces totally irrelevant frames. A similar problem is image collection summarization, where representative images are selected from a large set.

It was proposed in [40, 41] to use dictionary selection for video summarization. The algorithms are different. The first is based on optimization of a penalized error objective, the most used atoms being selected; the second is greedy and relies on a forward–backward principle: atoms are also removed from the dictionary, not only added as in Algorithm 5.2. Results on a large database show that the method from [41] is better that the state of the art. ■

Other applications of dictionary selections are akin to classification, for example that of photo albuming: the assignment of new photos to already created albums by means of sparse representations with a dictionary; such problems will be discussed in Chap. 8. Another selection problem, that of sensor placement, solved also with a greedy algorithm in [151], is in a sense dual to dictionary selection since it operates on the rows of the matrix $\boldsymbol{A}$ instead of columns.

5.8 Online DL

Online DL algorithms start from the assumption that the signals $\boldsymbol{Y}$ are not all available from the beginning, but arrive one by one or in small batches; also, they are especially appropriate when the signal model is time variant and hence the learned dictionary must adapt. Even when this is not the case, online DL algorithms can work by processing the columns of $\boldsymbol{Y}$ in a certain order, possibly with repetition. They are especially efficient when the number N of signals is large and so the batch algorithms discussed until now may have prohibitive complexity. Also, online algorithms need not store the whole signals matrix $\boldsymbol{Y}$ or the representation matrix $\boldsymbol{X}$, thus being memory efficient.

Let us assume that at time $t \in \mathbb{N}$ the online algorithm receives a signal $\boldsymbol{y}_t$ and the dictionary learned up to that time is $\boldsymbol{D}^{(t)}$. The purpose of the algorithm is to compute a sparse representation $\boldsymbol{x}_t$ and to update the dictionary to a new value $\boldsymbol{D}^{(t+1)}$. The nature of an online algorithm precludes the recomputation (or update) of the past representations, due to its excessive cost. So, once $\boldsymbol{x}_t$ is computed, it will always stay the same; however, $\boldsymbol{x}_t$ contributes to atoms updates not only at time t, but also at subsequent times; a forgetting factor φ helps diminishing the effect of past representations.

The typical structure of an online algorithm is that of alternating optimization. The sparse coding of the newly received signal $\boldsymbol{y}_t$ can be done by any method for computing sparse representations, for example with the call $\boldsymbol{x}_t$ = OMP($\boldsymbol{D}^{(t)}$, $\boldsymbol{y}_t$, s, 0), where s denotes the sparsity level, as usual. In the first online algorithm [129], the representations were computed as the solution of (1.9), but this is a matter of choice; like in the batch case, it is meaningful to match the sparse representation method used in learning with that used in the application for which the dictionary is designed.

The dictionary update step can be done in several ways, as detailed below. We stress again that low cost is an important constraint in choosing a method.

5.8.1 Online Coordinate Descent

Coordinate descent is used in [129], so it can be viewed as an online version of SGK [169] dictionary update, although the former paper precedes the latter. Using directly relation (3.24) to compute the new atom is not possible, due to the necessity to store all the signals. We store instead (at each time instant, so we drop the time index) the matrices

$$\boldsymbol{A} = \boldsymbol{X}\boldsymbol{X}^T, \quad \boldsymbol{B} = \boldsymbol{Y}\boldsymbol{X}^T, \tag{5.41}$$

whose size is constant (it does not depend on N). Revisiting relations (3.19)–(3.24), we can write the optimization of atom $\boldsymbol{d}_j^{(t+1)}$ while the other atoms are fixed at their current values as

$$\min \; \|\boldsymbol{Y} - \boldsymbol{D}\boldsymbol{X} + \boldsymbol{d}_j^{(t)}\boldsymbol{x}_j^T - \boldsymbol{d}_j^{(t+1)}\boldsymbol{x}_j^T\|_F^2 \tag{5.42}$$

Remind that here $\boldsymbol{x}_j^T$ is the jth row of the current matrix $\boldsymbol{X}$; so, $\boldsymbol{x}_j$ is a column vector obtained by transposing the row. Note also that the matrix $\boldsymbol{F}$, that appears also in (5.7), is in fact $\boldsymbol{Y} - \boldsymbol{D}\boldsymbol{X} + \boldsymbol{d}_j^{(t)}\boldsymbol{x}_j^T$. The solution is relation (3.24), written using (5.41) as follows

$$\boldsymbol{d}_j^{(t+1)} = (\boldsymbol{Y} - \boldsymbol{D}\boldsymbol{X} + \boldsymbol{d}_j^{(t)}\boldsymbol{x}_j^T)\boldsymbol{x}_j/(\boldsymbol{x}_j^T\boldsymbol{x}_j) = \boldsymbol{d}_j^{(t)} + \frac{1}{a_{jj}}(\boldsymbol{b}_j - \boldsymbol{D}\boldsymbol{a}_j). \tag{5.43}$$

The online DL algorithm using coordinate descent is presented as Algorithm 5.3, which describes the operations at time instant t. The update of the matrices (5.41) is made in (5.46) using the current signal and its representation using the current dictionary. The forgetting factor φ has the purpose of reducing the contribution of past signals, like in all typical online algorithms. To explain (5.46), we note that at time t, the representation error modified by the forgetting factor is

$$\sum_{\ell=1}^{t} \varphi^{t-\ell} \left\| \boldsymbol{y}_\ell - \boldsymbol{D}\boldsymbol{x}_\ell \right\|^2 \triangleq \|\boldsymbol{Y}_t - \boldsymbol{D}\boldsymbol{X}_t\|_F^2, \tag{5.44}$$

where

$$\boldsymbol{X}_t = \boldsymbol{X}\boldsymbol{\Phi}_t^{1/2}, \ \boldsymbol{Y}_t = \boldsymbol{Y}\boldsymbol{\Phi}_t^{1/2}, \ \boldsymbol{\Phi}_t = \text{diag}(\varphi^{t-1}, \dots, \varphi, 1).$$

It results that

$$\boldsymbol{X}_{t+1}\boldsymbol{X}_{t+1}^T = [\boldsymbol{X}\ \boldsymbol{\xi}] \begin{bmatrix} \varphi\boldsymbol{\Phi}_t & 0 \\ 0 & 1 \end{bmatrix} \begin{bmatrix} \boldsymbol{X}^T \\ \boldsymbol{\xi}^T \end{bmatrix} = \varphi\boldsymbol{X}_t\boldsymbol{X}_t^T + \boldsymbol{\xi}\boldsymbol{\xi}^T, \tag{5.45}$$

which explains the first relation from (5.46); the second results similarly. Of course, one can use $\varphi = 1$, but only for a definite number of iterations, since the elements of the matrices $\boldsymbol{A}$ and $\boldsymbol{B}$ would grow too large.

Algorithm 5.3 omits the initialization of the matrices $\boldsymbol{A}$ and $\boldsymbol{B}$, which is certainly important, because these matrices cannot be simply initialized with zero. Since the vectors $\boldsymbol{x}_j$ are sparse, if $\boldsymbol{A}$ would start from zero, some of the diagonal elements a_{jj} would be zero for a number of iterations, which would make impossible the computation of the update (5.43). So, we need to initialize $\boldsymbol{A}$ with a positive definite matrix, for example $\boldsymbol{A} = c\boldsymbol{I}$, where c is a constant, e.g., $c = 1$. For such an initialization, some of the atoms would become zero in the first iteration (check!) if $\boldsymbol{B} = 0$. So, $\boldsymbol{B}$ has to be initialized with some nonzero matrix; a possibility is to take $\boldsymbol{B} = \boldsymbol{D}$. Due to the forgetting factor, the bias due to initialization wears off in time.

We evaluate now the complexity of Algorithm 5.3. Operation (5.46) has a low computational cost, due to the sparsity of the representation vector $\boldsymbol{\xi}$. However, the dictionary update is quite costly, since it needs $O(mn^2)$ operations (a matrix–vector multiplication for each atom); the matrices $\boldsymbol{A}$ and $\boldsymbol{B}$ are not sparse. Even though, compared with batch algorithms, the factor N has disappeared from the complexity, the online DL is not really effective although based on possibly the simplest idea, that of coordinate descent. A way of reducing complexity is to update only part of the atoms at each time moment. This is basically equivalent with processing the signals in batches and updating all the atoms after each batch.

If several signals, say ν, are processed at time t, the modifications are minor. Only relation (5.46) is affected, although it formally keeps the same expression. However, now $\boldsymbol{\eta}$ is an $m \times \nu$ matrix with the signals on its columns and similarly $\boldsymbol{\xi}$ is an $n \times \nu$ matrix of corresponding representations. The dictionary update (5.43) is not affected, since it works directly with the matrices $\boldsymbol{A}$ and $\boldsymbol{B}$. So, the complexity is about ν times smaller; this would amount to updating only n/ν atoms in each iteration of Algorithm 5.3. Of course, reducing complexity by processing small batches or updating part of the atoms has the downside of a slower convergence,

Algorithm 5.3: Online DL Based on Coordinate Descent, at Time t

Data: current dictionary $\boldsymbol{D} \in \mathbb{R}^{m \times n}$
current signal $\boldsymbol{\eta} \in \mathbb{R}^m$
current product matrices $\boldsymbol{A} \in \mathbb{R}^{n \times n}$, $\boldsymbol{B} \in \mathbb{R}^{m \times n}$
forgetting factor φ
Result: updated dictionary $\boldsymbol{D}$ and product matrices $\boldsymbol{A}$, $\boldsymbol{B}$

1 Sparse coding: compute sparse representation $\boldsymbol{\xi}$ of signal $\boldsymbol{\eta}$
2 Update product matrices

$$\boldsymbol{A} \leftarrow \varphi \boldsymbol{A} + \boldsymbol{\xi}\boldsymbol{\xi}^T, \quad \boldsymbol{B} \leftarrow \varphi \boldsymbol{B} + \boldsymbol{\eta}\boldsymbol{\xi}^T \tag{5.46}$$

3 **for** $j = 1$ **to** n **do**
4 Update atom: $\boldsymbol{d}_j \leftarrow \boldsymbol{d}_j + \frac{1}{a_{jj}}(\boldsymbol{b}_j - \boldsymbol{D}\boldsymbol{a}_j)$
5 Normalize: $\boldsymbol{d}_j \leftarrow \boldsymbol{d}_j / \|\boldsymbol{d}_j\|$

if we measure "time" by the number of signals that are processed, since the descent steps are made at every ν signals.

5.8.2 *RLS DL*

The coordinate descent method presented in the previous section does not compute, at time t, the optimal dictionary corresponding to the current representations. In principle, optimality could be achieved if many sweeps of coordinates would be made, instead of a single one, as in Algorithm 5.3 (which is unpractical, since it would increase even more the complexity).

In order to emulate a Recursive Least Squares (RLS) algorithm for DL, a recursive relation should be created from the MOD dictionary update (3.31). Such a relation would provide optimality at each time. Let us denote $\boldsymbol{Y}$ and $\boldsymbol{X}$ the current signal matrix (size $m \times t$) and representation matrix (size $n \times t$); at time t, signal $\boldsymbol{\eta}$ is received and its representation $\boldsymbol{\xi}$ is computed (the notation is like in Algorithm 5.3). With the notation (5.41), the MOD-optimal dictionary from (3.31) is

$$\boldsymbol{D}^{(t)} = \boldsymbol{B}\boldsymbol{A}^{-1}. \tag{5.47}$$

Taking (5.46) into account, the next dictionary satisfies the relation (we ignore the forgetting factor for the moment)

$$\boldsymbol{D}^{(t+1)}(\boldsymbol{A} + \boldsymbol{\xi}\boldsymbol{\xi}^T) = \boldsymbol{B} + \boldsymbol{\eta}\boldsymbol{\xi}^T. \tag{5.48}$$

The RLS DL algorithm was developed in [177], where the explicit computation of $\boldsymbol{D}^{(t+1)}$ is made via the matrix inversion lemma. Writing $\boldsymbol{D}^{(t+1)} = \boldsymbol{D}^{(t)} + \boldsymbol{\Delta}$ and using (5.47) in (5.48) we obtain

$$\boldsymbol{\Delta} = (\boldsymbol{\eta} - \boldsymbol{D}^{(t)}\boldsymbol{\xi})\boldsymbol{\xi}^T(\boldsymbol{A} + \boldsymbol{\xi}\boldsymbol{\xi}^T)^{-1}.$$

Denoting $\boldsymbol{r} = \boldsymbol{\eta} - \boldsymbol{D}^{(t)}\boldsymbol{\xi}$ the residual of the sparse representation of the current signal and

$$\boldsymbol{u} = \boldsymbol{A}^{-1}\boldsymbol{\xi}, \quad \alpha = \frac{1}{1 + \boldsymbol{\xi}^T\boldsymbol{A}^{-1}\boldsymbol{\xi}} = \frac{1}{1 + \boldsymbol{\xi}^T\boldsymbol{u}},$$

and using the matrix inversion lemma

$$(\boldsymbol{A} + \boldsymbol{\xi}\boldsymbol{\xi}^T)^{-1} = \boldsymbol{A}^{-1} - \frac{\boldsymbol{A}^{-1}\boldsymbol{\xi}\boldsymbol{\xi}^T\boldsymbol{A}^{-1}}{1 + \boldsymbol{\xi}^T\boldsymbol{A}^{-1}\boldsymbol{\xi}},$$

it results that (notice that $\boldsymbol{\xi}^T\boldsymbol{A}^{-1}\boldsymbol{\xi} = (1-\alpha)/\alpha$)

$$\boldsymbol{\Delta} = \alpha\boldsymbol{r}\left(\frac{1}{\alpha}\boldsymbol{\xi}^T\boldsymbol{A}^{-1} + \boldsymbol{\xi}^T\boldsymbol{A}^{-1}\boldsymbol{\xi}\boldsymbol{\xi}^T\boldsymbol{A}^{-1}\right) = \alpha\boldsymbol{r}\boldsymbol{u}^T.$$

The above operations are assembled in Algorithm 5.4. We have also introduced the forgetting factor, whose presence in lines 3 and 4 is the consequence of replacing $\boldsymbol{A}$ with $\varphi\boldsymbol{A}$ and $\boldsymbol{B}$ with $\varphi\boldsymbol{B}$ in (5.48). The reader is invited to include the forgetting factor in the above derivation of the algorithm.

The initialization of $\boldsymbol{A}^{-1}$ is a standard issue in RLS algorithms: until gathering n input signals, the matrix $\boldsymbol{A}$ is not invertible. Of course, computing the inverse at time n or later is not feasible in the online philosophy, due to the large number of operations concentrated at a single moment. The typical simple solution is to initialize $\boldsymbol{A}^{-1} = c\boldsymbol{I}$, where c is a large constant. A more sophisticated scheme, which provides $\boldsymbol{A}^{-1}$ exactly once $\boldsymbol{A}$ becomes invertible, but with the expense of extra operations, can be found in [182].

It is remarkable that the number of operations is small: the costlier operations are matrix–vector multiplications and rank-1 updates. The overall complexity is only $O(n^2)$ per iteration, which is clearly less than the complexity of Algorithm 5.3. No matrix multiplication or inversion is involved as one might suspect by looking at (5.48). The main reason is the rank-1 structure of the above dictionary update. Note also that there is no need to store the product matrix $\boldsymbol{B}$.

Remark 5.11 A well-known problem of the standard form of the RLS algorithm is the lack of numerical stability due to the update of $\boldsymbol{A}^{-1}$ based on the matrix inversion lemma. The usual solution is to compute instead the Cholesky factorization

$$\boldsymbol{A} = \boldsymbol{L}\boldsymbol{L}^T \tag{5.49}$$

Algorithm 5.4: RLS DL, at Time t

Data: current dictionary $\boldsymbol{D} \in \mathbb{R}^{m \times n}$
current signal $\boldsymbol{\eta} \in \mathbb{R}^m$
current inverse matrix $\boldsymbol{A}^{-1} \in \mathbb{R}^{n \times n}$
forgetting factor φ
Result: updated dictionary $\boldsymbol{D}$ and inverse matrix $\boldsymbol{A}^{-1}$

1 Sparse coding: compute sparse representation $\boldsymbol{\xi}$ of signal $\boldsymbol{y}$
2 Compute representation residual: $\boldsymbol{r} = \boldsymbol{\eta} - \boldsymbol{D\xi}$
3 Compute auxiliary variables: $\boldsymbol{u} = \varphi^{-1}\boldsymbol{A}^{-1}\boldsymbol{\xi}$, $\alpha = \frac{1}{1+\boldsymbol{\xi}^T\boldsymbol{u}}$
4 Update inverse: $\boldsymbol{A}^{-1} \leftarrow \varphi^{-1}\boldsymbol{A}^{-1} - \alpha \boldsymbol{u}\boldsymbol{u}^T$
5 Update dictionary: $\boldsymbol{D} \leftarrow \boldsymbol{D} + \alpha \boldsymbol{r}\boldsymbol{u}^T$
6 Normalize dictionary

---

and to update the Cholesky factor based on the relation

$$\boldsymbol{A} + \boldsymbol{\xi}\boldsymbol{\xi}^T = [\boldsymbol{L}\ \boldsymbol{\xi}] \begin{bmatrix} \boldsymbol{L}^T \\ \boldsymbol{\xi}^T \end{bmatrix}. \tag{5.50}$$

(Note the similarity, but also the differences with respect to (3.1).) The update is computed by the orthogonal triangularization of the matrix $\begin{bmatrix} \boldsymbol{L}^T \\ \boldsymbol{\xi}^T \end{bmatrix}$, whose size is $(n+1)\times n$ and which is upper triangular in the first n rows. The last row is eliminated using Givens rotations; one can take advantage of the sparsity of $\boldsymbol{\xi}$. The procedure is detailed in [80, Sec. 6.5.4] and requires at most $O(n^2)$ operations. This procedure replaces line 4 of Algorithm 5.4.

For computing $\boldsymbol{u}$ in line 3 of the algorithm, we use the Cholesky factorization and get the expression

$$\boldsymbol{u} = \varphi^{-1}\boldsymbol{A}^{-1}\boldsymbol{\xi} = \varphi^{-1}\boldsymbol{L}^{-T}\boldsymbol{L}^{-1}\boldsymbol{\xi},$$

whose computation requires the solution of two triangular systems. This is again an $O(n^2)$ computation. So, the complexity stays within the same order of magnitude with respect to the standard version. The algorithm modified as above is numerically stable and thus recommended in all practical applications. The use of orthogonal transformations is crucial for the numerical stability. ■

Example 5.12 We compare Algorithms 5.3 and 5.4 on a simple DL problem with $m = 16$, $n = 40$, $N = 1000$, $s = 4$. The data are generated synthetically, with a random dictionary and an SNR of 20 dB. The algorithms are run for more than N iterations by circularly feeding them with the signals. The RLS-DL in the form of Algorithm 5.4 is indeed unstable and stops converging after 2500–

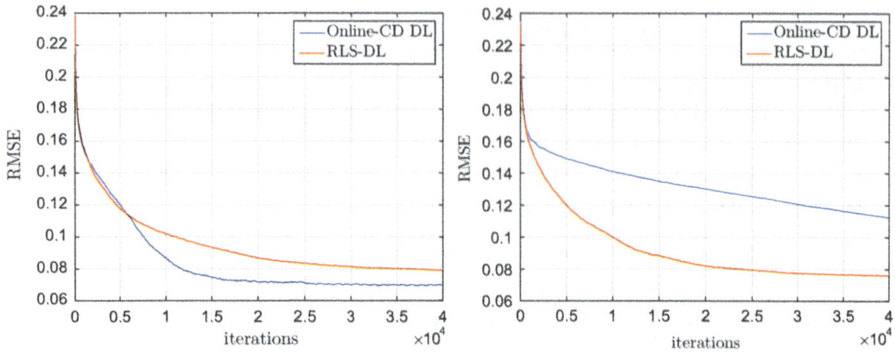

Fig. 5.1 RMSE given by online DL algorithms based on coordinate descent (blue) and RLS (red) for a forgetting factor $\varphi = 0.995$ (up) and $\varphi = 0.999$ (down)

3000 iterations; so, we report the results of the algorithm modified as presented in Remark 5.11. Figure 5.1 shows the evolution of the RMSE for the two algorithms for two forgetting factors, $\varphi = 0.995$ and $\varphi = 0.999$. The RMSE is averaged over ten runs with different data; in each run, both algorithms are initialized with the same random dictionary. For computing the RMSE, we use the definition (2.10), which is appropriate since the signals are used several times, hence we deal actually with a standard DL problem (not one in which the model may be indeed time variant). Note that computing the RMSE in such a way at each iteration is extremely expensive with respect to the complexity of the algorithm and is done here only for examination purposes. A first remark is that the convergence speed of RLS-DL is almost insensitive to the value of the forgetting factor. On the contrary, the coordinate descent algorithm is affected significantly when the forgetting factor is large. However, the RMSE attained by coordinate descent is lower, in these experiments. Although the results seem to give a small advantage to coordinate descent, we must not forget that RLS-DL has a much lower complexity and that it is also more robust with respect to the forgetting factor values. ■

5.9 DL with Incomplete Data

Training signals are often noisy. Until now, we mostly assumed Gaussian noise, the only exception being that from Sect. 5.5, where impulsive noise is treated. Another type of noise, possibly more disruptive, is that where the signals are incomplete. A typical example is that of an image where some pixels are simply missing.

So, instead of $\boldsymbol{Y}$ we have $\boldsymbol{M} \odot \boldsymbol{Y}$, where the elements of the known mask matrix $\boldsymbol{M}$ are zero or one and $\odot$ is the elementwise product. A zero in $\boldsymbol{M}$ means the respective signal element is not available; the corresponding zero in $\boldsymbol{M} \odot \boldsymbol{Y}$

carries thus no meaning. Taking the mask into consideration, the objective of the DL problem becomes

$$\|\boldsymbol{M} \odot (\boldsymbol{Y} - \boldsymbol{D}\boldsymbol{X})\|_F, \tag{5.51}$$

meaning that the missing elements are simply ignored. One may assume data are missing with probability ρ, independently of their position, but the standard DL algorithms can be adapted to (5.51) without any specific assumption.

Somewhat surprisingly, DL with incomplete data has received relatively little attention and was used in particular setups [87, 208, 226]. We present here the adaptation of SGK and AK-SVD to the incomplete data case, discussing both the sparse coding and dictionary update steps.

An algorithm for OMP with missing data was proposed in [35], providing recovery guarantees. Interestingly, it computes the support of the representation by just applying standard OMP to $\boldsymbol{M} \odot \boldsymbol{Y}$, i.e., considering also the zeros resulted from the application of the mask; only the final representation coefficients are computed differently from (1.14), using the naturally available estimate of ρ, which is the zeros density in $\boldsymbol{M}$. It is also tempting to adapt OMP by completely ignoring the missing elements (and the corresponding rows of the dictionary), hence running standard OMP for shorter vectors, which considerably decreases the execution time; we call masked OMP this interpretation of the algorithm. In the tests reported in [59], masked OMP was the best on images but otherwise it gave similar results with missing-data OMP, so it seems to bring the best trade-off between quality and complexity. Both algorithms were clearly superior to standard OMP applied with no modification to $\boldsymbol{M} \odot \boldsymbol{Y}$.

The AK-SVD dictionary update adaptation was presented sketchily in [87], although the idea was mentioned without details in previous work. With incomplete data, the objective of (3.21) becomes

$$\left\|\boldsymbol{M} \odot (\boldsymbol{F} - \boldsymbol{d}\boldsymbol{x}^T)\right\|_F. \tag{5.52}$$

Denoting $\mathcal{M} \subset \mathbb{N}^2$ the indices of available data in $\boldsymbol{Y}$, the LS problem corresponding to (5.52) is

$$d_i x_j = f_{ij}, \quad (i, j) \in \mathcal{M}. \tag{5.53}$$

If $\boldsymbol{x}$ is fixed, then the optimal atom has the elements

$$d_i = \frac{\sum_{(i,j)\subset\mathcal{M}} f_{ij} x_j}{\sum_{(i,j)\subset\mathcal{M}} x_j^2}, \quad i = 1 : m. \tag{5.54}$$

We thus obtain the incomplete data version of (3.22), which defines SGK. The numerators in (5.54) are the elements of the vector $(\boldsymbol{M} \odot \boldsymbol{F})\boldsymbol{x}$. The denominators are the norms of the compressed representations corresponding to the available signals that use d_i in their representation. Normalizing the atom (5.54) gives the incomplete data version of the AK-SVD dictionary update formula (3.28).

To complete the AK-SVD adaptation we conclude similarly that, when the atom is fixed in (5.53), the optimal representation coefficients are

$$x_j = \frac{\sum_{(i,j)\subset \mathcal{M}} f_{ij} d_i}{\sum_{(i,j)\subset \mathcal{M}} d_i^2}, \quad j = 1 : n. \tag{5.55}$$

The numerators are the elements of the vector $(\boldsymbol{M} \odot \boldsymbol{F})^T \boldsymbol{d}$. This is the equivalent of the vector $\boldsymbol{F}^T \boldsymbol{d}$ from (3.37).

Example 5.13 We run the incomplete data version of AK-SVD, combined with masked OMP, on a set of $N = 8000$ distinct image 8×8 patches, with 10% of the pixels missing, on randomly selected positions; the dictionary has $n = 128$ atoms and the sparsity level is $s = 6$. The evolution of the RMSE is shown in Fig. 5.2. The RMSE is computed as the error (5.51) divided to $\|\boldsymbol{M}\|_0^{1/2}$, respecting the notion of error per signal element. The quality of the designed dictionary $\boldsymbol{D}$ is evaluated on the full data by computing the RMSE (2.10), where the representation matrix $\boldsymbol{X}$ is that determined by the incomplete data. Naturally, when extending the use of the dictionary $\boldsymbol{D}$ and the representation matrix, designed with incomplete data, to the full matrix, the error is larger, since the missing elements of $\boldsymbol{Y}$ did not participate to the DL process. However, the error still has acceptable values, which explains the

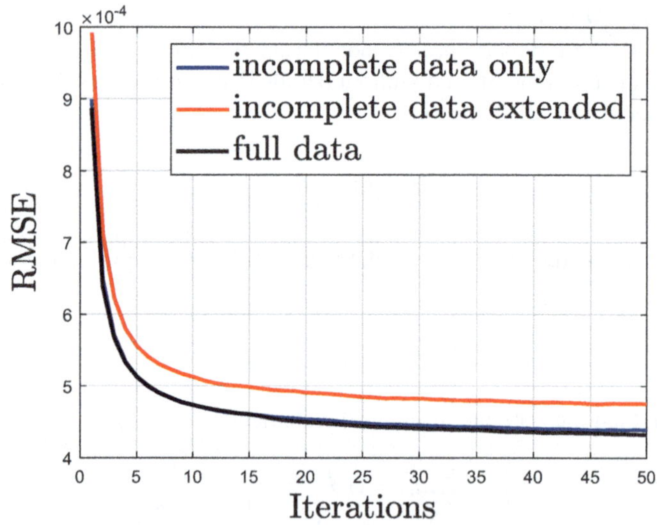

Fig. 5.2 RMSE evolution in AK-SVD on 10% missing data (blue) and on full data with the dictionary designed for incomplete data (red). For comparison, the RMSE of standard AK-SVD, run as usual on full data (black)

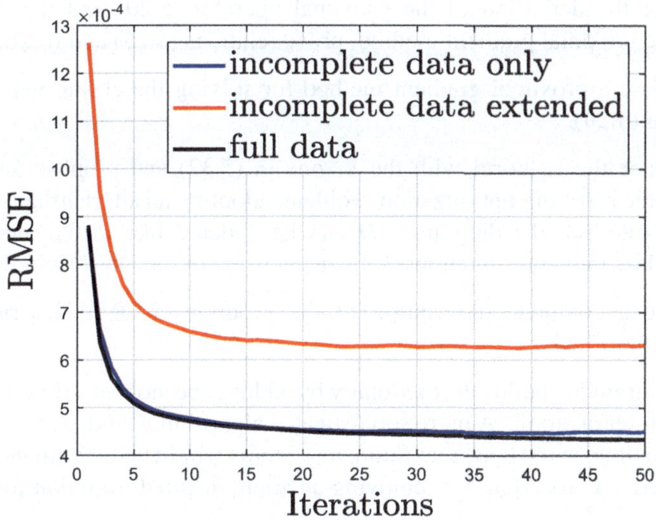

Fig. 5.3 Same curves as in Fig. 5.2, but for 30% missing data

use of the sparse representations in inpainting. Note that inpainting involves more refined techniques than simply extending the sparse representation, as explained in Sect. 2.4.2. Figure 5.3 shows the situation where the missing rate is 30%. The RMSE degrades accordingly.

For comparison, the RMSE of the standard AK-SVD run on the full data $\boldsymbol{Y}$ is also shown in both figures. The RMSE is only slightly better than for the incomplete data version, showing that in both cases the learning process can achieve the same performance on the available data. ■

Problems

P 5.1 It was noticed in Example 5.2 that for some training signals (columns of $\boldsymbol{Y}$), the representation with the optimal dictionary produced by the APrU Algorithm 5.1 has no coefficients at all (there are no nonzeros in the respective column of $\boldsymbol{X}$). How could the algorithm be modified such that each column of $\boldsymbol{X}$ has at least $s_{\min}$ nonzeros?

Similarly, the number of nonzeros can be very large. How can it be bounded to a value $s_{\max}$?

P 5.2 Investigate the coordinate descent approach for solving the DL problem with 0-norm penalty (5.27), following the lines of the APrU algorithm. Does that amount to replacing soft with hard thresholding? Find the exact formula for updating the representations that would replace (5.11).

P 5.3 Using the definition of the proximal operator (5.20) and the formulation (5.24) of the proximal gradient method, prove relations (5.25) and (5.26).

P 5.4 Propose a proximal gradient method for solving the elastic net regularized DL problem (5.29).

P 5.5 Replace the ℓ_0 norm with the ℓ_1 one in (5.32) and propose an algorithm for solving the resulting optimization problem, adopting an alternating optimization approach. Note that the dictionary $\boldsymbol{D}$ can be updated like in the methods from Sect. 4.4, while the representations $\boldsymbol{X}$ by using some of the ideas from this chapter.

P 5.6 Write a complete description of the group-level OMP described in the paragraph after (5.34).

P 5.7 Algorithm 5.2 builds the dictionary by adding one atom at a time via selection from a given set of atoms. A more sophisticated algorithm would also remove atoms from time to time, in the hope that future new atoms will fit better with the remaining atoms. Propose a procedure for removing an atom, inspired from that for adding an atom.

P 5.8 The optimal representations in the regularized AK-SVD algorithm are computed as shown in Proposition 4.4. In the incomplete data case, the objective becomes

$$\left\| \boldsymbol{M} \odot (\boldsymbol{F} - \boldsymbol{d}\boldsymbol{x}^T) \right\|_F^2 + \mu \left\| \boldsymbol{x} \right\|^2.$$

Derive the optimal representation $\boldsymbol{x}$, thus extending (5.55), and write the incomplete data version of regularized AK-SVD.

P 5.9 An alternative to the algorithms from Sect. 5.9 for DL with incomplete data is to use an ℓ_1 penalty, like in (5.4), thus solving the convex problem

$$\min_{\boldsymbol{D},\boldsymbol{X}} \quad \|\boldsymbol{M} \odot (\boldsymbol{Y} - \boldsymbol{D}\boldsymbol{X})\|_F^2 + \lambda \|\boldsymbol{X}\|_{1,1}$$

where $\boldsymbol{M}$ is the mask of available data. Adapt Algorithm 5.1 (APrU) to this objective, inspired from the relations defining AK-SVD with incomplete data.

Discuss the choice of the value of λ in APrU with full and incomplete data. Assuming that you have an oracle for the value in the former case, how would you choose λ in the latter?

P 5.10 Propose a task-driven approach to fill missing pixels from an image. Assume that you have a database of images from the same family as those with missing pixels. Compare with the direct DL approach from Section 5.9 and discuss advantages, drawbacks and differences.

Chapter 6
Optimizing Dictionary Size

Abstract Until now the number of atoms was an input parameter of the DL algorithms. Its choice was left to the user, leading usually to a trial and error approach. We discuss here possible ways to optimize the number of atoms. The most common way to pose the DL problem is to impose a certain representation error and attempt to find the smallest dictionary that can ensure that error. The algorithms solving this problem use the sparse coding and dictionary update ideas of the standard algorithms, but add and remove atoms during the DL iterations. They start either with a small number of atoms, then try to add new atoms that are able to significantly reduce the error, or with a large number of atoms, then remove the less useful ones; the growing strategy seems more successful and is likely to have the lowest complexity. Working on a general DL structure for designing dictionaries with variable size, we present some of the algorithms with best results, in particular Stagewise K-SVD and DLENE (DL with efficient number of elements); the first serves also as basis for an initialization algorithm that leads to better results than the typical random initializations. We present the main ideas of a few other methods, insisting on those based on clustering, in particular on the mean shift algorithm. Finally, we discuss how OMP can be modified to reduce the number of atoms without impeding too much on the quality of the representation.

6.1 Introduction: DL with Imposed Error

All the algorithms presented until now assume that the number n of atoms is an input data. Thinking strictly at the DL process, this is a convenient assumption that defines a better posed problem. However, in the larger context of an application where the dictionary will be used for sparse representations, this is in fact the manifestation of a trial and error approach where the designer picks a value for n that seems suited to the application. In case of unsatisfactory results, n is later increased or decreased based on the actual results obtained with that dictionary, and the whole design process is restarted. Without focusing on a specific application, we discuss here methods that not only design the dictionary, but also choose the number of atoms.

B. Dumitrescu, P. Irofti, *Dictionary Learning Algorithms and Applications*,
https://doi.org/10.1007/978-3-319-78674-2_6

There are two naturally contradictory requirements regarding the size of a dictionary. A larger number of atoms will almost certainly lead to smaller representation errors for the optimized dictionary. However, the number of atoms influences the representation speed (and that of the DL process): OMP and other representation algorithms need more time as n increases; moreover, as the size increases, the coherence properties worsen, hence finding the best sparse representation becomes more difficult. Finally, a too large number of atoms presents the classic danger of overfitting the training data; the resulted dictionary may become less representative for other data from the same family. A convenient trade-off is to try to obtain the smallest dictionary that does the desired job, but transforming this desire into an optimization problem is not necessarily immediate.

Example 6.1 Figure 6.1 illustrates the RMSE attained by AK-SVD after 50 iterations when designing dictionaries with $m = 64$, $N = 8192$, and several values of n and s, starting from random initializations, with training data that are vectorized 8×8 patches of some images, as described in Sect. 2.3.1; each initial dictionary for some n is a submatrix of all initial dictionaries with larger n; totally random initializations make the curves a bit less smooth. A similar experiment, with more algorithms but fewer values of n, was reported in Fig. 3.8. (Note that the RMSE has here smaller values because the patches can partially overlap, while there they were completely disjoint; also, fewer images are used here for training.) It is visible that, as the dictionary size grows, the decrease of the error is smaller. Although not the case in this experiment, it also may happen that, due to the nonconvexity of the DL

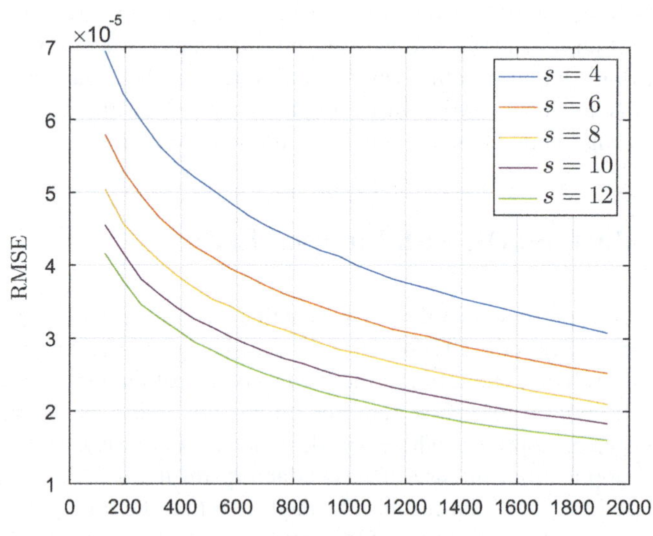

Fig. 6.1 RMSE attained for $64 \times n$ dictionaries designed with AK-SVD with 50 iterations

problem and to the difficulty to find a good approximation of the global minimum, the error increases even though the dictionary is larger.

In this example, it is almost clear that a dictionary size of about 2000 is certainly an overkill, and that taking n in the low hundreds is more likely to be appropriate, but choosing the "right" size is certainly not easy. ■

The probably simplest way to pose the DL problem such that size is considered in the optimization process is to impose a given error level ε and find the minimum number of atoms n for which a dictionary satisfies the error. So, we replace the basic DL problem (2.2) with

$$\begin{aligned} \min_{n,\boldsymbol{D},\boldsymbol{X}} \quad & n \\ \text{s.t.} \quad & \|\boldsymbol{Y} - \boldsymbol{D}\boldsymbol{X}\|_F \le \varepsilon \\ & \|\boldsymbol{x}_\ell\|_0 \le s, \ \ell = 1:N \\ & \|\boldsymbol{d}_j\| = 1, \ j = 1:n \end{aligned} \tag{6.1}$$

The solution is, however, much more difficult than for (2.2), due to the discrete objective. The naive approach would be to completely design dictionaries, starting with a tentative n and then decreasing or increasing it if the error constraint is satisfied or not, respectively. We dismiss such an approach due to its high complexity. Ideally, the number of atoms should be varied during a single run of a DL algorithm until it attains its optimal value.

Finding an appropriate value ε is also a significant issue. A relatively good starting point is to select a desired RMSE (2.10), denoted here ς, and set $\varepsilon = \sqrt{mN}\varsigma$ in (6.1); of course, in many applications the quality measure is not directly the error, so setting an average error level is also a matter of trial and error (or a possibly difficult optimization process).

The algorithms that attempt solving (6.1) are based on diverse heuristics, some inspired from optimization techniques that seem unrelated to dictionary learning. We will present several strategies that proved successful, without giving the algorithms in all their intricacies. Another important drive behind size optimization is the hope to obtain algorithms that produce a final error smaller than that given by a standard DL algorithm for the same n. So, the underlying idea is that a standard DL algorithm, even knowing the optimal size, can be beaten by a more complex one for which n is a variable. We will see that this is possible, even though at the price of a higher complexity.

6.2 A General Size-Optimizing DL Structure

If size optimization is desired during the DL iterations, we can modify the DL basic alternate optimization Algorithm 3.1 by inserting operations for adding and removing atoms, obtaining Algorithm 6.1. Let us first discuss its main ingredients, besides the usual sparse coding and dictionary update operations.

Initialization Unlike in standard DL, the size n_0 of the initial dictionary is inherently different from that of the final one, hence we need to choose it. Obviously, the choice is the result of a strategy. There are two possible approaches that proved successful. One can start with a small dictionary and then grow it up to the (hopefully) optimal size; the reasoning behind this strategy is that a small dictionary can be easily well trained, although it will give a large representation error; adding new atoms will then reduce the error; if the growing process is sufficiently slow, then one can hope that the new atoms, despite having fewer update iterations, will be trained as well as the old ones.

The reverse strategy is to start with a large dictionary and then prune it to the optimal size. The presumed advantage is that the many initial atoms cover all useful directions and so the learning process will only slightly modify them, while the useless atoms are removed gradually. The obvious drawback is the complexity of handling large dictionaries.

The most successful strategies start with a small dictionary. Interestingly, an intermediary strategy that starts with a medium sized dictionary has not been tried, most likely due to the similarity with the usual DL algorithms.

When the initial dictionary is large, the methods for generating the initial dictionary are in principle the same as discussed in Chap. 3: random atoms or a random selection of signals. When n_0 is small, one may try more targeted initializations, like those described in the next sections.

Adding and Removing Atoms In each iteration of the DL algorithm we can add some atoms and remove others; the order of these operations is not relevant. We denote α the number of added atoms and β the number of removed atoms; either of these numbers can be zero, depending on the strategy.

A strategy is defined by its answers to the following questions:

1. how many atoms do we add/remove in each iteration?
2. how do we generate the new atoms?
3. which atoms do we remove?

Before describing some successful strategies, let us examine a few quick answers to the above questions.

1. The simplest decision is to keep α and β fixed and hence add or remove a constant number of atoms in each iteration. A more complicated approach is to let α and β depend on the context.
2. Adding random atoms or random signals, like at initialization, seems not a good strategy, since it is difficult to argue that such an approach has some advantage. The computation of new atoms should be based on ideas more refined than sheer randomness. For example, one can build atoms that are suited for the worst represented signals; the new atoms will likely be placed in zones where the atom density is very low, but where there are enough signals to warrant the introduction

---

Algorithm 6.1: DL with Size Optimization

Data: signals set $\boldsymbol{Y} \in \mathbb{R}^{m \times N}$
    sparsity s
    desired error ε
    initial dictionary number of atoms n_0
    maximum number of iterations K
Result: trained dictionary $\boldsymbol{D}$

Initialize with dictionary $\boldsymbol{D} \in \mathbb{R}^{m \times n_0}$
for $k = 1$ **to** K **do**
    if *more atoms needed* **then**
        add_atoms
    if *fewer atoms needed* **then**
        remove_atoms
    Sparse coding: compute sparse representations $\boldsymbol{X}$ with current $\boldsymbol{D}$
    Dictionary update: compute new dictionary $\boldsymbol{D}$; the matrix $\boldsymbol{X}$ may be changed or not
    Exit with current dictionary if stopping criteria are met
Optional: final adjustment of n

---

of a new atom and thus reduce representation error. An opposite strategy is to replace a very useful atom (which appears in the representation of many signals) by several atoms; otherwise said, we increase atoms density where signals are dense.

3. We have already tackled in Sect. 3.9 the obvious case of atom removal. When an atom is not used, then it is a clear candidate for elimination, no matter the context. Generalizing, we can remove atoms that are seldom used, i.e., they appear in the representation of few signals. Another viewpoint is to merge atoms that have similar directions, thus looking for redundancies.

Going towards more quantitative descriptions, the importance of the atoms may be measured in at least two ways. Let $\boldsymbol{x}_j^T$ be the jth row of the representation matrix $\boldsymbol{X}$ and $\mathcal{I}_j = \{\ell \mid x_{j\ell} \neq 0\}$ the positions of nonzero elements on this row. The power of the representations using atom $\boldsymbol{d}_j$ can be defined as

$$p_{j,2} = \|\boldsymbol{x}_j^T\|_2^2 = \sum_{\ell \in \mathcal{I}_j} x_{j\ell}^2 \tag{6.2}$$

and emphasizes the atoms that have large coefficients. The number of nonzero coefficients

$$p_{j,0} = \|\boldsymbol{x}_j^T\|_0 = |\mathcal{I}_j| \tag{6.3}$$

emphasizes the atoms that are present in many representations, regardless of their actual weight in the representation. Sorting the atoms in decreasing order based upon one of these measures gives a list in which the first atoms are very important and so probably very useful, while the last atoms are almost useless. So, if we want to reduce the number of atoms, the last atoms in this list are the most likely candidates for removing.

Sparse Coding This step is done as usual, using the full dictionary that includes the new atoms.

Atom Update The dictionary update step is performed as usual, with the preferred algorithm. We still can make a fine distinction between the new and old atoms. All atoms need update, but we can run more dictionary update iterations for the new atoms, such they recover the deficit in training with respect to the other atoms. This is possible in all DL algorithms, see problem **P** 6.2.

Stopping Criteria If we attempt to solve (6.1), a good moment to stop the DL iterations is when all constraints are satisfied. However, this is meaningful when a growing strategy is adopted. When starting with many atoms, the constraints are naturally satisfied, so it is more appropriate to exit the loop when the constraints are no longer satisfied, keeping the dictionary and the number of atoms from the previous iteration.

Final Operations We leave the option to fine-tune the number of atoms after the DL algorithm. For example, with reference to (6.1), one can try to remove atoms as long as the constraints are still satisfied, while updating the remaining atoms. One can see these final operations as other DL iterations in which a different strategy is applied.

6.3 Stagewise K-SVD

Stagewise K-SVD [160] has a strategy based on growing the dictionary, with K-SVD [4] as underlying DL algorithm. We will describe it following the general structure from Algorithm 6.1.

Initialization A set of $n_0 = s$ random signals. The main reason for this choice is that at least s atoms are needed to reach a sparsity level s. Technically, one can start with fewer atoms, since OMP can work with any number of atoms and one can easily impose a sparsity level that is not larger than n, but this may imply tuning the code, since this is an unusual situation. Of course, it is important that s is always a small number; it is likely that a slightly larger n_0 would do as well.

Removing Atoms Stagewise K-SVD adopts a very simple strategy: in each iteration it removes $\beta = 1$ atom, namely the least used in terms of the power (6.2). So, even though the strategy is founded on growing, always removing the least used atom ensures that useless or almost useless atoms never survive.

Adding Atoms Let us first consider an auxiliary problem. Let $\boldsymbol{W}$ be a subset of signals, i.e., a submatrix of $\boldsymbol{Y}$. We want to select α atoms that represent the main directions covered by the signals. They can be obtained by computing the singular value decomposition of $\boldsymbol{W} = \boldsymbol{U}\boldsymbol{\Sigma}\boldsymbol{V}^T$ and taking as atoms the first α left singular vectors, i.e., the first α columns of $\boldsymbol{U}$. Since this is in fact a selection of the principal directions, let us denote PR_DIR$(\boldsymbol{W}, \alpha)$ this procedure.

Stagewise K-SVD builds a set $\boldsymbol{W}$ of the worst represented signals. Let

$$\boldsymbol{r}_\ell = \boldsymbol{y}_\ell - \boldsymbol{D}\boldsymbol{x}_\ell$$

be the residual when representing the ℓth signal (now $\boldsymbol{x}_\ell$ is the ℓth column of $\boldsymbol{X}$). We can sort the signals in decreasing order of the relative residual norms $\|\boldsymbol{r}_\ell\| / \|\boldsymbol{y}_\ell\|$. The first signals in this list are the worst represented: they have the largest residuals (relatively to the signals norms). It seems natural to add atoms that help representing better these signals, since it should be easier to decrease large errors. Let $\boldsymbol{W}$ be the set of worst represented signals; in [160], their number is taken as 5% of the total number N of signals; of course, this is not a magical number, but it was good enough in the performed experiments; it is only important that $\boldsymbol{W}$ contains a small fraction of all signals. In each iteration, Stagewise K-SVD adds α atoms computed with PR_DIR$(\boldsymbol{W}, \alpha)$.

Atom Update In order to reduce complexity, a small integer parameter ν is introduced, with the following significance. At each νth iteration, the full dictionary update algorithm (K-SVD in [160]) is applied; otherwise, only the atoms added in that iteration are updated. As mentioned in the previous section, the sparse coding step is always performed with the whole dictionary.

Stopping Criterion Stagewise K-SVD stops when the error constraint is satisfied in (6.1).

Final Operations In fact, the iterative process continues by removing one atom per iteration until the error becomes larger than the bound ε. (One can see this as putting $\alpha = 0$.) Then, the last atom removal is undone and the final dictionary is obtained.

Algorithm 6.2 summarizes the above operations. Dictionary updates are performed with K-SVD only because this was the option in the original version of Stagewise K-SVD, but in principle any algorithm could be used. The complexity clearly depends on α and ν; if the dictionary grows faster (α is larger), then the error decreases faster, hence the total number of iterations is smaller; if full dictionary updates are used more often (ν is smaller), then the complexity is clearly larger, since updating n atoms takes considerably more time than updating α atoms. So, $\alpha = 2$ and $\nu = 1$ give the highest complexity; on the other side, these parameter values are likely to give the best dictionaries, since a slow growth leaves enough time to the atoms to reach (nearly) optimal directions. Compared to plain K-SVD, although the number of iterations is not significantly larger, the complexity

Algorithm 6.2: Stagewise K-SVD

Data: signals set $\boldsymbol{Y} \in \mathbb{R}^{m \times N}$
sparsity s
desired error ε
number α of added atoms
number ν of iterations for full K-SVD
maximum number of iterations K

Result: trained dictionary $\boldsymbol{D}$

1 Initialize with dictionary $\boldsymbol{D} \in \mathbb{R}^{m \times s}$
2 Sparse coding: $\boldsymbol{X} = \text{OMP}(\boldsymbol{D}, \boldsymbol{Y}, s, 0)$
3 **for** $k = 1$ **to** K **do**
4   Remove from $\boldsymbol{D}$ the least used atom based on (6.2)
5   Build set $\boldsymbol{W}$ of 5% worst represented signals
6   Add new atoms: $\boldsymbol{D} \leftarrow [\boldsymbol{D}\ \text{PR\_DIR}(\boldsymbol{W}, \alpha)]$
7   Sparse coding: $\boldsymbol{X} = \text{OMP}(\boldsymbol{D}, \boldsymbol{Y}, s, 0)$
8   **if** $k \bmod \nu = 0$ **then**
9   Update all atoms with K-SVD dictionary update
10   **else**
11   Update only the α new atoms with K-SVD
12   **if** $\|\boldsymbol{Y} - \boldsymbol{D}\boldsymbol{X}\|_F \leq \varepsilon$ **then**
13   Exit for loop
14 **if** $k < K$ **then**
15   **while** $\|\boldsymbol{Y} - \boldsymbol{D}\boldsymbol{X}\|_F \leq \varepsilon$ **do**
16   Store current dictionary: $\boldsymbol{D}_{\text{prev}} \leftarrow \boldsymbol{D}$
17   Remove from $\boldsymbol{D}$ the least used atom based on (6.2)
18   Perform sparse coding and K-SVD dictionary update on $\boldsymbol{D}$ and $\boldsymbol{X}$
19   Restore previous result: $\boldsymbol{D} \leftarrow \boldsymbol{D}_{\text{prev}}$

of Stagewise K-SVD is higher due also to the SVD operations implied by the computation of new atoms (step 6 of Algorithm 6.2).

Example 6.2 An experiment detailed in [160] shows that Stagewise K-SVD is able to outperform standard K-SVD. The latter algorithm is run on $N = 11{,}000$ image patches, with $m = 64$, $s = 6$ and $n = 441$. The attained error is used as stopping tolerance for Stagewise K-SVD, run with several values of α and ν. The optimal number of atoms reached by Stagewise K-SVD is shown in Table 6.1. The results are clearly better than for K-SVD, even for large values like $\alpha = 6$, $\mu = 5$. Interestingly, for these parameter values, Stagewise K-SVD is also faster than K-SVD, the SVD computations from step 6 of Algorithm 6.2 being more than compensated by the lower number of atom updates. However, when $\alpha = 2$ and $\nu = 1$, i.e., when the best

Table 6.1 Number of atoms obtained by Stagewise K-SVD in the experiment described in Example 6.2

	ν				
α	1	2	3	4	5
2	175	197	212	222	219
3	189	197	205	211	226
4	194	225	243	226	258
5	199	224	238	239	246
6	199	228	244	241	318

number of atoms is obtained as expected, the execution time is 4.5 times larger than for K-SVD.

Other experiments from [160] show that, running K-SVD with several values of n, the optimal number of atoms obtained by Stagewise K-SVD with the same error level ranges from $0.71n$ when $n = 128$ to $0.4n$ when $n = 512$; naturally, the benefits are more visible for large n, when the optimization problem is harder and it is more likely that less useful atoms appear in the dictionary. Also, the results of K-SVD are sensitive to the initialization, while those of Stagewise K-SVD are practically independent of the initialization. Hence, Stagewise K-SVD has to be used only once, while standard algorithms possibly need more initializations. Thus, its higher complexity becomes a smaller drawback. An additional benefit is that the mutual coherence is lower than for K-SVD; this is a natural outcome, due to the lower number of atoms, but also to the explicit consideration of the worst represented signals, which adds atoms that are very different from the existing ones. ■

6.4 An Initialization Method

Stagewise K-SVD is the source of inspiration for initializing a DL algorithm. As mentioned in Sect. 3.9, initialization is usually an extremely quick operation, with a random character. However, since DL algorithms are time consuming, we could spend more effort on initialization in the hope to find better dictionaries; an initialization method should be only relatively cheap, its complexity being that of a few iterations of the DL algorithm. In this section we return to the usual DL setup, in which the number n of atoms is fixed.

We describe here the initialization method from [162]. The underlying idea is to grow a dictionary that is simultaneously adapted to the training signals but still has enough variety to be able to cover all interesting directions. The method starts with an initial dictionary $\boldsymbol{D}$ that is incoherent; its size is $m \times n_0$, with $n_0 < n$. It can be an already available incoherent frame (see Chap. 4), built analytically, or, in the simplest case where $n_0 = m$, the unit matrix. The method has two stages and can be seen as a quick and simple DL algorithm.

Stage 1 (Training of a Small Incoherent Dictionary) Compute representations matrix $\boldsymbol{X}$ for the initial dictionary and repeat the following operations for a fixed number of iterations.

(1a) Optimize orthogonal matrix $\boldsymbol{Q}$ such that the error $\|\boldsymbol{Y} - \boldsymbol{Q}\boldsymbol{D}\boldsymbol{X}\|_F$ is minimized. This optimization is detailed in Remark 4.15. Set the new dictionary: $\boldsymbol{D} \leftarrow \boldsymbol{Q}\boldsymbol{D}$.
(1b) Compute representations $\boldsymbol{X}$ for the current $\boldsymbol{D}$.
(1c) Eliminate the atoms $\boldsymbol{d}_j$ for which the power (6.2) of the associated coefficients is below a given threshold.

Stage 2 (Increase Dictionary Size) Repeat the following operations.

(2a) Like in Stagewise K-SVD, select the (say) 5% worst reconstructed atoms, in terms of the reconstruction error $\|\boldsymbol{y}_\ell - \boldsymbol{D}\boldsymbol{x}_\ell\|/\|\boldsymbol{y}_\ell\|$, and form the matrix $\boldsymbol{W}$ with them. Compute the singular value decomposition $\boldsymbol{W} = \boldsymbol{U}\boldsymbol{\Sigma}\boldsymbol{V}^T$ and add to the dictionary the first α left singular vectors (the first α columns of $\boldsymbol{U}$), which correspond to the largest α singular values.
(2b) Compute representations $\boldsymbol{X}$ for the new (extended) $\boldsymbol{D}$.
(2c) If the number of atoms exceeds n, then eliminate the atoms $\boldsymbol{d}_j$ with smallest power (6.2) such that n atoms remain, and stop.

Stage 1 aims to train a small incoherent dictionary that is adapted to the given training signals. This dictionary is already small from the beginning and then the less used atoms are pruned.

Stage 2 grows the dictionary to the desired size by adding atoms that are suited to represent the currently worst represented signals. Of course, the coherence of the dictionary increases during stage 2, but one may hope that not by much, as one adds each time a group of orthogonal atoms. Unlike Stagewise K-SVD, this method makes no attempt to update the atoms; once an atom is added to the dictionary, it stays unchanged. We can take much larger values for α than in Stagewise K-SVD, such that the target number of atoms is reached in few iterations; the simulations in [162] were done with $\alpha = 20$.

The complexity of the algorithm is clearly much less than that of K-SVD, since stage 1 operates with smaller matrices and both stages have a small number of iterations. The choice of the parameters involved in the algorithm is relatively robust. It is reported in [162] that, with this initialization, either K-SVD reaches the same error level faster or a smaller error in the same running time, compared with K-SVD initialized in the standard way. So, the effort put into initialization pays off and the whole procedure is much faster (up to more than ten times faster) than Stagewise K-SVD. However, for the same number of atoms, the error levels attained by Stagewise K-SVD are better. The user can thus tune the trade-off complexity/quality to the application at hand.

6.5 An Atom Splitting Procedure

The dictionary size optimization can be posed in other ways than (6.1). For example, in [131], it is proposed to let the number of nonzeros on the columns of $\boldsymbol{X}$ be variable, like in the DL approaches from the first part of Chap. 5. To obtain a well-posed problem, the representation error must be bounded and likewise the average number of nonzeros. Like in (5.2), we define a bound ς for the RMSE and we impose it for each signal; this makes easy the choice of the sparse coder: OMP with error stopping. We also impose a bound $\bar{s}$ for the average number of nonzeros, thus obtaining the problem

$$\begin{aligned} \min_{n,\boldsymbol{D},\boldsymbol{X}} \quad & n \\ \text{s.t.} \quad & \|\boldsymbol{y}_\ell - \boldsymbol{D}\boldsymbol{x}_\ell\| \le \sqrt{m}\varsigma, \ 1 \le \ell \le N \\ & \|\boldsymbol{X}\|_0 \le N\bar{s} \\ & \|\boldsymbol{d}_j\| = 1, \ j = 1:n \end{aligned} \tag{6.4}$$

The algorithm from [131], called DLENE (DL with Efficient Number of Elements) is based on the underlying idea that, since the error constraint in (6.4) is satisfied automatically by using the error-based OMP, the DL process reduces the number of nonzeros when the dictionary size is constant. When $\|\boldsymbol{X}\|_0$ stagnates or even increases, it is time to add new atoms. The characteristics of the algorithm are detailed below, for each of the specific steps of Algorithm 6.1.

Initialization DLENE starts with only two atoms. One has equal elements and acts as a DC model; it is permanently kept in the dictionary without modification; this is useful in image processing. The second atom is the average of all input signals.

Removing Atoms An atom $\boldsymbol{d}_j$ is eliminated if the number (6.3) of nonzero coefficients is lower than a threshold, defined as a fraction of the current number of nonzeros (per atom)

$$p_{j,0} < \theta \frac{\|\boldsymbol{X}\|_0}{n}. \tag{6.5}$$

The coefficient θ should be small enough such that only obviously unnecessary atoms are culled. The value used in [131] is $\theta = 0.035$.

Adding Atoms The creation of new atoms is based on splitting an existing atom into two atoms that differ only in a single element. Let us denote $\boldsymbol{d}_j$ the atom to be split and assume that the ith element will be transformed into two values. (We will explain later how i and j are chosen.) Let $\boldsymbol{F}$ be the error matrix (3.19) with the contributions of all atoms but $\boldsymbol{d}_j$ removed; remind that $\mathcal{I}_j$ is the set of nonzero positions on row j of $\boldsymbol{X}$.

The original contribution of [131] is to replace $\boldsymbol{d}_j$ with two new atoms, $\boldsymbol{d}_j^+$ and $\boldsymbol{d}_j^-$ that are identical with $\boldsymbol{d}_j$ with the exception of the ith element; the conditions

$$\boldsymbol{d}_{ij}^+ \geq 0, \quad \boldsymbol{d}_{ij}^- < 0 \tag{6.6}$$

are also imposed; they are meant to enhance diversity in the two atoms, which are otherwise identical. The new atoms will be normalized after the values $\boldsymbol{d}_{ij}^+$ and $\boldsymbol{d}_{ij}^-$ are computed. The coefficient $x_{i\ell}$ has also to be "split"; since we want to preserve the total number of nonzero, the coefficient will be allocated to one of the new atoms, the other atom having a zero coefficient in that position. Let $\mathcal{I}_j^+$ be the set of positions allocated to $\boldsymbol{d}_j^+$ (or, otherwise said, the set of signals whose sparse representations contain $\boldsymbol{d}_j^+$); define $\mathcal{I}_j^-$ similarly. Looking only at the ith atom element, its contribution to the representation error is

$$\sum_{\ell \in \mathcal{I}_j} (f_{i\ell} - d_{ij} x_{j\ell})^2 .$$

After splitting, the error becomes

$$\sum_{\ell \in \mathcal{I}_j^+} \left(f_{i\ell} - d_{ij}^+ x_{j\ell} \right)^2 + \sum_{\ell \in \mathcal{I}_j^-} \left(f_{i\ell} - d_{ij}^- x_{j\ell} \right)^2 . \tag{6.7}$$

We have to find the sets $\mathcal{I}_j^+$ and $\mathcal{I}_j^-$, and the values d_{ij}^+ and d_{ij}^- such that (6.7) is minimized.

The problem is hard to solve exactly and is related to what is known in quantization as cell splitting. It is proposed in [131] to solve it as follows. The sets of indices are defined by

$$\ell \in \begin{cases} \mathcal{I}_j^+, & \text{if } \operatorname{sign}(f_{i\ell}) = \operatorname{sign}(x_{j\ell}) \\ \mathcal{I}_j^-, & \text{otherwise} \end{cases} \tag{6.8}$$

Since the signs of d_{ij}^+ and d_{ij}^- are predetermined, the above choice is meant to most likely (although not surely) decrease the error in (6.7), moving towards zero (with the possibility of overshooting) the individual errors. Once the sets (6.8) are defined, the optimal values (for these sets!) of the new coefficients result by simple scalar optimization and are

$$d_{ij}^+ = \frac{\sum_{\ell \in \mathcal{I}_j^+} f_{i\ell} x_{j\ell}}{\sum_{\ell \in \mathcal{I}_j^+} x_{j\ell}^2}, \quad d_{ij}^- = \frac{\sum_{\ell \in \mathcal{I}_j^-} f_{i\ell} x_{j\ell}}{\sum_{\ell \in \mathcal{I}_j^-} x_{j\ell}^2}. \tag{6.9}$$

Note that the choice (6.8) indeed ensures the sign assumption (6.6). The index set splitting could be iteratively improved, but the gain is most likely not worth the computation cost. We answer now shortly a few questions defining the full algorithm.

How do we choose the index i? The best index is that minimizing the error (6.7) when (6.8) and (6.9) are used for splitting the atom. We need to check all indices from 1 to m.

How do we choose the atom? Here there is no unique best answer, like at the previous question. The minimum requirement, effectively used in DLENE, is that both new atoms satisfy the threshold condition (6.5), i.e., neither will be immediately eliminated. Other more cautious approaches could be used, like splitting the atom(s) with the most energy, but they have not been tested.

When do we split atoms? This is again a question with multiple possibly good answers. The approach in DLENE is to split atoms when the average number of nonzeros does not decrease in ten iterations (the trend is computed by linear least squares). This means that, for the current dictionary size, the algorithm cannot reduce $\|X\|_0$ any longer and the dictionary size must be increased. Such an approach, combined with the liberal condition for splitting atoms, makes the dictionary size increase in large leaps well spaced in time. On the contrary, the decreases are small and may occur at any time.

Atom update is made with the AK-SVD algorithm.

Stopping Criterion When the sparsity bound in (6.4) is satisfied, DLENE can be stopped. However, we should ensure that a certain number of iterations has passed from the last increase of n, to let all atoms be updated a number of times and also to allow the less used atoms be eliminated.

Comparisons DLENE is compared with Stagewise K-SVD in [131]. DLENE seems to have a slight advantage in terms of dictionary size, for similar RMSEs and average number of nonzeros, although there are cases where Stagewise K-SVD is better. The execution times of the two methods are also similar. More investigations are necessary, but it may be possible that at least some of the advantage of DLENE comes from the free distribution of the nonzeros, while Stagewise K-SVD works with fixed sparsity level per signal. In any case, both methods are better than other size optimization algorithms, based on decreasing a large dictionary.

6.6 Clustering as a DL Tool

Clustering is the operation of grouping vectors according to some nearness criterion. Given a set of ν vectors z_i, $i = 1 : \nu$, we want to group them into p clusters $\mathcal{C}_j$, $j = 1 : p$, where p is given or, more generally, should also be found. Each vector z_i is associated with a single cluster $\mathcal{C}_j$, which we denote $i \in \mathcal{C}_j$. A cluster may have a "center," which is a representative vector for the cluster and is usually not one of the vectors z_i. Clustering has no unique result; the obtained clusters usually depend on the choice of some parameters. In DL, clustering has been used at least in [72, 134, 152] for several operations, but especially in connection with dictionary size optimization.

There are many clustering algorithms and it is beyond our topic to discuss and compare them here. For the sake of example, we describe shortly the *mean shift* algorithm, without asserting that it is better than other algorithms; it certainly has an intuitive support and basically consists of a single formula. Mean shift was used only in [152], while [134] appealed to competitive agglomeration and [72] to subtractive clustering.

The underlying idea is that the vectors z_i are samples of an unknown distribution, for which we adopt a non-parametric model. The associated probability density function has the form

$$P(z) = \frac{1}{\nu} \sum_{i=1}^{\nu} K(z - z_i), \tag{6.10}$$

where $K(\cdot)$ is a kernel. The notion of kernel may have several meanings, depending on the (applied) mathematics field where it occurs. Here we use the definition commonly met in statistics. There will be a related but different one in Chap. 9, belonging to machine learning. A kernel is a nonnegative function, typically symmetric, that has large values in the neighborhood of zero and smaller values as the argument grows. The kernels from (6.10) have multivariate arguments, but the useful situation is that where they actually depend only on the magnitude of the vector argument:

$$K(\zeta) = k\left(\frac{\|\zeta\|^2}{h^2}\right). \tag{6.11}$$

The parameter h is a measure of the width of the kernel around the origin. Since kernels are themselves probability density functions, they should integrate to 1; however, we ignore the normalization constants, since they are not important as long as a single kernel is used in (6.10). Two important examples are the Gaussian, $k(u) = \exp(-u)$, and the Epanechnikov kernel

$$k(u) = \begin{cases} 1 - u, & \text{if } u \le 1 \\ 0, & \text{otherwise} \end{cases} \tag{6.12}$$

Note the square in (6.11) when computing $K(\zeta)$, which explains why $k(u)$ needs to be defined only for nonnegative u. Figure 6.2 shows the shapes of the Gaussian and Epanechnikov kernels (6.11) when $\zeta \in \mathbb{R}^2$, for $h = 1$; if we take a signal processing view and see the kernels as windows, it is clear that the Gaussian kernel is wider.

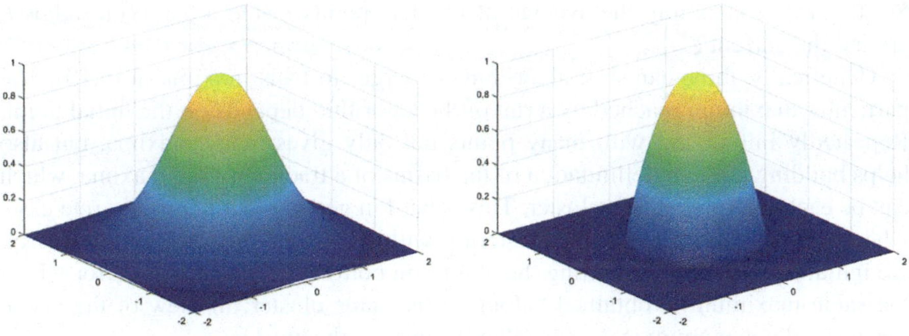

Fig. 6.2 The Gaussian (left) and Epanechnikov (right) kernels in two variables

Armed with this information, we can now describe the mean shift algorithm, which searches the maxima of (6.10), which are natural cluster centers. The gradient of (6.10) is

$$\frac{\partial P(z)}{\partial z} = \frac{2}{\nu h^2} \sum_{i=1}^{\nu} k'(z - z_i)(z - z_i). \tag{6.13}$$

When the gradient is zero, stationary points are obtained, including maxima. However, this equation is not easy to solve; in particular, an analytic solution is not foreseeable. The mean shift algorithm simply transforms the equation into the iteration

$$z \leftarrow \frac{\sum_{i=1}^{\nu} k'(z - z_i) z_i}{\sum_{i=1}^{\nu} k'(z - z_i)}. \tag{6.14}$$

The new z is obtained as a weighted average of the given vectors z_i, with weights given by the kernel derivative computed in the current z. The value $k'(z - z_i)$ can be seen as a similarity measure between z and z_i. For the Gaussian kernel, the iteration is

$$z \leftarrow \frac{\sum_{i=1}^{\nu} \exp(-\|z - z_i\| / h^2) z_i}{\sum_{i=1}^{\nu} \exp(-\|z - z_i\| / h^2)}. \tag{6.15}$$

For the Epanechnikov kernel, the iteration is even simpler. Denoting $\mathcal{N}(z)$ the indices of data vector z_i for which $\|z - z_i\| \le h$, the mean shift algorithm is defined by

$$z \leftarrow \frac{\sum_{i \in \mathcal{N}(z)} z_i}{|\mathcal{N}(z)|}. \tag{6.16}$$

So, the new z is simply the average of the data points inside a ball with radius h around the current z.

Generically, the mean shift algorithm converges to local maxima of (6.10). The particular maximum reached by a run of the algorithm depends on the initial point. Repeatedly initializing with many points not only gives many maxima, but also helps building a rough delimitation of the basins of attraction of the maxima, which can be each interpreted as a cluster. This is hard in general, but in our discrete case, where we have ν data vectors, clustering with the mean shift algorithm is easy: we initialize with each z_i, letting the algorithm converge. The vectors z_i for which the same maximum is obtained belong to the same cluster; in view of the above formulas, the maximum is the (weighted) center of the cluster.

It is clear that the clusters depend on the parameter h. If h is very small, then there are many maxima, hence many clusters. As h approaches zero, then each z_i has its own cluster. On the contrary, when h grows, the number of clusters decreases, eventually all data vectors being grouped in a single cluster. So, although the mean shift algorithm produces itself the number p of clusters, this number depends on the choice of h, which is entirely the user's.

Let us now see how clustering can be put to work in the diverse DL stages, especially those related to optimizing the number of atoms. We keep the discussion at a general level, since the details may be quite intricate and it is hard to recommend a best option.

Initialization Using clustering on the signals y_ℓ, $\ell = 1 : N$, can show what signals are representative and thus can be used for a good initialization of the dictionary. For example, we can take from each cluster a number of signals that is proportional with the size of the cluster. If there are many clusters (i.e., we have used a small value for h), we can simply take the cluster centers as initial atoms.

The clustering method from [134], named Enhanced K-SVD, does even more: it decides the dictionary size during the initialization process. The clustering is more refined: atoms are considered as cluster centers and the signals may be associated with several clusters, with different degrees of membership, mimicking sparse coding in the sense that a signal is associated with only a few atoms. The clustering process is evolutionary; by changing the atoms and the degrees of membership, some atoms become useless and can be removed. Once this initialization process has ended, the number of atoms is fixed and a standard DL algorithm (K-SVD, in fact) is run as usual.

Adding Atoms Remind the Stagewise K-SVD algorithm, where the new atoms were chosen by using the worst represented signals for computing the most significant principal directions using the SVD decompositions. Clustering provides an alternative. The worst represented signals could be input into a clustering algorithms and the centers of the most populated clusters could be used as new atoms. In [72] a somewhat different strategy is used. Clustering is applied not directly on signals but on the residuals of sparse coding obtained for representations with only $s/2$ nonzero coefficients. Thus, the effect of important atoms is removed and the new atoms are selected between those well modeling these residuals. Of course, it is needed to check if the new atoms are not in the vicinity of already existing atoms and remove them in this case.

Such operations should be separated by a number of regular DL iterations, such that the new atoms have the time to be trained.

Removing Atoms Unlike the approaches sketched above, the method from [152] uses clustering directly on the atoms, without looking at the signals. If we start with a large number of atoms, DL iterations will tend to produce atoms that are highly correlated. If two atoms are near each other, it is unlikely that they will be both used in the sparse representation of the same signal. (Remind that in OMP the residual is orthogonal on the selected atoms. So, if an atom is selected, the residual will be almost orthogonal on the nearby atoms, which reduces their chances to be selected in a further OMP iteration.) Also, replacing an atom with a nearby one will only slightly increase the representation error.

The method from [152] runs the mean shift algorithm using (6.16) (with $h = 0.5$) with the atoms as entries. The atoms from the same cluster are replaced with the center of the cluster. A collateral effect is that the new dictionary has good coherence properties. Hence, the method not only reduces the number of atoms to an efficient one, but can also be seen as an alternative to the algorithms from Chap. 4. Again, since dictionary update tends to change atoms rather smoothly, it takes a few iterations for some atoms to cluster together; so, the mean shift algorithm has to be run only from time to time; its computational burden is not significant.

6.7 Other Methods

We present here other methods that attempt the optimization of the dictionary size. The first two modify the DL objective in order to attain this purpose.

In [210] the objective is

$$\|\boldsymbol{Y} - \boldsymbol{D}\boldsymbol{X}\|_F^2 + \lambda \|\boldsymbol{X}\|_{1,1} + \gamma \|\boldsymbol{D}\|_{2,1}, \tag{6.17}$$

where

$$\|\boldsymbol{D}\|_{2,1} = \sum_{j=1}^{n} \left(\sum_{i=1}^{m} d_{ij}^2 \right)^{1/2} = \sum_{j=1}^{n} \|\boldsymbol{d}_j\|_2 .$$

The first penalty term induces sparsity in the representation, like in Sect. 5.1, while the second penalty is an atom sparsity enforcer. Of course, the problem (6.17) cannot have atom normalization constraints. Since the penalty is a sum of atom norms, it is likely that some of the norms will tend to zero, as typical in ℓ_1 minimization. The method from [210] is based on majorization techniques, see Sect. 5.3, and atom norms are bounded (not fixed). When an atom norm becomes small, the atom can be removed. Note that a small atom cannot force large representation coefficients (without modifying the error term of the objective), due to the presence of the $\|X\|_{1,1}$ term which, besides sparsifying the representation, keeps also under control the magnitude of the coefficients.

Another modification with similar purpose, but now acting on the representations, was proposed in [59], where the objective is

$$\frac{1}{2}\|Y - DX\|_F^2 + \frac{\mu_1}{2}\|X\|_F^2 + \mu_2 \left\|X^T\right\|_{2,1}. \tag{6.18}$$

The first penalty term is simply regularization, like in Sect. 4.2. The second enforces sparsity on the rows of X. If the entire row j of X is (almost) zero, then the atom d_j is (practically) unused and can be removed from the dictionary. The objective (6.18) is easily amenable to an AK-SVD dictionary update. The atom update is exactly like in AK-SVD. The representation update is slightly more complicated, but still straightforward, see problem **P** 6.3.

In a totally different approach, Bayesian learning and an Indian Buffet Process (IBP) are combined in [45] to find the optimal dictionary size. The algorithm starts with a single atom and grows the dictionary, leaving open the possibility of removing atoms. The IBP prior favors the reuse of the most used atoms, thus clustering the signal representations towards the most successful atoms; at the same time, the use of new atoms is also encouraged.

Worth mentioning is also [73], where the atoms form non-disjoint groups. Only atoms from the same group are used for representing a signal. This is somewhat akin to clustering, in the sense that each signal is associated with a group of atoms. Usual dictionary update is run on each group, but removing and adding atoms in a group are based on heuristic rules.

Another angle of attack is proposed in [150], where Minimum Description Length (MDL) is employed to decide the sparsity level and the size of the dictionary. To implement the MDL principle, a virtual coder is built to find the dictionary that provides the shortest code of the training signals. A drawback is the exhaustive search over all possible numbers of atoms, unlike almost all algorithms mentioned in this chapter, which try to adapt n during the DL process.

Dictionary size optimization is a problem not yet fully solved and there is no extensive comparison between existing algorithms that would give a ranking, at least for some categories of signals. Although a few heuristics showed good behavior of the algorithms built around them, our impression is that there are many possibilities to explore and that the field is still open.

6.8 Size-Reducing OMP

We have seen several methods for reducing the number of atoms, all occurring outside the sparse coding stage (if the DL method contained such a stage), which is assumed to stay unchanged. Of course, the representation matrix X produced by sparse coding contains important information for seeing which atoms are more or less useful. However, we may wonder how sparse coding could be altered to make the representation matrix even more meaningful. We show here how OMP can be modified to accelerate the process of selecting the most relevant atoms.

We remind that standard OMP represents a signal by greedily selecting atoms that have large projections on the signal. Like in Sect. 1.3, we denote $\boldsymbol{y} \in \mathbb{R}^m$ a signal, $\boldsymbol{x} \in \mathbb{R}^n$ its sparse representation and $\boldsymbol{e} = \boldsymbol{y} - \boldsymbol{D}\boldsymbol{x}$ the residual. In each iteration, OMP adds to the representation the atom $\boldsymbol{d}_j$ for which the projection $|\boldsymbol{r}^T\boldsymbol{d}_j|$ is the largest.

OMP treats the atoms equally. However, we could consider that some atoms are more likely than others to appear in a representation. So, we associate to atom $\boldsymbol{d}_j$ the probability $p(\boldsymbol{d}_j)$, not yet specified, quantifying how likely is the atom to enter a representation. It was proposed in [68] to use Bayes' rule and thus change the OMP selection criterion by selecting the atom that maximizes

$$p(\boldsymbol{d}_j|\boldsymbol{e}) \sim p(\boldsymbol{r}|\boldsymbol{d}_j) \cdot p(\boldsymbol{d}_j), \tag{6.19}$$

where the probability of the residual given the atom can be naturally taken proportional with the projection, i.e.

$$p(\boldsymbol{e}|\boldsymbol{d}_j) \sim |\boldsymbol{e}^T\boldsymbol{d}_j|. \tag{6.20}$$

This choice leads to a weighted OMP. Another possibility is given in [64]; however, it needs an estimation of the variance of the noise. An additive rule is proposed in [170] instead of (6.19), where a logit term proportional to $\log[p(\boldsymbol{d}_j)/(1 - p(\boldsymbol{d}_j))]$ is added to the projection; again, the variance of the noise is needed.

If we want to use the weighted OMP for accelerating the elimination of some atoms by their simple lack of use, a natural measure of the probability of an atom would be given by the weight of its coefficients with respect to the whole current representation matrix $\boldsymbol{X}$. So, if $\boldsymbol{x}_j^T$ the jth row of matrix $\boldsymbol{X}$, it was proposed in [59] to use (6.19) with prior probabilities

$$p(\boldsymbol{d}_j) \sim \psi(\|\boldsymbol{x}_j^T\|^2/\|\boldsymbol{X}\|_F^2), \tag{6.21}$$

where ψ is a function that allows adaptation to various situations. For example, ψ could be the identity. For standard OMP, ψ is constant.

In the beginning of the DL process, assuming that the dictionary has many atoms, we start with the identity function in (6.21). This will favor atoms that contribute in reducing the error for many signals, even though they may be not optimal. However, as the DL process advances and the number of atoms ideally settles, we want OMP to work in the standard mode, where it tries to find the combination of atoms that minimizes the error. So, denoting k the iteration number in the alternate DL process (like in Algorithm 3.1), the idea from [59] is to use in (6.21) the function

$$\psi(\xi) = \xi + \frac{ck}{n}, \tag{6.22}$$

where $c \geq 0$ is a parameter. As k grows, the values of $\psi(\xi)$ tend to have similar values for all $\xi \in [0, 1]$. It is also possible to switch, at a certain iteration or depending on other factors, from (6.22) to the constant function, which means switching from weighted to standard OMP.

The weighted OMP can be used together with a dictionary update based on an objective like (6.18), effectively cutting the number of atoms and producing a better RMSE than a standard DL algorithm run with the same number of atoms. Of course, the computation cost is higher, especially due to the larger initial dictionary size. Another drawback is the extra parameters that may need tuning for obtaining the best results.

Problems

P 6.1 It was mentioned in Example 6.1 that, on the same training data $\boldsymbol{Y}$, an increase of the number n of atoms does not necessarily lead to a reduction of the representation error. What initialization should we use to be (almost) sure that the error decreases? (Remind that there is no totally fail-safe procedure, since the sparse coding algorithm is not guaranteed to find a better solution even if that solution exists.)

P 6.2 Modify an algorithm that updates the atoms sequentially, like K-SVD, such that the number of atom updates tends to be the same for all atoms, regardless the iteration where they were added to the dictionary. For example, associate an update counter c_j with atom j, increased each time when the atom is updated; at iteration k, if $c_j = k-1$, update the atom once, as usual; if $c_j < k-1$, update the atom twice (or even more times). Organize the atom sweeps such that an atom is not updated the second time before another atom is updated. Alternatively to a fixed order, randomly generate atom indices such that the same goal is attained.

P 6.3 To derive an AK-SVD-like algorithm for minimizing (6.18), assume that the dictionary $\boldsymbol{D}$ and the representations $\boldsymbol{X}$ are fixed, with the exception of i) atom $\boldsymbol{d}$, and ii) corresponding row $\boldsymbol{x}^T$. Show that in case i) we obtain (3.28) and that in case ii) the optimal representation is

$$\boldsymbol{x} = \max\left(0, \frac{\|\boldsymbol{F}^T\boldsymbol{d}\| - \mu_2}{1+\mu_1}\right) \cdot \frac{1}{\|\boldsymbol{F}^T\boldsymbol{d}\|}\boldsymbol{F}^T\boldsymbol{d}. \tag{6.23}$$

The matrix $\boldsymbol{F}$ is, as usual, the error after removing the contributions of all other atoms, like in (5.7).

P 6.4 Dictionary selection solves the DL problem by choosing n atoms out of n_a candidates, as explained in Sect. 5.7. Combine this idea with atom removing to obtain a selection algorithm that solves the DL problem with imposed error (6.1). Examine possible variants of the algorithm when starting from a void dictionary or, on the contrary, from a dictionary containing all candidates.

Chapter 7
Structured Dictionaries

Abstract Endowing the dictionary with a structure may be beneficial by better modeling certain signals and by speeding up the representation and learning processes, despite losing some of the freedom of a general dictionary. We study here several unrelated types of structures and present DL algorithms adapted to the respective structures. *Sparse* dictionaries assume that the atoms are sparse combinations of the columns of a matrix, usually those of a square transform. This is equivalent to a factorization of the dictionary as a product between a dense and a sparse matrix or, generalizing the concept, a product of several sparse matrices. This structure can be seen as the ultimate approach to parsimony via sparsity. Dictionaries made of *orthogonal blocks* have several appealing properties, including better incoherence. Of particular interest is the case where a single block is used for the sparse representation, thus making sparse coding extremely fast because of its simplicity and parallelism. *Shift invariant* dictionaries bring the advantage of being insensitive to the way a long signal is cut into smaller patches for processing. They also have fast representation algorithms based on FFT. *Separable* dictionaries work with 2D signals without vectorization; a pair of dictionaries is used instead of a single one. The representation is more economic and may be better suited to image processing. The concept can be generalized to more than two dimensions, working with tensors; we present a few theoretical notions that pave the way to a tensor DL. Finally, *composite* dictionaries have two components: one is learned off-line, as usual, but the other directly on the set of signals to be processed. This slows the processing, but can bring extra quality.

7.1 Short Introduction

Until now, the dictionaries we have designed had no special structure, in the sense that the atoms were free to assume any position in the m-dimensional space. The only exception was Chap. 4, where incoherence and regularization conditions implicitly limited the freedom and did not allow atoms too close one from another, but without other special constraints.

B. Dumitrescu, P. Irofti, *Dictionary Learning Algorithms and Applications*,
https://doi.org/10.1007/978-3-319-78674-2_7

Structured dictionaries are built with explicit constraints on atoms or groups of atoms, endowing them with a special shape. Unlike the fixed dictionaries discussed in Sect. 1.6, e.g., based on DCT, the dictionaries of interest here still have a part that can be learned; so, they are in between fixed and unconstrained dictionaries.

Among the many possibilities, we will discuss in this chapter about

- sparse dictionaries, whose atoms have themselves a sparse representation;
- dictionaries made of orthogonal blocks;
- shift invariant dictionaries, that aim to model with the same accuracy any shifted version of a signal;
- separable dictionaries, that economically model 2D (or higher dimensional) non-vectorized signals.

To these, we add the so-called composite dictionaries, characterized by combining parts obtained by training off- and on-line.

There are various reasons for thinking at structured dictionaries.

- The underlying structure of the signals can be better modeled, hence the structured dictionary is more suited to that class of signals than a general dictionary.
- It is easier to obtain incoherence or incoherence is (at least partially) structurally guaranteed.
- The trade-off between the attained representation error and the complexity of finding it is better than with general dictionaries. In particular, finding sparse representations is cheaper when compared with unstructured dictionaries with the same size.
- The training of the dictionary is less complex than with a general dictionary.

It is not necessarily possible to achieve all these goals with a single dictionary structure, but even improvement in one of them may be acceptable if advances are obtained in a critical direction for the application at hand.

7.2 Sparse Dictionaries

We present in this section several dictionary constructions that use sparsity, either in the atoms themselves or, more interestingly, in atom representations that mimic the sparse representation of signals.

7.2.1 *Double Sparsity*

The dictionary has the form

$$\boldsymbol{D} = \boldsymbol{\Phi}\boldsymbol{A}, \tag{7.1}$$

where $\boldsymbol{\Phi}$ is a fixed dictionary, for example based on DCT or a wavelet transform, and $\boldsymbol{A}$ is a matrix with sparse columns. For convenience, we assume the constraint $\|\boldsymbol{a}_j\|_0 \leq s_A$, meaning that each atom of $\boldsymbol{D}$ is the linear combination of a few columns of $\boldsymbol{\Phi}$. The number of columns of $\boldsymbol{\Phi}$ may have any relation with the number n of atoms. The model (7.1) was proposed in [157] and named *doubly sparse*, since it is based on two sparse representations, one for atoms and the other (the usual one) for signals.

The dictionary form (2.2) can be seen as a trade-off between a fixed transform, which may have good representation qualities for a class of signals but cannot be fully adapted to them, and a free dictionary. One may hope that, by reducing the number of parameters (the matrix $\boldsymbol{A}$ has fewer nonzero elements than the dictionary $\boldsymbol{D}$), the dictionary optimization process will be more likely to find a better local optimum, in less time.

The DL problem (2.2) has now the form

$$
\begin{aligned}
\min_{\boldsymbol{A},\boldsymbol{X}} \quad & \|\boldsymbol{Y} - \boldsymbol{\Phi}\boldsymbol{A}\boldsymbol{X}\|_F^2 \\
\text{s.t.} \quad & \|\boldsymbol{x}_\ell\|_0 \leq s,\ \ell = 1:N \\
& \|\boldsymbol{a}_j\|_0 \leq s_A,\ j = 1:n \\
& \|\boldsymbol{\Phi}\boldsymbol{a}_j\| = 1,\ j = 1:n
\end{aligned}
\tag{7.2}
$$

Note that the normalization constraint is imposed on the columns of $\boldsymbol{D}$, as usual, since $\boldsymbol{d}_j = \boldsymbol{\Phi}\boldsymbol{a}_j$.

We present the adaptation of AK-SVD for solving (7.2). Mimicking the developments from Sects. 3.3.2 and 3.5, we denote simply $\boldsymbol{d} = \boldsymbol{\Phi}\boldsymbol{a}$ the atom that is currently optimized, while all the others are fixed, and $\boldsymbol{x}^T$ the (reduced) vector of representations corresponding to that atom. Defining the (reduced) error matrix $\boldsymbol{F}$ like in (3.19), the optimal atom can be found by solving a problem similar to that from Proposition 3.4, namely

$$
\begin{aligned}
\min_{\boldsymbol{a}} \quad & \left\|\boldsymbol{F} - \boldsymbol{\Phi}\boldsymbol{a}\boldsymbol{x}^T\right\|_F^2 \\
\text{s.t.} \quad & \|\boldsymbol{a}\|_0 \leq s_A \\
& \|\boldsymbol{\Phi}\boldsymbol{a}\| = 1
\end{aligned}
\tag{7.3}
$$

This problem is somewhat more intricate than (3.27), due to the more complicated objective and, especially, to the sparsity constraint on $\boldsymbol{a}$. Transforming the objective of (7.3) like in (3.29) and taking the norm constraint into account, we obtain

$$
\begin{aligned}
\left\|\boldsymbol{F} - \boldsymbol{\Phi}\boldsymbol{a}\boldsymbol{x}^T\right\|_F^2 &= \|\boldsymbol{F}\|_F^2 - 2\boldsymbol{x}^T\boldsymbol{F}^T\boldsymbol{\Phi}\boldsymbol{a} + \|\boldsymbol{x}\|^2 \\
&= \|\boldsymbol{F}\|_F^2 + \|\boldsymbol{F}\boldsymbol{x} - \boldsymbol{\Phi}\boldsymbol{a}\|^2 - \|\boldsymbol{F}\boldsymbol{x}\|^2 - 1 + \|\boldsymbol{x}\|^2 .
\end{aligned}
\tag{7.4}
$$

Only the second term depends on $\boldsymbol{a}$, hence (7.3) is equivalent to

$$\begin{aligned} \min_{\boldsymbol{a}} \quad & \|\boldsymbol{F}\boldsymbol{x} - \boldsymbol{\Phi}\boldsymbol{a}\|_F^2 \\ \text{s.t.} \quad & \|\boldsymbol{a}\|_0 \le s_A \\ & \|\boldsymbol{\Phi}\boldsymbol{a}\| = 1 \end{aligned} \tag{7.5}$$

Since the angle between the vectors $\boldsymbol{\Phi}\boldsymbol{a}$ and $\boldsymbol{F}\boldsymbol{x}$ must be minimum regardless of the norm constraint, the solution can be computed in two steps: first the objective is minimized under the sparsity constraint only, then the solution is normalized such that the norm constraint is satisfied. The first step can be implemented by calling OMP via

$$\boldsymbol{a} = \text{OMP}(\boldsymbol{\Phi}, \boldsymbol{F}\boldsymbol{x}, s, 0).$$

Once the optimal atom has been computed, the optimal corresponding representation is given by Proposition 3.8:

$$\boldsymbol{x} = \boldsymbol{F}^T \boldsymbol{\Phi}\boldsymbol{a}. \tag{7.6}$$

The overall dictionary update stage is shown in Algorithm 7.1. Its complexity is somewhat larger than that of AK-SVD, due to the replacement of a simple matrix–vector multiplication with the OMP call from step 5. However, due to the smaller number of parameters, convergence is faster. Experiments in [157] show that the double sparsity approach is very robust in image representation, in the sense that the trained dictionary behaves very well on validation images not used for training.

7.2.2 Greedy Selection

A less common situation is that where the atoms are sparse, which may be seen as taking $\boldsymbol{\Phi} = \boldsymbol{I}$ in (7.1); of course, sparse atoms are appropriated where the signals themselves are sparse, a situation that seldom appears; for example, speech signals inherently contain silent portions; so, at least some sparse atoms would be useful in the dictionary, if not all of them. A very simple method for building a dictionary that contains many sparse atoms was proposed in [103]. It is closer to the selection methods from Sect. 5.7 than to the usual DL approach by optimization. The idea is to take the first atom as the sparsest signal from the set $\boldsymbol{Y}$. Then, the contribution of the chosen atom is removed from all the other signals. The second atom is the sparsest modified signal (which can be seen as a residual), etc. The whole process resembles the Matching Pursuit algorithm, with the difference that atom selection is different.

Algorithm 7.1: Double Sparsity AK-SVD Dictionary Update

Data: signals set $\boldsymbol{Y} \in \mathbb{R}^{m \times N}$
fixed dictionary $\boldsymbol{\Phi}$
current sparse dictionary $\boldsymbol{A}$
representation matrix $\boldsymbol{X} \in \mathbb{R}^{n \times N}$
sparsity level s

Result: updated sparse dictionary $\boldsymbol{A}$

Compute error $\boldsymbol{E} = \boldsymbol{Y} - \boldsymbol{\Phi A X}$
for $j = 1 : n$ **do**
  Let $\mathcal{I}_j$ be the set of indices of nonzeros on row j of $\boldsymbol{X}$
  Modify error: $\boldsymbol{F} = \boldsymbol{E}_{\mathcal{I}_j} + \boldsymbol{\Phi a}_j \boldsymbol{X}_{j,\mathcal{I}_j}$
  Update sparse atom: $\boldsymbol{a}_j = \text{OMP}(\boldsymbol{\Phi}, \boldsymbol{F X}_{j,\mathcal{I}_j}^T, s, 0)$
  Normalize atom: $\boldsymbol{a}_j \leftarrow \boldsymbol{a}_j / \left\| \boldsymbol{\Phi a}_j \right\|$
  Update representation: $\boldsymbol{X}_{j,\mathcal{I}_j}^T = \boldsymbol{F}^T \boldsymbol{\Phi a}_j$
  Recompute error: $\boldsymbol{E}_{\mathcal{I}_j} = \boldsymbol{F} - \boldsymbol{\Phi a}_j \boldsymbol{X}_{j,\mathcal{I}_j}$

Algorithm 7.2: Greedy Adaptive Dictionary

Data: signals set $\boldsymbol{Y} \in \mathbb{R}^{m \times N}$
dictionary size n

Result: dictionary $\mathcal{D} \in \mathbb{R}^{m \times n}$

Initialize dictionary $\mathcal{D} = \emptyset$, residuals $\boldsymbol{E} = \boldsymbol{Y}$, candidate indices $\mathcal{I} = 1 : N$
for $j = 1 : n$ **do**
  Find index of sparsest residual: $\ell = \arg\min_{i \in \mathcal{I}} \|\boldsymbol{e}_i\|_1 / \|\boldsymbol{e}_i\|_2$
  Set new atom: $\boldsymbol{d}_j = \boldsymbol{e}_\ell / \|\boldsymbol{e}_\ell\|$, $\mathcal{I} \leftarrow \mathcal{I} \setminus \{\ell\}$
  for $i \in \mathcal{I}$ **do**
    Update residuals: $\boldsymbol{e}_i \leftarrow \boldsymbol{e}_i - \boldsymbol{e}_i^T \boldsymbol{d}_j \cdot \boldsymbol{e}_\ell$

Let $\boldsymbol{E}$ be the residuals matrix, initialized with $\boldsymbol{Y}$. As sparsity measure of a residual, say $\boldsymbol{e}_\ell$, the ratio $\|\boldsymbol{e}_\ell\|_1 / \|\boldsymbol{e}_\ell\|_2$ is used. Once the residual with the lowest sparsity measure is chosen, denote it again $\boldsymbol{e}_\ell$, and is normalized and appended to the dictionary, all the other residuals are updated via

$$\boldsymbol{e}_i \leftarrow \boldsymbol{e}_i - \boldsymbol{e}_i^T \boldsymbol{e}_\ell \cdot \boldsymbol{e}_\ell, \ \forall i \neq \ell,$$

i.e., by subtracting their projection on the new atom, in Matching Pursuit style.

The procedure, named Greedy Adaptive Dictionary (GAD) in [103], continues until a preset number of atoms is reached or the norm of the residuals becomes smaller than a threshold. The former version is presented in Algorithm 7.2.

It is clear that the sparsity level of the atoms is likely to increase as the process evolves. At some point, the atoms are no longer sparse, as illustrated in [103] for speech signals. The complexity is obviously much smaller than that of standard DL algorithms.

7.2.3 Multi-Layer Sparse DL

A generalization of the double sparsity dictionary (7.1) was proposed in [115]: the dictionary is a product of sparse transforms and the optimization criterion is thus

$$\begin{aligned} \min_{\boldsymbol{D}_i} \quad & \left\| \boldsymbol{Y} - \prod_{i=1}^{p} \boldsymbol{D}_i \right\|_F^2 \\ \text{s.t.} \quad & \boldsymbol{D}_i,\ i = 1 : p,\ \text{have sparse columns} \end{aligned} \tag{7.7}$$

So, $\boldsymbol{D}_i, i = 1 : p-1$, can be seen as dictionaries, while $\boldsymbol{D}_p$ serves as representation matrix. The underlying motivation is the possibility to express many standard transforms as the product of sparse matrices; the most prominent example is the butterfly structure of the FFT, where each transform in the product has two nonzero entries per column. Instead of a fixed structure, like for a usual transforms, the only constraint on the product dictionary is the sparsity of each factor, which can be enforced explicitly or by adding to the objective of (7.7) the sum

$$\lambda \sum_{i=1}^{p} g(\boldsymbol{D}_i), \tag{7.8}$$

where $g(\boldsymbol{D}_i)$ is a sparsity enforcing function, for example the norm (5.5), and λ a trade-off parameter. The latter strategy is adopted in [115], using a block coordinate descent method working on a dictionary $\boldsymbol{D}_i$ at a time. Good image denoising results are reported, using $p = 5$, a value close to the logarithm of $m = 64$; the sizes of the factor dictionaries are 64×64, with $\boldsymbol{D}_4$ being 64×256; of course, $\boldsymbol{D}_5$ is $256 \times N$. So, there are many possibilities regarding the sizes, but a natural choice seems sufficient.

7.2.4 Multiscale Dictionaries

The multiscale dictionaries proposed in [128] can be seen as a particular case of sparse dictionaries and also a step towards shift invariance, a topic further discussed in Sect. 7.4. The dictionary has the structure

$$D = \left[\begin{array}{c|cc|cccc} & D_2 & & D_3 & & & \\ D_1 & & & & D_3 & & \\ & & D_2 & & & D_3 & \\ & & & & & & D_3 \end{array}\right]. \tag{7.9}$$

The atoms from D_1 have length m and are not constrained, hence they are like in any regular dictionary. The atoms from D_2 have length $m/2$ and they occupy either the first or the last $m/2$ positions of a full-sized atom, the remainder of the elements being zero. Similarly, the atoms from D_3 have length $m/4$ and fill a quarter of a full atom, etc.

Learning a dictionary like (7.9) requires minor changes in the usual alternate optimization setup. OMP can be run as usual; taking into account the structure will reduce the number of operations, without changing the principle of the algorithm. Dictionary update is discussed in a more general case in Sect. 7.4.2; to anticipate, let us look at the atoms of D_2; they are used in the representations of two distinct (non-overlapping) parts of the signals—the two halves; updating the atom in SGK or AK-SVD style would give a different value for each half; the optimal update is a weighted average of these two values; see problem **P 7.2** for details.

7.3 Orthogonal Blocks

We are motivated to train dictionaries because they can sparsely represent signals via specialized vectors but, historically, the first "dictionaries" were built from a collection of fixed basis representing popular matrix transforms such as Fourier, discrete cosine or wavelets.

In fact, in the context of sparse representations, the term dictionary was first coined by Donoho [53] when studying if there is any advantage of representing a signal y through a combined set of vectors resulted from concatenating two orthogonal basis $\Phi \in \mathbb{R}^{m \times m}$ and $\Psi \in \mathbb{R}^{m \times m}$. The matrix $D = [\Phi \; \Psi]$ was called a dictionary and the first theoretical guarantees on sparse representations revolved around this structure. For example, Proposition 1.14(a) was first proved for such a dictionary. Note that, for this dictionary, the mutual coherence (1.25) is efficiently computed by skipping the trivial inner-products within the same basis

$$M = \max_{1 \le i,j \le m} |\phi_i^T \psi_j|. \tag{7.10}$$

The bound (1.30) was later improved in [63]

$$\|x\|_0 < \frac{1}{M} \tag{7.11}$$

where it was also shown that the solutions of ℓ_0 and its ℓ_1 relaxation coincide for

$$\|\boldsymbol{x}\|_0 < \frac{\sqrt{2}-0.5}{M} \approx \frac{0.9142}{M}. \tag{7.12}$$

The result, belonging to the same family as Proposition 1.19, was generalized in [84] for unions of $L \geq 2$ orthogonal bases

$$\|\boldsymbol{x}\|_0 < \left(\frac{1}{2} + \frac{1}{2(L-1)}\right)\frac{1}{M} \tag{7.13}$$

and for ℓ_1 equivalence the stricter bound becomes

$$\|\boldsymbol{x}\|_0 < \left(\sqrt{2} - 1 + \frac{1}{2(L-1)}\right)\frac{1}{M}. \tag{7.14}$$

Given these theoretical guarantees and going back to our initial drive, a natural idea is to adapt DL to orthogonal bases, thus generalizing orthogonal transforms. This structured DL problem is

$$\begin{aligned} \min_{Q_1,\dots,Q_L,X} \quad & \|\boldsymbol{Y} - [\boldsymbol{Q}_1\ \boldsymbol{Q}_2\ \dots\ \boldsymbol{Q}_L]\,\boldsymbol{X}\|_F^2 \\ \text{s.t.} \quad & \|\boldsymbol{x}_i\|_0 \leq s,\ i = 1:N \\ & \boldsymbol{Q}_j{}^T\boldsymbol{Q}_j = \boldsymbol{I}_m,\ j = 1:L \end{aligned} \tag{7.15}$$

The fact that we store the entire basis $\boldsymbol{Q}_j$ in the dictionary, instead of only a few key vectors for the data set at hand, has the advantage of being somewhat future-proof in that it is less likely to overfit on the training set. This comes at the cost of dictionary size, which does not necessarily impact performance, as we will see.

7.3.1 Orthogonal Basis Training

When dealing with dictionaries structured as a union of orthogonal bases, learning is performed separately for each block. Dedicated algorithms differentiate themselves through the representation strategy, while dictionary training seems to be performed ubiquitously by the 1ONB algorithm, which solves the DL problem (2.2) for a dictionary that is an orthogonal matrix $\boldsymbol{Q} \in \mathbb{R}^{m \times m}$, using the typical alternate optimization adapted to the orthogonal structure. (For historical reasons, the abbreviation ONB comes from "orthonormal bases" [117].) We present its instructions in Algorithm 7.3 and proceed with the details in the following paragraphs.

The signals $\boldsymbol{Y}$ can be exactly represented by dictionary $\boldsymbol{Q}$ because it is also a full vector basis, $\boldsymbol{Y} = \boldsymbol{Q}\boldsymbol{X}$, thus dense representations can be directly computed via $\boldsymbol{X} = \boldsymbol{Q}^T\boldsymbol{Y}$ (step 3). This is of course insufficient because we require sparse solutions.

Algorithm 7.3: 1ONB

Data: signals set $\boldsymbol{Y} \in \mathbb{R}^{m \times N}$
target sparsity level s
number of training rounds R
Result: updated dictionary $\boldsymbol{Q}$, representation matrix $\boldsymbol{X}$

```
1 Initialization: Let Q = U, taken from the SVD Y = UΣV^T
2 for r = 1 : R do
3 |   Update representation: X = Q^T Y
4 |   for j = 1 : N do
5 |   |   Sparsify representation by thresholding: x_j = SELECT(x_j, s)
6 |   Update dictionary: Q = VU^T, using the SVD XY^T = UΣV^T
```

Sparse coding is implemented in 1ONB in step 5 by thresholding all but the first s largest (in absolute value) coefficients; due to orthogonality, this operation is optimal, see Problem 7.3; there is no need of OMP or other representation algorithm. Selecting the largest entries from vector $\boldsymbol{x} \in \mathbb{R}^m$ can be achieved through multiple methods. The straightforward and naive way would be through in-place sorting: find the absolute maximum value and swap it with the top position, repeat s times by performing the search on the remaining entries from $\boldsymbol{x}$. Even though this totals to $O(sm)$ instructions, which is asymptotically inefficient, in our case s and m are small enough that it makes little difference compared to the fastest (but more involved) partial sorting algorithm [30] that performs the operation in $O(m)$ steps. No matter the algorithm, we denote this thresholding operation as SELECT$(\boldsymbol{x}, s)$ and apply it for all representation vectors (step 4).

In the update step, using the new matrix $\boldsymbol{X}$, the dictionary is refined (step 6) by using the product of the resulting orthogonal matrices from the SVD computation of $\boldsymbol{X}\boldsymbol{Y}^T$. This orthogonal $\boldsymbol{Q}$ minimizes $\|\boldsymbol{Y} - \boldsymbol{Q}\boldsymbol{X}\|_F$ (the problem is called Procrustes orthogonalization [83]), as shown in Proposition 4.14.

Results from [117] indicate that, with proper initialization (step 1), good results can be reached by just a few iterations ($R \leq 5$ in step 2). A good starting point is the left singular vectors matrix of the SVD decomposition of the given data set

$$\boldsymbol{Y} = \boldsymbol{U}\boldsymbol{\Sigma}\boldsymbol{V}^T \Rightarrow \boldsymbol{Q}_0 = \boldsymbol{U}. \tag{7.16}$$

This is in fact a principal component reduction of the input signal.

The complexities of the required computations for each 1ONB iteration are $O(m^2N)$ for the matrix multiplication in the representations update, $O(smN)$ for the N thresholding operations, and $O(m^3)$ the dictionary update consisting of SVD and matrix multiplication. This sums up to approximately $O(Rm^2N)$ operations.

Algorithm 7.4: UONB

Data: signals set $\boldsymbol{Y} \in \mathbb{R}^{m \times N}$
current dictionary $\boldsymbol{D} = [\ \boldsymbol{Q}_1 \ldots \boldsymbol{Q}_L\ ] \in \mathbb{R}^{m \times Lm}$
representation matrix $\boldsymbol{X} \in \mathbb{R}^{Lm \times N}$
number of training rounds R
Result: updated dictionary $\boldsymbol{D}$

1 Compute error $\boldsymbol{E} = \boldsymbol{Y} - \boldsymbol{D}\boldsymbol{X}$
2 **for** $j = 1 : L$ **do**
3 Modify error: $\boldsymbol{F} = \boldsymbol{E}_{\mathcal{I}_j} + \boldsymbol{Q}_j \boldsymbol{X}_{\mathcal{J},\mathcal{I}_j}$
4 Update basis: $\boldsymbol{Q}_j = \boldsymbol{V}\boldsymbol{U}^T$ using the SVD $\boldsymbol{X}_{\mathcal{J},\mathcal{I}_j}\boldsymbol{F}^T = \boldsymbol{U}\boldsymbol{\Sigma}\boldsymbol{V}^T$
5 Update representations: $\boldsymbol{X} = \text{OMP}(\boldsymbol{D}, \boldsymbol{Y}, s, 0)$
6 Recompute error: $\boldsymbol{E} = \boldsymbol{Y} - \boldsymbol{D}\boldsymbol{X}$

7.3.2 *Union of Orthonormal Bases*

The Union of Orthonormal Basis (UONB) algorithm [117] was the first proposed DL method for this type of dictionary structure. Its design follows that of the K-SVD family. Representations are computed by performing OMP for each signal $\boldsymbol{y}$ in the usual manner, choosing from all the atoms of the union $\boldsymbol{D}$. Note that this approach ignores the special dictionary structure. We describe the dictionary update routine in Algorithm 7.4.

The initial approximation error is computed in step 1. Based on it, the orthogonal matrices $\boldsymbol{Q}_j$ are successively updated in the loop from step 2. Let $\mathcal{I}_j$ be the set of the indices of the signals using (at least an atom from) the current basis in their representation and let $\mathcal{J}$ be the row indices set of matrix $\boldsymbol{X}$ corresponding to basis j. Then, we can write the signals using basis $\boldsymbol{Q}_j$ as $\boldsymbol{Y}_{\mathcal{I}_j}$ and the associated block from the representation matrix as $\boldsymbol{X}_{\mathcal{J},\mathcal{I}_j}$. The representation error for signals $\mathcal{I}_j$ becomes

$$\boldsymbol{E}_{\mathcal{I}_j} = \boldsymbol{Y}_{\mathcal{I}_j} - \boldsymbol{D}\boldsymbol{X}_{\mathcal{I}_j} = \boldsymbol{Y}_{\mathcal{I}_j} - \sum_{k \neq j} \boldsymbol{Q}_k \boldsymbol{X}_{\mathcal{K},\mathcal{I}_j} - \boldsymbol{Q}_j \boldsymbol{X}_{\mathcal{J},\mathcal{I}_j}, \tag{7.17}$$

where $\mathcal{K}$ contains the row indices from $\boldsymbol{X}$ corresponding to the block $\boldsymbol{Q}_k$. Fixing all the other orthogonal blocks, the matrix

$$\boldsymbol{F} = \boldsymbol{Y}_{\mathcal{I}_j} - \sum_{k \neq j} \boldsymbol{Q}_k \boldsymbol{X}_{\mathcal{K},\mathcal{I}_i} \tag{7.18}$$

is the fixed part of the error, computed for each orthogonal basis in step 3. Following the same rationale as behind 1ONB, we use $\boldsymbol{F}$ and the associated sparse representation $X_{\mathcal{J},\mathcal{I}_j}$ to update the dictionary block Q_j (step 4) via Procrustes approximation (4.32). Next, due to the structure agnostic representation strategy of UONB, the entire matrix X has to be updated via OMP (step 5) as the new basis Q_j might trigger a change in the global sparsity pattern which could in turn modify set $\mathcal{I}_j$. Finally, the representation error E is updated in step 6 and the loop proceeds to update the subsequent dictionary basis.

When estimating the total number of instructions necessary for updating a single basis via UONB, we can ignore the multiplications for computing matrices E and F because of the much more complex operations required by SVD for Procrustes orthogonalization $O(|\mathcal{J}|^2 m + m^3)$ [80] and OMP for representation update $O(Nms(Lm + s^2))$. Updating the entire dictionary rounds up to $O(LNms(Lm + s^2) + N^2 m + Lm^3)$.

7.3.3 Single Block Orthogonal DL

Unlike UONB, single block orthogonal algorithm (SBO) [161] takes advantage of the special dictionary structure by constraining each data item from Y to use a single block Q_j for its representation. While this may seem a hard constraint, we will see that it has clear benefits in terms of speed, both in representation and in learning. Another significant difference is that SBO iteratively increases the dictionary size, adapting each new basis to the current context in order to improve the approximation error.

We remind from Sect. 7.3.1 that the s-sparse representation x of a signal y using an orthogonal matrix Q_j can be computed by a partial sorting function through the call $x = \text{SELECT}(Q_j^T y, s)$.

Proposition 7.1 *Given a union of L orthogonal blocks, the best basis j to represent a given signal y is picked by computing the energy of the resulting representation coefficients and selecting the block where the energy is highest. More precisely, if*

$$x^i = SELECT(Q_i^T y, s) \tag{7.19}$$

is the representation using block Q_i, and $E_{x^i} = \|x^i\|^2$ is its energy, then the best orthogonal block is given by

$$j = \arg\max_{i=1:L} (E_{x^i}). \tag{7.20}$$

Proof Indeed, denoting $\mathcal{S}$ the support of the representation x and $\xi = Q^T y$, and taking into account that multiplication with an orthogonal matrix Q preserves the norm of a vector, the sparse representation error with this matrix is

$$\|y - Qx\| = \left\| Q^T y - x \right\| = \sum_{k \notin S} \xi_k^2 = \|y\|^2 - \|x\|^2 .$$

We have used the equalities $\|y\| = \|\xi\|$ and $\xi_k = x_k$, for $k \in S$. Hence, the error is minimum when the energy of the sparse representation is maximum. ■

Following this method, each data item from Y is represented by each orthogonal basis from D in $O(Lm^2)$ operations. Thresholding each candidate solution x^i takes $O(Lm)$ operations as discussed around Algorithm 7.3. The final steps of computing the energy and finding the basis maximizing it need $O(Ls)$ and $O(L)$ operations, respectively, and so are insignificant in comparison. In total, for all N signals of data set Y, the representation stage rounds up to $O(LNm^2)$ operations, which is a clear improvement over OMP as used in the UONB algorithm.

Note that SBO representation preserves the nice properties of OMP, ensuring that each selected atom is orthogonal to the existing set, and at the same time provides significant reductions in execution times. Obviously, OMP approximation quality is superior as it has access to atoms from the entire dictionary, an edge that SBO attempts to compensate through larger dictionaries that provide a wider palette of bases to choose from.

The complete set of SBO instructions is summarized in Algorithm 7.5. SBO has two big stages. The first is for initialization, where L_0 orthogonal bases are trained via 1ONB based on disjunct subsets $\mathcal{P}_j$ of the signal set Y (steps 1–4). The subsets need not cover the whole data set; the size of each subset should be at least a small multiple of m. With the initial dictionary formed with these L_0 orthogonal blocks, we compute the sparse representations X (steps 5–7) by choosing the optimal basis from the existing union for each data item as earlier described. Even though typical values for L_0 are small (e.g., around 5) this warm-up stage is important as it determines the quality of the resulting dictionary.

The second stage, starting with step 8, comprises the proper SBO iterations, in which the union is enlarged by one block at a time and all bases are trained on all signals. Each new basis Q_j is trained on the $\mathcal{W}$ signal subset corresponding to the worst existing representations (steps 9–10); this operation is similar in spirit to the addition of new atoms in Stagewise K-SVD Algorithm 6.2 form Sect. 6.3; the percentage of 5% for the size of $\mathcal{W}$ is only an indication of magnitude and can be varied upon necessities. Using the extended dictionary from step 11, the representation process is repeated in steps 12–14. Before completing the iteration, the signals $\mathcal{I}_b$ using basis b in their representation are used to refine Q_b through a few 1ONB rounds (step 16). This process is repeated for each union member (step 15). This strategy ensures that the signals are reallocated to bases several times and each base is several times adapted to the signals that it represents.

Due to its improved representation, the complexity of SBO is dictated by the dictionary learning phase unlike standard algorithms from Chap. 3, where representation may have a significant weight in the complexity. Representation amounts to $O(NL_0m^2)$ operations during initialization and then $O(Njm^2)$ for iteration j in step 8 totaling to $O(NL^2m^2)$ operations. There are L 1ONB calls

Algorithm 7.5: SBO

Data: signals set $\boldsymbol{Y} \in \mathbb{R}^{m \times N}$
number of orthogonal blocks L
number initial orthogonal blocks L_0
number of training rounds R
Result: dictionary $\boldsymbol{D} = [\ \boldsymbol{Q}_1 \ldots \boldsymbol{Q}_L\ ] \in \mathbb{R}^{m \times Lm}$

`Initialization`

1 Randomly generate sets $\mathcal{P}_j$, $j = 1 : L_0$, of distinct signal indices
2 **for** $j = 1 : L_0$ **do**
3   Train new basis: $\boldsymbol{Q}_j = 1\text{ONB}(\boldsymbol{Y}_{\mathcal{P}_j}, s, R)$
4   Extend dictionary: $\boldsymbol{D} = [\ \boldsymbol{Q}_1 \ldots \boldsymbol{Q}_{j-1}\ \boldsymbol{Q}_j\ ]$
5 **for** $k = 1 : N$ **do**
6   Choose basis: $b = \arg\max_{i=1:L}(E_{\boldsymbol{x}_k^i})$
7   Compute representation: $\boldsymbol{x}_k = \boldsymbol{x}_k^b$

`Iterations`

8 **for** $j = L_0 + 1 : L$ **do**
9   Find the index set $\mathcal{W}$ of the worst 5% represented signals
10   Train new basis: $\boldsymbol{Q}_j = 1\text{ONB}(\boldsymbol{Y}_{\mathcal{W}}, s, R)$
11   Update dictionary: $\boldsymbol{D} = [\ \boldsymbol{Q}_1 \ldots \boldsymbol{Q}_{j-1}\ \boldsymbol{Q}_j\ ]$
12   **for** $k = 1 : N$ **do**
13   Choose basis: $b = \arg\max_{i=1:L}(\boldsymbol{E}_{\boldsymbol{x}_k^i})$
14   Update representation: $\boldsymbol{x}_k = \boldsymbol{x}_k^b$
15   **for** $b = 1 : j$ **do**
16   Refine basis: $\boldsymbol{Q}_b = 1\text{ONB}(\boldsymbol{Y}_{\mathcal{I}_b}, s, R)$

for the entire union with an associated cost of $O(RLm^3)$ and $L(L+1)/2$ more for refinement during step 15. The learning phase thus rounds up to $O(RL^2m^3)$ and overall SBO requires $O(RL^2m^3 + NL^2m^2)$ operations.

Remark 7.2 (Improvements via Parallelism) In taking advantage of the dictionary structure, SBO offers the opportunity for parallelization due to the decoupled nature of its data manipulations. Indeed in [97] P-SBO is proposed, where P stands for parallel.

The representation stage is expressed through the map-reduce paradigm. Signal-basis pairs are formed that produce the energy of the resulting sparse representation $E_{\boldsymbol{x}^i}$ from (7.20). We reduce the list, for each signal in $\boldsymbol{Y}$, to the element with the largest energy leading to the choice of a single representation block.

For dictionary refinement, each instance of steps 2 and 15 in Algorithm 7.5 is concurrently executed and further down, the operations within each 1ONB call are also parallelized.

P-SBO also modifies dictionary expansion (steps 9–11) to permit the simultaneous addition of $\tilde{j} > 1$ orthogonal bases. For this, the worst represented signal set $\mathcal{W}$ can be evenly split and the new $\tilde{j}$ basis can be trained in parallel as discussed above. This speeds up the DL process by shrinking the number of iterations required to build the final union of L orthogonal bases from $L - L_0$ to $\lceil (L - L_0)/\tilde{j} \rceil$.

A rather surprising empirically proven fact of this modification is that the representation error is also improved. The effect is visible and growing as $\tilde{j}$ is increased until it reaches the point where the training set $\mathcal{W}$ becomes too small for properly training the orthogonal bases and so the error starts to slightly depreciate. The result is consistent across different sizes of the worst reconstruction set $\mathcal{W}$. ■

Remark 7.3 (Relation Between SBO and [171]) A similar result to SBO is presented in [171], where the L orthogonal blocks are seen as the means through which a signal $\boldsymbol{y}$ is classified into one of L classes. The dictionaries are trained similarly to SBO through Procrustes orthogonalization (4.32) with the notable difference that the dictionary size is fixed across DL as in the UONB case.

The classification (or representation) strategy uses the criterion

$$j = \arg\min_{i=1:L} \left(\left\| \boldsymbol{y} - \boldsymbol{Q}_i \boldsymbol{x}^i \right\|^2 + \lambda \left\| \boldsymbol{x}^i \right\|_0 \right) \tag{7.21}$$

to assign data item $\boldsymbol{y}$ to class (or orthogonal basis) j. SBO has the advantage here because it uses only the error in its criterion (7.20) without having to use cross-validation or similar techniques to set the value of the extra parameter λ in (7.21); however, SBO sets the sparsity level s to a fixed value, while that resulting from (7.21) is variable. With fixed basis $\boldsymbol{Q}_i$, the sparsification process of $\boldsymbol{x}^i$ is also slightly different because it solves

$$\boldsymbol{x}^i = \arg\min_{\boldsymbol{x}} \left(\| \boldsymbol{y} - \boldsymbol{Q}_i \boldsymbol{x} \|^2 + \lambda \| \boldsymbol{x} \|_0 \right) \tag{7.22}$$

by hard-thresholding the entries of $\boldsymbol{\xi} = \boldsymbol{Q}_i^T \boldsymbol{y}$ with $\sqrt{\lambda}$

$$x_k^i = \begin{cases} \xi_k, & \text{if } |\xi_k| \geq \sqrt{\lambda} \\ 0, & \text{if } |\xi_k| < \sqrt{\lambda} \end{cases}, \quad k = 1 : m. \tag{7.23}$$

The criteria from (7.21) and (7.22) are conceptually closer to OMP with error stopping than to SBO, whose sparsity level is fixed. ■

7.4 Shift Invariant Dictionaries

The standard DL problem is formulated for signals that have the same length. In most cases, these signals are obtained by simply cutting blocks from a long signal, which is the actual output of the source that we want to model. The obvious example is that of images, that are divided into smaller patches for processing, the main reason being the computational requirements. Splitting the long signal is often done without any analysis of the signal, in equal adjacent blocks. Hence, significant patterns may appear in arbitrary positions in the blocks. A possible and partial cure is to allow overlapping blocks in the training set, with the price of increasing its dimension; however, considering all possible blocks is unrealistic; for example, a single 512×512 image would generate 505×505 different 8×8 patches.

Even when the signals have naturally the same length (e.g., a heart beat sampled with a fixed number of points), significant information may appear in several (if not all) shifted versions and explicit alignment may be impossible due to the actually unknown shape of the sought pattern.

All the above point to the necessity of formulating a shift invariant version of the DL problem, explicitly addressing the issue. In this section, we will discuss two solutions for obtaining shift invariance. One is to transfer the burden of modeling the shift to the dictionary, which, instead of containing only independent atoms, it will also include all their (circularly) shifted versions. So, no matter the position of pattern in the signal, the dictionary will contain an atom aligned with it.

The second is to just reformulate the DL problem, such that it is expressed for a single long signal, without cutting it into blocks. The atoms still have a small length, like in the standard problem, but they can be shifted in any position, such that the resulting overall sparse representation has a small error. Such an approach is also named convolutional sparse coding and can have several flavors, of which we will present only the basics. The two approaches are obviously related, although technically they lead to quite different algorithms.

7.4.1 Circulant Dictionaries

An explicit solution to shift invariance is to work with a dictionary that, together with an atom, contains all its m rotations. That is, the dictionary is a concatenation of circulant matrices. Although apparently we increase the size of the dictionary from n to mn, there are fast algorithms that highly alleviate the complexity issues. Our presentation is based on [146, 164], which take advantage of the relation between circulant matrices and the DFT.

Remark 7.4 Let $\boldsymbol{P}$ be the permutation matrix that circularly down-shifts a vector by multiplication. Let $\boldsymbol{c} \in \mathbb{R}^m$ be a vector; then, the circulant matrix defined by $\boldsymbol{c}$ is

$$\boldsymbol{C} = [\boldsymbol{c}\ \boldsymbol{P}\boldsymbol{c}\ \boldsymbol{P}^2\boldsymbol{c}\ \dots\ \boldsymbol{P}^{m-1}\boldsymbol{c}].$$

The eigenvalue decomposition of a circulant matrix can be written with the DFT matrix $\boldsymbol{H}$ as (remind that $\boldsymbol{H}^T\boldsymbol{H} = m\boldsymbol{I}$)

$$\boldsymbol{C} = \left(\frac{1}{\sqrt{m}}\boldsymbol{H}^H\right)\text{diag}(\boldsymbol{H}\boldsymbol{c})\left(\frac{1}{\sqrt{m}}\boldsymbol{H}\right). \tag{7.24}$$

So, the multiplication of a circulant matrix with a vector can be done using the FFT:

$$\boldsymbol{C}\boldsymbol{y} = \text{IFFT}(\text{FFT}(\boldsymbol{c}) \circ \text{FFT}(\boldsymbol{y})), \tag{7.25}$$

where $\circ$ stands for element-wise multiplication. ■

We describe now the changes required by SGK for learning a dictionary having the form

$$\boldsymbol{D} = [\boldsymbol{C}_1\ \boldsymbol{C}_2\ \ldots\ \boldsymbol{C}_n], \tag{7.26}$$

where $\boldsymbol{C}_i$, $i = 1 : n$, are circulant matrices. The dictionary has mn atoms, but only n are independent; hence we stick to our notation with n being the number of atoms.

Remark 7.5 Let us examine first the sparse coding step. We note that OMP is shift invariant if the dictionary (7.26) is used, meaning that if an atom $\boldsymbol{d}_j$ is chosen when the vector $\boldsymbol{y}$ is represented, then the atom $\boldsymbol{P}^q\boldsymbol{d}_j$ is chosen when the vector $\boldsymbol{P}^q\boldsymbol{y}$ is represented. This can be proved by induction; we show only the first step. The first selected atom is that $\boldsymbol{d}_j$ for which the absolute inner product $|\boldsymbol{y}^T\boldsymbol{d}_j|$ is the largest. For $\boldsymbol{P}^q\boldsymbol{y}$, the shifted version of the same atom is selected, because $(\boldsymbol{P}^q\boldsymbol{y})^T\boldsymbol{P}^q\boldsymbol{d}_j = \boldsymbol{y}^T\boldsymbol{d}_j$. (Note that $\boldsymbol{P}$ is an orthogonal matrix.) The residual of the representation is similarly shifted for $\boldsymbol{P}^q\boldsymbol{y}$, etc.

So, the behavior of OMP in the presence of a dictionary made of circulant matrices is exactly what we should expect, i.e., OMP is robust to shifts of the input signals.

The adaptation of the OMP algorithm to the dictionary (7.26) is left to the reader, see problem **P 7.5**. ■

Since we focus on SGK, the dictionary update is based on the optimization of a single atom, while the others are fixed. However, due to the structure (7.26) of the dictionary, we have to optimize a full circulant matrix $\boldsymbol{C}_j$, while keeping the others fixed. Splitting the representation matrix $\boldsymbol{X}$ on n block rows (each with m rows) and paralleling relations (3.18)–(3.20), we express the representation error as

$$\|\boldsymbol{Y} - \boldsymbol{D}\boldsymbol{X}\|_F = \left\|\boldsymbol{Y} - \sum_{i \neq j}\boldsymbol{C}_i\boldsymbol{X}_i - \boldsymbol{C}_j\boldsymbol{X}_j\right\|_F, \tag{7.27}$$

denote the fixed part

$$F = \left[Y - \sum_{i \neq j} C_i X_i \right]_{\mathcal{I}_j}, \tag{7.28}$$

where $\mathcal{I}_j$ is the set of indices of signals whose representations contain at least an atom from C_j, and we end up with the optimization problem

$$\min_{C_j} \quad \left\| F - C_j X_{j,\mathcal{I}_j} \right\|_F^2 \tag{7.29}$$

Denoting

$$\Sigma = \text{diag}(H c_j) \triangleq \text{diag}(\sigma_1, \ldots, \sigma_m) \tag{7.30}$$

the diagonal matrix from (7.24) (with C_j instead of C) and knowing that $H/\sqrt{m}$ is an orthogonal matrix, it results that

$$\left\| F - C_j X_{j,\mathcal{I}_j} \right\|_F = \left\| m^{-1/2} H F - m^{-1/2} \Sigma H X_{j,\mathcal{I}_j} \right\|_F \triangleq \left\| \tilde{F} - \Sigma \tilde{X} \right\|_F. \tag{7.31}$$

So, the problem has been simplified: the variable is now a diagonal matrix. Moreover, the problem can be decoupled on the rows of the matrices:

$$\left\| \tilde{F} - \Sigma \tilde{X} \right\|_F^2 = \sum_{i=1}^{m} \left\| \tilde{f}_i^T - \sigma_i \tilde{x}_i^T \right\|^2. \tag{7.32}$$

Hence, each variable can be computed separately by solving the corresponding least squares problem, whose solution is

$$\sigma_i = \frac{\tilde{x}_i^H \tilde{f}_i}{\|\tilde{x}_i\|^2}. \tag{7.33}$$

Note that most of the values σ_i are complex. However, they should appear in conjugate pairs, since they are generated in (7.30) through a DFT transform of a real vector. This indeed happens in (7.33), since $\tilde{F}$ and $\tilde{X}$ are also obtained through DFT transforms of real matrices.

Once the values σ_i are obtained with (7.33), the updated atom c_j defining the current circulant matrix C_j is obtained through IFFT, see (7.30). The atom has to be normalized, since the procedure does not constrain the norm.

Updating also the representations, like in AK-SVD, means minimizing (7.29) with $X_{j,\mathcal{I}_j}$ as variable. A solution is possible, but it is complicated by the fact that the matrix $X_{j,\mathcal{I}_j}$ is sparse. So, the objective of (7.29) should be split on columns, keeping only the relevant elements from each column, whose values are computed by solving a least squares problem for each column. Updating the representations was not attempted in [146, 164].

Although the dictionary (7.26) has m times more atoms than a usual dictionary, the complexity of the dictionary update is only $O(\log m)$ times larger, since the multiplications with circulant matrices that appear in (7.27) and (7.31) are performed like in (7.25) via the FFT.

7.4.2 Convolutional Sparse Coding

We now work directly with a long signal $\overline{y} \in \mathbb{R}^{\overline{m}}$ without extracting blocks (or patches) from it. For simplicity, we assume the signal is one-dimensional, but extension to 2D is straightforward (although somewhat tedious to implement). We still assume that the dictionary has n atoms of length m, denoted as usual $\boldsymbol{d}_j$, $j = 1 : n$. The signal has a convolutional sparse representation if it can be written as

$$\overline{y} = \sum_{j=1}^{n} \boldsymbol{d}_j * \boldsymbol{x}_j, \tag{7.34}$$

where $*$ denotes convolution between two vectors and the vectors $\boldsymbol{x}_j$ are sparse. Denoting $\overline{\boldsymbol{d}}_{tj}$ a vector of length $\overline{m}$ having the atom $\boldsymbol{d}_j$ as subvector starting with position t and zeros elsewhere, relation (7.34) has the form

$$\overline{y} = \sum_{j=1}^{n} \sum_{t} \overline{\boldsymbol{d}}_{tj} x_{tj} = \overline{\boldsymbol{D}}\overline{\boldsymbol{x}}. \tag{7.35}$$

So, the signal is represented with the help of copies of the (short) atoms, placed where necessary, in any position $t = 1 : \overline{m} - m + 1$. The full dictionary $\overline{\boldsymbol{D}} = [\overline{\boldsymbol{D}}_1 \ \ldots \ \overline{\boldsymbol{D}}_n]$ (never formed explicitly) looks like in Fig. 7.1; each matrix $\overline{\boldsymbol{D}}_j$ is a convolution matrix formed with the extended atoms $\overline{\boldsymbol{d}}_{tj}$ for all possible values of t, representing all possible shifts of the atom $\boldsymbol{d}_j$; one can also include the shifts for t going up to $\overline{m}$, the atoms being shifted circularly in the first rows of $\overline{\boldsymbol{D}}_j$, but this is a matter of choice and is not visible in the figure. Imagine the length of the signal $\overline{y}$ much larger than shown in the figure; the number of nonzeros in the representation vector $\overline{\boldsymbol{x}}$ will also be large, proportionally with the length of the vector. The representation setup from (7.35) is traditionally called convolutional sparse coding (CSC), see [86, 125, 185, 204] for some of the early approaches and [206] for a thorough bibliographic coverage.

Similarly with the standard problem (2.2), we can formulate the CSC DL problem as

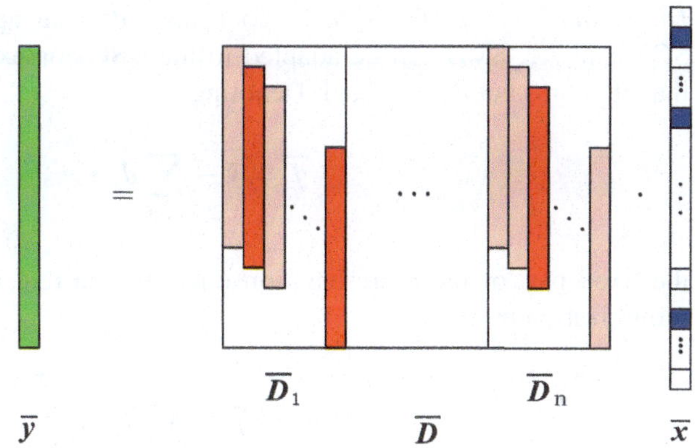

Fig. 7.1 Convolutional sparse representation of a (long) signal. The used atoms are red and the nonzero coefficients are blue. The unused atoms are pink

$$
\begin{aligned}
\min_{D,\overline{x}} \quad & \left\|\overline{y} - \overline{D}\overline{x}\right\|^2 \\
\text{s.t.} \quad & \|\overline{x}\|_0 \leq S \\
& \left\|d_j\right\| = 1, \; j = 1 : n
\end{aligned}
\tag{7.36}
$$

Note that the length m atoms are variables of the problem, hence the number of dictionary variables is the same as in (2.2). The normalization constraint is also the same and has the same purpose. The parameter S is no longer the sparsity level, but the value $Sm/\overline{m}$ could be seen as the average sparsity level (5.1). (We could define $N = \overline{m}/m$ as the number of signals when $\overline{y}$ is split into blocks of size m.)

The problem (7.36) can be solved with the standard alternate optimization approach. We discuss below the characteristic aspects of the two main stages.

Sparse Coding OMP can be easily adapted to CSC. A basic operation in OMP is the multiplication of the dictionary with the residual. The convolution structure of $\overline{D}$ allows the use of FFT for these multiplications. However, the complexity is larger than in the case where short signals are used and where the sparse coding can be made independently on each signal. Here we have a single OMP instance on a huge signal. A possible bottleneck is the computation of the least squares solution for each support, where the structure of the dictionary is of little help. A cheaper solution is to avoid orthogonalization and use Matching Pursuit, as proposed in [125] with good results.

Alternatively to (O)MP one can appeal to the Basis Pursuit problem

$$
\min_{\overline{x}} \quad \left\|\overline{y} - \overline{D}\overline{x}\right\|^2 + \lambda \|\overline{x}\|_1 \tag{7.37}
$$

The obstacle raised by the huge size of the problem can be overcome by taking advantage of the particular form of the subproblems that arise when ADMM is used [16, 206]. The complexity per iteration can be reduced to as much as $O(\overline{m}n \log \overline{m})$.

Dictionary Update The (block) coordinate descent approach that is the base of SGK and AK-SVD can be adapted to the CSC context. Let us assume, as usual, that all atoms but $\boldsymbol{d}_j$ are fixed. Denoting

$$\overline{\boldsymbol{f}} = \overline{\boldsymbol{y}} - \sum_{i \neq j} \boldsymbol{d}_i * \boldsymbol{x}_i, \tag{7.38}$$

the fixed part of the representation error, the current atom is the solution of the minimization problem

$$\min_{\boldsymbol{d}_j} \left\| \overline{\boldsymbol{f}} - \sum_t \overline{\boldsymbol{d}}_{tj} x_{tj} \right\|^2. \tag{7.39}$$

This is a least squares problem that has a simple solution when the copies of the atom $\boldsymbol{d}_j$ do not overlap in the extended atoms $\overline{\boldsymbol{d}}_{tj}$ for which x_{tj} is nonzero. Denote $\mathcal{S} = \{t \mid x_{tj} \neq 0\}$ the support of $\boldsymbol{x}_j$; non-overlapping means that the difference between any two values from $\mathcal{S}$ is at least m (in absolute value). Denoting $\boldsymbol{f}_t$ the length m subvector of $\overline{\boldsymbol{f}}$ starting at position t, the objective of (7.39) becomes

$$\sum_{t \in \mathcal{S}} \| \boldsymbol{f}_t - \boldsymbol{d}_j x_{tj} \|^2 \tag{7.40}$$

and the optimal atom is (before normalization)

$$\boldsymbol{d}_j = \sum_{t \in \mathcal{S}} x_{tj} \boldsymbol{f}_t. \tag{7.41}$$

This is a simple weighted average of the length m parts of $\overline{\boldsymbol{f}}$ that correspond to the shifts of the atom.

If there is overlapping between the copies of $\boldsymbol{d}_j$, then the least squares solution is more difficult to compute. A way around is proposed in [125] by splitting the vectors $\boldsymbol{f}_t$.

If we aim to update also the representation coefficients in AK-SVD style, this can again be done by solving a least squares problem.

Remark 7.6 All the developments in this section can be easily adapted to the situation where the atoms have different lengths. We have adopted the equal length case only for simplicity, but one can go towards a multiresolution representation by allowing the length to vary. ■

7.5 Separable Dictionaries

A compromise that can be reached when dealing with data blocks drawn from a larger multidimensional signal, as described in Sect. 7.4, is to work directly with the blocks, or patches, without vectorization. This has the advantage of preserving vicinity-based patterns that might be lost in the standard column-vector approach.

This multidimensional approach requires a dedicated dictionary for each facet. Take an image for example. One data item is now represented as a 2D block which means its sparse representation is a low-rank matrix. This reminds us of the SVD decomposition that represents the dense matrix $\boldsymbol{A}$ via the sparse diagonal matrix $\boldsymbol{\Sigma}$

$$\boldsymbol{A} = \boldsymbol{U}\boldsymbol{\Sigma}\boldsymbol{V}^T,$$

where the matrices $\boldsymbol{U}$ and $\boldsymbol{V}$ are orthogonal (and can be seen as relatives of the orthogonal dictionaries from Sect. 7.3, although this is of no importance here). The fact that $\boldsymbol{\Sigma}$ is diagonal implies that atom j from $\boldsymbol{U}$ is used exclusively with atom j from $\boldsymbol{V}$, a constraint that we would like to loosen in the DL case. A generalization that relaxes the orthogonality constraint on the dictionaries and the diagonal form of the representation matrix, but keeps the low-rank representation principle is to approximate an image patch $\boldsymbol{Y} \in \mathbb{R}^{m \times m}$ with

$$\boldsymbol{Y} = \boldsymbol{D}_1 \boldsymbol{X} \boldsymbol{D}_2^T = \sum_i \sum_j x_{ij} \boldsymbol{d}_{1i} \boldsymbol{d}_{2j}^T, \tag{7.42}$$

where $\boldsymbol{X}$ is sparse, $\boldsymbol{D}_1 \in \mathbb{R}^{m \times n_1}$ is the left dictionary and $\boldsymbol{D}_2 \in \mathbb{R}^{m \times n_2}$ is the right dictionary. Figure 7.2 illustrates relation (7.42); the position (i, j) of a nonzero element of the representation matrix $\boldsymbol{X}$ corresponds to a unique $(\boldsymbol{d}_{1i}, \boldsymbol{d}_{2j})$ atom combination of the two dictionaries. So, atom $\boldsymbol{d}_{1i}$ may form bonds with any other right-hand side atom and the same can happen for atom $\boldsymbol{d}_{2j}$. Note also that (7.42) shows that when the representation matrix $\boldsymbol{X}$ is sparse, then the 2D signal is indeed modeled as the sum of a few rank-1 matrices.

Given now N matrices $\boldsymbol{Y}_k \in \mathbb{R}^{m \times m}$, $k = 1 : N$, which are the training 2D signals in their genuine form, and target sparsity $s \in \mathbb{N}$, the DL problem becomes

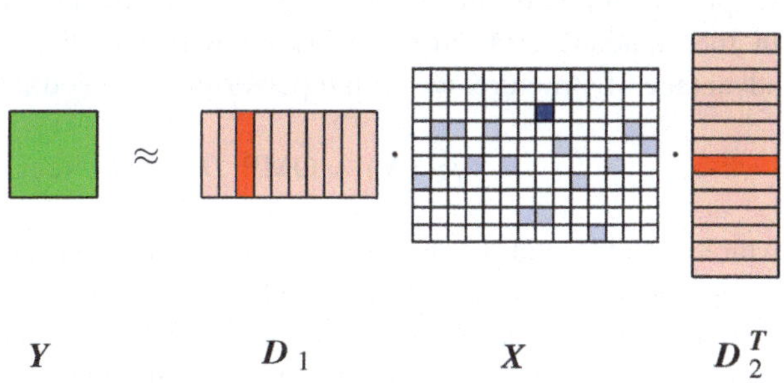

Fig. 7.2 Sparse representation of a 2D signal. The nonzero coefficients are blue. The bright blue element corresponds to the bright red atoms pair

$$
\begin{aligned}
\min_{D_1, D_2, X} \quad & \sum_{k=1}^{N} \|Y_k - D_1 X_k D_2^T\|_F^2 \\
\text{s.t.} \quad & \|X_k\|_0 \le s, \; k = 1 : N \\
& \|d_{1i}\|_2 = 1, \; i = 1 : n_1 \\
& \|d_{2j}\|_2 = 1, \; j = 1 : n_2
\end{aligned}
\tag{7.43}
$$

The first constraint says that the matrix X_k has at most s nonzero elements and the other two constraints impose the usual normalization of the atoms. Since

$$
\|Y_k - D_1 X_k D_2^T\|_F = \|\text{vec}(Y_k) - (D_2 \otimes D_1)\text{vec}(X_k)\|_2, \tag{7.44}
$$

where $\otimes$ is the Kronecker product and the operator vec$(\cdot)$ stacks the columns of the matrix argument, problem (7.43) is in fact the standard DL problem with dictionary

$$
D = D_2 \otimes D_1, \tag{7.45}
$$

which is also the reason why D is called separable. Note that D has $n_1 n_2$ atoms of size m^2, hence its size is much larger than that of D_1 or D_2, witnessing its highly structured form.

Although the bi-dimensional case will be our central focus in the following sections, most of the time generalizations to any number of dimensions are possible if not trivial.

7.5.1 2D-OMP

Even though we can use the formulations (7.44) and (7.45) to perform sparse coding via the standard methods from Chap. 1, we will show that exploiting the separable structure permits significant complexity reductions while maintaining identical representation quality.

Given that OMP is our preferred representation algorithm, we present its adaptation for two-dimensional signals, called 2D-OMP [70], in Algorithm 7.6. In the standard OMP form, working with the dictionary (7.45) without taking advantage of structure, we would perform sparse coding via

$$
\text{vec}(X) = \text{OMP}(D_2 \otimes D_1, \text{vec}(Y), s) \tag{7.46}
$$

which would lead us to the vectorized version of our 2D sparse signal in $O(m^2 s(n_1 n_2 + s^2))$ operations (see Sect. 1.3 for details).

Let us analyze each operation required by OMP and discuss its adaptation and performance in the 2D case. Atom selection, step 3 in Algorithm 1.1, has to be done in pairs, resulting in a total of $n_1 n_2$ choices that corresponds to the number of atoms in the dictionary $\boldsymbol{D}_2 \otimes \boldsymbol{D}_1$. The selection comes down to computing the atom-residual correlation vector

$$\boldsymbol{c} = \boldsymbol{D}^T \boldsymbol{e} = (\boldsymbol{D}_2 \otimes \boldsymbol{D}_1)^T \text{vec}(\boldsymbol{E}) = \text{vec}(\boldsymbol{C}) \tag{7.47}$$

where $\boldsymbol{E}$ and $\boldsymbol{C}$ are the 2D residual and correlation matrix. This task requires $O(m^2 n_1 n_2)$ instructions. If we take advantage of the separable structure, we can rewrite (7.47) as

$$\boldsymbol{C} = \boldsymbol{D}_1^T \boldsymbol{E} \boldsymbol{D}_2 \tag{7.48}$$

whose absolute largest entry dictates atom selection as follows: the row index will correspond to the atom from $\boldsymbol{D}_1$ and the column index to the atom from $\boldsymbol{D}_2^T$ (step 3 in Algorithm 7.6). The complexity of (7.48) becomes $O(mn_1 n_2)$, an improvement by a factor of m over (7.47).

In step 4, the support $\mathcal{S}$ is updated with the current (i, j) pair. Later on we denote with $\mathcal{I}$ and $\mathcal{J}$ the set of all atoms selected from the left-hand side dictionary and, respectively, from the right-hand side; these sets are ordered and may include repetitions. For computing the coefficients by solving a least squares problem, we need the restriction of the Kronecker dictionary (7.45) to the selected atoms, denoted $\boldsymbol{A}$. Unfortunately, the Kronecker structure is of little help and we store the full atoms, as shown in step 5, where the new atom

$$\boldsymbol{d} = \boldsymbol{d}_{2j} \otimes \boldsymbol{d}_{1i} \tag{7.49}$$

is appended to the dictionary, where $\boldsymbol{d}_{1i}$ and $\boldsymbol{d}_{2j}$ are the selected atoms of the current iteration. The solution of the least squares problem is

$$\boldsymbol{\xi} = (\boldsymbol{A}^T \boldsymbol{A})^{-1} \boldsymbol{A}^T \text{vec}(\boldsymbol{Y}), \tag{7.50}$$

and the relation between it and the 2D representation $\boldsymbol{X}$ results by rewriting (7.42) as

$$\sum_{(i,j)\in\mathcal{S}} x_{ij} \boldsymbol{d}_{1i} \boldsymbol{d}_{2j}^T = \boldsymbol{D}_{1\mathcal{I}} \text{diag}(\boldsymbol{\xi}) \boldsymbol{D}_{2\mathcal{J}}^T, \tag{7.51}$$

which gives

$$x_{i_k, j_k} = \xi_k, \quad (i_k, j_k) \in \mathcal{S},\ k = 1 : s. \tag{7.52}$$

Overall, the complexity of computing $\boldsymbol{X}$ remains the same as that of OMP: $O(m^2 s^2)$. Finally, the residual is updated by step 7 to be used in the following iteration. This operation is efficiently performed in $O(m^2 s)$ instructions, matching standard OMP. It follows that, although 2D-OMP provides identical representations

Algorithm 7.6: 2D Orthogonal Matching Pursuit

Data: dictionaries $\boldsymbol{D}_1 \in \mathbb{R}^{m\times n1}$, $\boldsymbol{D}_2 \in \mathbb{R}^{m\times n_2}$
signal $\boldsymbol{Y} \in \mathbb{R}^{m\times m}$
sparsity level s
stopping error ε
Result: representation support $\mathcal{S}$, solution $\boldsymbol{X}$

Initialize $\mathcal{S} = \emptyset$, $\boldsymbol{A} = \emptyset$, $\boldsymbol{E} = \boldsymbol{Y}$
while $|\mathcal{S}| < s$ and $\|\boldsymbol{E}\|_F > \varepsilon$ **do**
  Find new index: $(i, j) = \arg\max_{(i,j)\notin\mathcal{S}} |\boldsymbol{d}_{1i}^T \boldsymbol{E}^T \boldsymbol{d}_{2j}|$
  Build new support: $\mathcal{S} \leftarrow \mathcal{S} \cup \{(i, j)\}$
  Update Kronecker dictionary: $\boldsymbol{A} = [\boldsymbol{A}\ \boldsymbol{d}_{2j} \otimes \boldsymbol{d}_{1i}]$
  Compute new coefficients: $\boldsymbol{\xi} = (\boldsymbol{A}^T\boldsymbol{A})^{-1}\boldsymbol{A}^T \text{vec}(\boldsymbol{Y})$
  Compute new residual: $\boldsymbol{E} = \boldsymbol{Y} - \boldsymbol{D}_{1\mathcal{I}} \text{diag}(\boldsymbol{\xi}) \boldsymbol{D}_{2\mathcal{J}}^T$

as OMP applied directly to the dictionary (7.45), its total complexity drops to $O(ms(n_1n_2 + ms^2))$ due to the separable structure.

7.5.2 SeDiL

SeDiL [92] approaches DL similarly in spirit to SimCO, through a geometrical interpretation of the optimization problem solved via gradient descent methods, only this time learning is performed on smooth Riemannian submanifolds.

Let us start from the generic submanifold $\boldsymbol{M} = \mathbb{R}^{m\times n_1} \times \mathbb{R}^{m\times n_2} \times \mathbb{R}^{n_1\times n_2\times N}$, and let $f : \boldsymbol{M} \to \mathbb{R}$ be the DL differentiable cost function such that

$$\boldsymbol{S} = \min_{\mathcal{S}\in\boldsymbol{M}} f(\mathcal{S}) \tag{7.53}$$

where $\boldsymbol{S} = \{\boldsymbol{D}_1, \boldsymbol{D}_2, \boldsymbol{X}\}$ is the end result of the DL process (7.43).

We start defining the cost function f by matching the objective function of (7.43) that we normalize with the number of samples N

$$f(\mathcal{S}) = f(\boldsymbol{D}_1, \boldsymbol{D}_2, \boldsymbol{X}) = \frac{1}{2N}\sum_{k=1}^{N} \|\boldsymbol{Y}_k - \boldsymbol{D}_1\boldsymbol{X}_k\boldsymbol{D}_2^T\|_F^2 \tag{7.54}$$

and we note that, in this brute form, the final solution $\boldsymbol{S}$ would most probably lack all of the desired properties represented as constraints in (7.43). To that end, we start reshaping (7.53) towards our original goal by first making the observation that column normalized matrices have a product of spheres manifold structure

$$\mathcal{D}(m,n) = \{\boldsymbol{D} \in \mathbb{R}^{m\times n} | \mathrm{tr}(\boldsymbol{D}^T\boldsymbol{D}) = n\}. \tag{7.55}$$

Thus we can obtain atom normalization by restricting the cost function domain to $\boldsymbol{M} = \mathcal{D}(m, n_1) \times \mathcal{D}(m, n_2) \times \mathbb{R}^{n_1\times n_2\times N}$. Now we are still left with the sparsity constraint—or lack thereof. SeDiL obtains sparsity through regularization, thus modifying f. In order to maintain smoothness, the familiar ℓ_0-norm is replaced with the overall sparsity measure

$$g(\boldsymbol{X}) = \sum_{k=1}^{N}\sum_{i=1}^{n_1}\sum_{j=1}^{n_2} \ln(1 + \rho\,|x_{ijk}|^2) \tag{7.56}$$

where x_{ijk} is the (i, j)-entry of $\boldsymbol{X}_k$ and $\rho \in \mathbb{R}$ is a weighting parameter. The new objective function (denoted also f) becomes

$$f(\mathcal{S}) = f(\boldsymbol{D}_1, \boldsymbol{D}_2, \boldsymbol{X}) = \frac{1}{2N}\sum_{k=1}^{N} \|\boldsymbol{Y}_k - \boldsymbol{D}_1\boldsymbol{X}_k\boldsymbol{D}_2^T\|_F^2 + \frac{\lambda}{N} g(\boldsymbol{X}) \tag{7.57}$$

where we have added the sparsity regularization along with its associated tuning parameter λ. We now have enough to satisfy the pursuit in (7.43), but SeDiL takes things one step further and also promotes coherence. Again, the standard coherence function (1.25) is replaced by an alternative differentiable mutual coherence measure

$$r(\boldsymbol{D}) = -\sum_{1\le i\le j\le n} \ln(1 - (\boldsymbol{d}_i^T\boldsymbol{d}_j)^2) \tag{7.58}$$

such that the final objective function remains smooth

$$f(\mathcal{S}) = \frac{1}{2N}\sum_{k=1}^{N} \|\boldsymbol{Y}_k - \boldsymbol{D}_1\boldsymbol{X}_k\boldsymbol{D}_2^T\|_F^2 + \frac{\lambda}{N} g(\boldsymbol{X}) + \kappa r(\boldsymbol{D}_1) + \kappa r(\boldsymbol{D}_2), \tag{7.59}$$

where κ is the coherence tuning parameter.

To solve (7.59) and exploit the manifold structure, SeDiL authors make use of the geometric conjugate gradient method [2], the details of which do not make the object of this book. Nonetheless, we sketch out the main ingredients needed in order to gain intuition should the reader decide to perform an in-depth study of this approach. First, let $\boldsymbol{T}_{\mathcal{S}}$ represent the tangent space at point $\mathcal{S} \in \boldsymbol{M}$ and let $\mathcal{T} \in \boldsymbol{T}_{\mathcal{S}}$ be the tangent vector at $\mathcal{S}$, then the Riemannian gradient $\mathcal{G}(f(\mathcal{S}))$ belongs to $\boldsymbol{T}_{\mathcal{S}}$. During gradient descent, the new solution $\mathcal{S}^{k+1}$ is computed by following the smooth curve $\mathcal{C}$ starting from point $\mathcal{S}^k \in \boldsymbol{M}$ towards the search direction $\mathcal{H}^k \in \boldsymbol{T}_{\mathcal{S}^k}$ with step size α

$$\mathcal{S}^{k+1} = \mathcal{C}(\mathcal{S}^k, \mathcal{H}^k, \alpha). \tag{7.60}$$

The step size α can be computed via standard line search methods. Once a new solution $\mathcal{S}^{k+1}$ is available, we can compute the new gradient $\mathcal{G}(f(\mathcal{S}^{k+1}))$ and update the search direction $\mathcal{H}^{k+1}$ as a linear combination between the former search $\mathcal{H}^k$ and the current gradient.

7.6 Tensor Strategies

When dealing with separable dictionaries we can take a few steps closer to home by rearranging the 2D training signals $\boldsymbol{Y}_k$ from (7.43) as slices of a third-order tensors, as suggested in Fig. 7.3. This paves the way towards various factorization strategies that are more or less the generalization of the K-SVD family to the next dimension.

First let us define a few useful tensor operations. The Khatri–Rao product is the matching column Kronecker product of two matrices $\boldsymbol{A} \in \mathbb{R}^{p\times n}$ and $\boldsymbol{B} \in \mathbb{R}^{q\times n}$

$$\boldsymbol{A} \odot \boldsymbol{B} = \left[\boldsymbol{a}_1 \otimes \boldsymbol{b}_1 \; \boldsymbol{a}_2 \otimes \boldsymbol{b}_2 \cdots \boldsymbol{a}_n \otimes \boldsymbol{b}_n\right] \in \mathbb{R}^{pq\times n}. \tag{7.61}$$

The Hadamard product is the elementwise product of two identically shaped matrices $\boldsymbol{A}, \boldsymbol{B} \in \mathbb{R}^{n\times m}$

$$\boldsymbol{A} \circ \boldsymbol{B} = \begin{bmatrix} a_{11}b_{11} & a_{12}b_{12} & \dots & a_{1m}b_{1m} \\ a_{21}b_{21} & a_{22}b_{22} & \dots & a_{2m}b_{2m} \\ \vdots & & \ddots & \vdots \\ a_{n1}b_{n1} & a_{n2}b_{n2} & \dots & a_{nm}b_{nm} \end{bmatrix}. \tag{7.62}$$

The two products are tied by the following relationships:

$$(\boldsymbol{A} \odot \boldsymbol{B})^T(\boldsymbol{A} \odot \boldsymbol{B}) = (\boldsymbol{A}^T\boldsymbol{A}) \circ (\boldsymbol{B}^T\boldsymbol{B}) \tag{7.63}$$

and

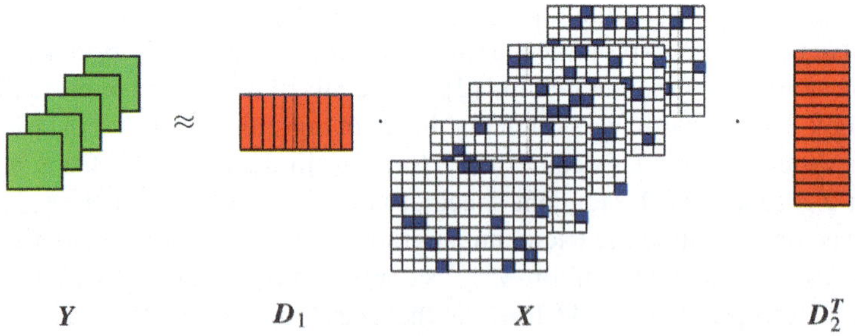

Fig. 7.3 Separable Dictionary Learning where patches and representations are N frontal slices of a third-order tensor. The nonzero coefficients of $\boldsymbol{X}$ are marked blue

$$(\boldsymbol{A} \odot \boldsymbol{B})^{\dagger} = [(\boldsymbol{A}^T \boldsymbol{A}) \circ (\boldsymbol{B}^T \boldsymbol{B})]^{\dagger} (\boldsymbol{A} \odot \boldsymbol{B})^T, \tag{7.64}$$

often used to reduce the total number of required multiplications.

Finally, let $\boldsymbol{T}$ be a tensor of size $I_1 \times I_2 \times \ldots \times I_n$. We denote $\boldsymbol{T}_{(k)}$ the mode-k matricization of a tensor $\boldsymbol{T}$ by taking the mode-k fibers $t_{i_1 i_2 \ldots i_{k-1} : i_{k+1} \ldots i_n}$ (the colon means that all values of that index are taken for building the fiber, while all other indices are fixed; fibers are for tensors what rows or columns are for a matrix) and arranging them as columns in $\boldsymbol{T}_{(k)}$. The mode-k tensor product with a matrix, between $\boldsymbol{T}$ and a matrix $\boldsymbol{M} \in \mathbb{R}^{J \times I_k}$, is a tensor of size $I_1 \times \ldots \times I_{k-1} \times J \times I_{k+1} \times \ldots \times I_n$ obtained by multiplying each mode-k fiber with $\boldsymbol{M}$. The mode-k product can be expressed via matrix multiplication through

$$(\boldsymbol{T} \otimes_k \boldsymbol{M})_{(k)} = \boldsymbol{M} \boldsymbol{T}_{(k)}. \tag{7.65}$$

7.6.1 CP Decomposition

One of the most popular tensor tools is the CANDECOMP/PARAFAC (CP) decomposition. In fact, it is so popular that it was first proposed in 1927 [93] and reinvented at least four more times since [111]. Regardless of nomenclature, given the order n tensor $\boldsymbol{T}$, CP decomposition is its factorization as a sum of rank-one tensor outer-products

$$\boldsymbol{T} = \sum_{r=1}^{R} \boldsymbol{\alpha}_r^{(1)} \boldsymbol{\alpha}_r^{(2)} \cdots \boldsymbol{\alpha}_r^{(n)}, \tag{7.66}$$

where R is the tensor rank. The element $t_{i_1 i_2 \ldots i_n}$ of $\boldsymbol{T}$ is given by

$$t_{i_1 i_2 \ldots i_n} = \sum_{r=1}^{R} \boldsymbol{\alpha}_{i_1 r}^{(1)} \boldsymbol{\alpha}_{i_2 r}^{(2)} \cdots \boldsymbol{\alpha}_{i_n r}^{(n)}. \tag{7.67}$$

If the vectors $\boldsymbol{\alpha}_r^{(i)}$ are normalized and grouped as columns in a matrix $\boldsymbol{A}^{(i)} \in \mathbb{R}^{I_i \times R}$, the (7.66) decomposition is denoted by

$$\boldsymbol{T} = [\![\boldsymbol{\lambda}; \boldsymbol{A}^{(1)}, \boldsymbol{A}^{(2)}, \ldots, \boldsymbol{A}^{(n)}]\!] = \sum_{r=1}^{R} \lambda_r \boldsymbol{\alpha}_r^{(1)} \boldsymbol{\alpha}_r^{(2)} \cdots \boldsymbol{\alpha}_r^{(n)}, \tag{7.68}$$

where $\boldsymbol{\lambda} \in \mathbb{R}^R$ is the vector whose entries are

$$\lambda_r = \prod_{r=1}^{R} \left\| \boldsymbol{\alpha}_r^{(i)} \right\|.$$

Then, it has been shown in [110] that

$$T_{(k)} = A^{(k)}\Lambda\left(A^{(n)} \odot \cdots \odot A^{(k+1)} \odot A^{(k-1)} \odot \cdots \odot A^{(1)}\right)^T, \tag{7.69}$$

where $\Lambda = \text{diag}(\lambda)$.

Relation (7.68) bears an obvious resemblance to the singular value decomposition for matrices. However, while the matrix SVD can always be computed exactly, the equality in (7.66) is hard to achieve in practice; Given a tensor T, the problem of finding $A^{(i)}$ and the rank R is NP-hard. Most applications settle for an approximation instead, but even so we are faced with challenges in choosing the number of components R. Assuming that R is given to us by some oracle, there are many procedures to choose from for computing the CP decomposition. In this book we will focus on the well-known alternating least squares (ALS) method [26, 91] that solves the problem exactly how the main DL algorithms do: through alternate optimization.

At step i, ALS keeps all $A^{(j)}$ with $i \neq j$ fixed and optimizes $A^{(i)}$. From (7.69) we can write this as the following optimization problem

$$\min_{A} \left\| T_{(i)} - A\left(A^{(n)} \odot \cdots \odot A^{(i+1)} \odot A^{(i-1)} \odot \cdots \odot A^{(1)}\right)^T \right\|_F, \tag{7.70}$$

where $A = A^{(i)}\Lambda$, whose least squares solution is

$$A = T_{(i)}\left[\left(A^{(n)} \odot \cdots \odot A^{(i+1)} \odot A^{(i-1)} \odot \cdots \odot A^{(1)}\right)^T\right]^{\dagger}. \tag{7.71}$$

Using (7.63) and (7.64) we can reduce the pseudo-inverse operations to

$$\left(A^{(n)} \odot \cdots \odot A^{(i+1)} \odot A^{(i-1)} \odot \cdots \odot A^{(1)}\right) P^{\dagger} \tag{7.72}$$

where

$$P = (A^{(1)T}A^{(1)}) \circ \cdots \circ (A^{(i-1)T}A^{(i-1)}) \circ (A^{(i+1)T}A^{(i+1)}) \circ \cdots \circ (A^{(n)T}A^{(n)}) \tag{7.73}$$

is an $R \times R$ matrix that is much easier to invert than the original matrix with $(I_n \cdots I_{i+1}I_{i-1} \cdots I_1) \times R$ elements.

The ALS instructions are put together in Algorithm 7.7. There are various techniques for initializing $A^{(i)}$ in step 1, a good starting point being the R left singular vectors of the corresponding $T_{(i)}$ matrix

$$A^{(i)} = U_{\{1,\dots,R\}}, \text{ where } T_{(i)} = U\Sigma V^T \text{ is the SVD}, \tag{7.74}$$

much like what 1ONB does in Algorithm 7.3. Keeping all other components fixed, step 4 updates $A^{(i)}$ as earlier described. Given that the vectors $A^{(j)}$, $i \neq j$, are

Algorithm 7.7: CP Decomposition via Alternating Least Squares

Data: tensor $\boldsymbol{T} \in \mathbb{R}^{I_1 \times I_2 \times \ldots \times I_n}$
approximation rank R
number of iterations K
Result: CP decomposition $\hat{\boldsymbol{T}} = [\![\boldsymbol{\lambda}; \boldsymbol{A}^{(1)}, \boldsymbol{A}^{(2)}, \ldots, \boldsymbol{A}^{(n)}]\!]$

```
Initialize A^(i) ∈ R^{I_i × r}, ∀i ∈ 1 : n
for k = 1 : K do
    for i = 1 : n do
        Update A^(i) as described around (7.71)
        Recompute λ_r = ‖α_r^(i)‖, ∀r ∈ 1 : R
        Normalize columns α_r^(i), ∀r ∈ 1 : R
```

already normalized, the norm vector $\boldsymbol{\lambda}$ is simply replaced by the new column norms in step 5, after which $\boldsymbol{A}^{(i)}$ is normalized in step 6. These operations are repeated for all n components by the loop at step 3 and the entire process is iterated until convergence is reached (or for a maximum number of iterations).

The bulk of ALS instructions is centered around computing $\boldsymbol{A}^{(i)}$, the normalization process being trivial in comparison. Efficiently solving (7.71) first requires us to compute $\boldsymbol{P}$ from (7.73). The Gram matrices take $O(R^2(I_1 + \ldots I_{i-1} + I_{i+1} + \cdots + I_n)$ operations and the Hadamard products an extra $O((n-1)R^2)$ instructions. Next, the Khatri–Rao products from (7.72) and their multiplication with $\boldsymbol{T}_{(i)}$ take $O(R^2 I_1 \cdots I_n)$ operations. Finally the least squares operation costs $O(R^3 I_i)$. If we assume that $I = I_1 = I_2 = \ldots = I_n$, the complexity of updating all n matrices is $O(n^2 R^2 I + nR^3 I)$.

7.6.2 CP Dictionary Update

Going back to Fig. 7.3, we can rewrite (7.43) in tensor form as

$$\min_{\boldsymbol{D}_1, \boldsymbol{D}_2, \boldsymbol{X}} \quad \|\boldsymbol{Y} - \boldsymbol{X} \otimes_1 \boldsymbol{D}_1 \otimes_2 \boldsymbol{D}_2\| \tag{7.75}$$

where $\boldsymbol{Y}$ is an $m \times m \times N$ tensor, $\boldsymbol{X}$ is an $n_1 \times n_2 \times N$ tensor and the tensor norm is the generalization of the Frobenius norm. Let us detail the expression of the objective. Denoting $\boldsymbol{Z} = \boldsymbol{X} \otimes_1 \boldsymbol{D}_1 \in \mathbb{R}^{m \times n_2 \times N}$, the mode-$k$ multiplication definition (7.65) says that the frontal slices of $\boldsymbol{Z}$ are obtained by multiplying $\boldsymbol{D}_1$ with the frontal slices of $\boldsymbol{X}$, i.e.,

$$\boldsymbol{Z}_{(1)} \stackrel{\Delta}{=} [\boldsymbol{Z}_1\ \boldsymbol{Z}_2\ \ldots\ \boldsymbol{Z}_N] = \boldsymbol{D}_1 \boldsymbol{X}_{(1)} = \boldsymbol{D}_1 [\boldsymbol{X}_1\ \boldsymbol{X}_2\ \ldots\ \boldsymbol{X}_N].$$

Recomposing the tensor gives

$$\boldsymbol{Z} = \left[\, \boldsymbol{D}_1\boldsymbol{X}_1 \middle| \boldsymbol{D}_1\boldsymbol{X}_2 \middle| \ldots \middle| \boldsymbol{D}_1\boldsymbol{X}_N \,\right]. \tag{7.76}$$

Denoting now $\hat{\boldsymbol{Y}} = \boldsymbol{Z} \otimes_2 \boldsymbol{D}_2 \in \mathbb{R}^{m\times m\times N}$, the operator $\otimes_2$ works with the rows of the frontal slices of $\boldsymbol{Z}$ and concatenates them as columns in $\boldsymbol{Z}_{(2)}$, giving

$$\hat{\boldsymbol{Y}}_{(2)} = \boldsymbol{D}_2\boldsymbol{Z}_{(2)} = \boldsymbol{D}_2[\boldsymbol{Z}_1^T\ \boldsymbol{Z}_2^T\ \ldots\ \boldsymbol{Z}_N^T] = \boldsymbol{D}_2[\boldsymbol{X}_1^T\ \boldsymbol{X}_2^T\ \ldots\ \boldsymbol{X}_N^T]\boldsymbol{D}_1^T.$$

It results that

$$\hat{\boldsymbol{Y}} = \left[\, \boldsymbol{D}_1\boldsymbol{X}_1\boldsymbol{D}_2^T \middle| \boldsymbol{D}_1\boldsymbol{X}_2\boldsymbol{D}_2^T \middle| \ldots \middle| \boldsymbol{D}_1\boldsymbol{X}_N\boldsymbol{D}_2^T \,\right], \tag{7.77}$$

which makes obvious the correspondence with Fig. 7.3.

In [95] the authors apply CP decomposition to solve (7.75) in AK-SVD fashion. Let $\boldsymbol{d}_1$ and $\boldsymbol{d}_2$ denote the fixed (i, j) atom pair from the two dictionaries and $\boldsymbol{x}_{a,b}$ the mode-3 fiber (or tube) crossing the N frontal slices of $\boldsymbol{X}$ at position (a, b)

$$\boldsymbol{x}_{a,b} = [x_{ab1}\ x_{ab2}\ \ldots\ x_{abN}]^T. \tag{7.78}$$

We shorten $\boldsymbol{x}_{i,j}$ as $\boldsymbol{x}$. Then the approximation error is written as

$$\begin{aligned} \boldsymbol{E} = \boldsymbol{Y} - \sum_{a\neq i}^{n_1}\sum_{b\neq j}^{n_2} \boldsymbol{x}_{a,b} \otimes \boldsymbol{d}_{2b} \otimes \boldsymbol{d}_{1a} - \boldsymbol{x} \otimes \boldsymbol{d}_2 \otimes \boldsymbol{d}_1 - \\ - \sum_{a\neq i}^{n_1} \boldsymbol{x}_{a,j} \otimes \boldsymbol{d}_2 \otimes \boldsymbol{d}_{1a} - \sum_{b\neq j}^{n_2} \boldsymbol{x}_{i,b} \otimes \boldsymbol{d}_{2b} \otimes \boldsymbol{d}_1 \end{aligned} \tag{7.79}$$

and the fixed matrix is similar to (3.19)

$$\boldsymbol{F} = \boldsymbol{Y} - \sum_{a\neq i}^{n_1}\sum_{b\neq j}^{n_2} \boldsymbol{x}_{a,b} \otimes \boldsymbol{d}_{2b} \otimes \boldsymbol{d}_{1a}, \tag{7.80}$$

but the variable part is more involved due to the $\boldsymbol{x}_{a,j}$ and $\boldsymbol{x}_{i,b}$ tubes. The solution proposed in [95] uses 2D-OMP to perform sparse representation and then performs a rank-1 CP decomposition using regularized ALS (RALS) [118] to compute the variables $\{\boldsymbol{d}_1, \boldsymbol{d}_2, \boldsymbol{x}, \boldsymbol{x}_{i,b}, \boldsymbol{x}_{a,j}\}$. RALS solves the same problem as plain ALS but adds an extra penalty term $\left\| A_k^{(i)} - A_{k+1}^{(i)} \right\|_F^2$ in (7.70), where k is the iteration number from step 2 of Algorithm 7.7. The dictionary update phase is repeated until no significant changes are detected in $\boldsymbol{d}_1$, $\boldsymbol{d}_2$, and $\boldsymbol{x}$.

In order to reduce the large number of variables in (7.79), the last two terms are canceled in [95] by using 2D rank-1 non-separable dictionary atoms, which are approximated by the outer product of two atoms $\boldsymbol{d}_{1i}$ and $\boldsymbol{d}_{2i}$ that do not individually participate in the construction of any other 2D atom. The $\boldsymbol{x}_{a,j}$ and $\boldsymbol{x}_{i,b}$ terms cease to exist and the error becomes

$$\boldsymbol{E} = \boldsymbol{F} - \boldsymbol{x} \otimes \boldsymbol{d}_2 \otimes \boldsymbol{d}_1. \tag{7.81}$$

Not surprisingly, the expression can be minimized through rank-1 tensor approximation of $\boldsymbol{F}$ via CP decomposition. If we decide to keep the pair of 1D atoms in separate dictionaries, $\boldsymbol{D}_1$ and $\boldsymbol{D}_2$ will have to be much larger than in the separable case as each will have to hold a total of $n_1 n_2$ atoms. The final dictionary is thus a tensor consisting of $n_1 n_2$ slices of size $m \times m$ which, unlike (7.45), is generated by the Khatri–Rao product

$$\boldsymbol{D} = \boldsymbol{D}_2 \odot \boldsymbol{D}_1. \tag{7.82}$$

and the s-sparse representation of $\boldsymbol{Y}_i$ is expressed as the linear combination of s slices from $\boldsymbol{D}$

$$\boldsymbol{Y}_i = \sum_{k \in \mathcal{S}} x_{ik} \boldsymbol{A}_k \approx \sum_{k \in \mathcal{S}} x_{ik} \otimes \boldsymbol{d}_{2k} \otimes \boldsymbol{d}_{1k}, \tag{7.83}$$

where $|\mathcal{S}| = s$ is the support computed via 2D-OMP and $\boldsymbol{A}_k$ are the approximated 2D atoms.

The non-separable dictionary update stage is presented in Algorithm 7.8. First the initial error $\boldsymbol{E}$ (7.75) is computed in step 1. For each atoms pair (step 2) the error is modified (step 3) to exclude its contribution in the final approximation. Following is the main operation of performing CP decomposition (step 4) to simultaneously update the atoms and their associated coefficients (step 5). This is very similar to K-SVD, that uses the singular value decomposition instead. The iteration ends by updating the error with fresh data in step 6.

The SuKro algorithm [46] generalizes the current design by writing the dictionary as a sum of Kronecker products

$$\boldsymbol{D} = \sum_{k=1}^{\alpha} \boldsymbol{D}_2^{(k)} \otimes \boldsymbol{D}_1^{(k)}. \tag{7.84}$$

In order to let the learning process choose the number of α products, SuKro rearranges the elements of $\boldsymbol{D}$ into a rank-α matrix $\tilde{\boldsymbol{D}}$ via the $\mathcal{R}$ operator [194]

$$\tilde{\boldsymbol{D}} = \mathcal{R}(\boldsymbol{D}) = \sum_{k=1}^{\alpha} \mathcal{R}(\boldsymbol{D}_2^{(k)} \otimes \boldsymbol{D}_1^{(k)}) = \sum_{k=1}^{\alpha} \text{vec}(\boldsymbol{D}_2^{(k)}) \text{vec}(\boldsymbol{D}_1^{(k)})^T \tag{7.85}$$

and extends the standard DL optimization problem (2.2) to include a convex rank penalty during the dictionary update stage

Algorithm 7.8: 2D Non-separable Approximation Dictionary Update

Data: tensor $\boldsymbol{Y} \in \mathbb{R}^{m \times m \times N}$,
sparsity level s,
number of iterations K

Result: dictionaries $\boldsymbol{D}_1 \in \mathbb{R}^{m \times n_1 n_2}$, $\boldsymbol{D}_2 \in \mathbb{R}^{m \times n_1 n_2}$

1 Compute error $\boldsymbol{E} = \boldsymbol{Y} - \boldsymbol{X} \otimes_1 \boldsymbol{D}_1 \otimes_2 \boldsymbol{D}_2$
2 **for** $j = 1 : n_1 n_2$ **do**
3 Modify error: $\boldsymbol{F} = \boldsymbol{E} + \boldsymbol{x}_j \otimes \boldsymbol{d}_{2j} \otimes \boldsymbol{d}_{1j}$
4 Compute CP rank-1 decomposition of $\boldsymbol{F}$
5 Update $\boldsymbol{d}_{1j}, \boldsymbol{d}_{2j}$ and $\boldsymbol{x}_j$
6 Recompute error: $\boldsymbol{E} = \boldsymbol{F} - \boldsymbol{x}_j \otimes \boldsymbol{d}_{2j} \otimes \boldsymbol{d}_{1j}$

$$\begin{aligned} \min_{\boldsymbol{D},\tilde{\boldsymbol{D}}} \quad & \|\boldsymbol{Y} - \boldsymbol{D}\boldsymbol{X}\|_F^2 + \lambda \left\|\tilde{\boldsymbol{D}}\right\|_* \\ \text{s.t.} \quad & \tilde{\boldsymbol{D}} = \mathcal{R}(\boldsymbol{D}) \end{aligned} \tag{7.86}$$

where $\|.\|_*$ is the nuclear norm

$$\|\boldsymbol{A}\|_* = \text{tr}(\sqrt{\boldsymbol{A}^T\boldsymbol{A}}) = \sum_{i=1}^{\min(m,n)} \sigma_i \tag{7.87}$$

with $\boldsymbol{A} \in \mathbb{R}^{m \times n}$. While sparse coding is performed by standard OMP, dictionary update uses the convex alternating method of multipliers (ADMM) [15] to solve (7.86) and leaves atom normalization as an extra final step. It is clear that for $\alpha = 1$ we have to deal with separable dictionaries discussed thus far, but note that SuKro differs from SeDiL and CP-based methods due to the different dictionary design. As the parameter λ is increased, so is the rank of $\tilde{\boldsymbol{D}}$ and implicitly the number of terms in (7.84). Fortunately, even for $\alpha > 1$, we are still able to exploit the separable structure when performing matrix multiplication.

Among other convex DL variations we mention KDRSDL [8] that moves the sparsity constraint as a penalty in the objective function and uses ADMM to learn both the representations and dictionary.

The joint multiple dictionary learning (JMDL) algorithm [74] is a K-SVD-like generalization of the separable DL problem to M-dimensional tensors. Each frontal slice $\boldsymbol{Y}_k$ is now an $m_1 \times m_2 \times \cdots \times m_M$ tensor and the associated representation $\boldsymbol{X}_k$ is an $n_1 \times n_2 \times \cdots \times n_M$ sparse tensor such that

$$\|\boldsymbol{Y}_k - \boldsymbol{X}_k \otimes_1 \boldsymbol{D}_1 \otimes_2 \boldsymbol{D}_2 \cdots \otimes_M \boldsymbol{D}_M\| \tag{7.88}$$

is minimized under sparsity level constraints. This is solved by generalizing 2D-OMP, termed N-BOMP in [74], to find an M sized index in step 4 of Algorithm 7.6 and extend the dictionary construction in step 5 to perform an M-term Kronecker product of the corresponding atoms. To avoid computing and updating all possible pairs that form among all M dictionaries, JMDL splits the atoms into frequency pattern lists. Each list $\mathcal{P}$ is processed sequentially until all atoms are updated. As a natural extension of (7.81), the atoms within $\mathcal{P}$ are simultaneously updated using the CP decomposition of the fixed error

$$\boldsymbol{F}_{\mathcal{P}} = \boldsymbol{Y}_{\mathcal{I}} - \boldsymbol{X}_{I_1, I_2, \ldots, I_M, \mathcal{I}} \otimes_1 \boldsymbol{D}_1 \otimes_2 \boldsymbol{D}_2 \cdots \otimes_M \boldsymbol{D}_M \otimes_{M+1} \boldsymbol{I}, \tag{7.89}$$

where $\mathcal{I}$ is the set of frontal slices in $\boldsymbol{Y}$ and $\boldsymbol{X}$ using the atoms from $\mathcal{P}$ in their representation and $I_{1:M} \notin \mathcal{P}$. Given a threshold t, the set $\mathcal{P}$ is built using the FP-tree [90] algorithm such that $|\mathcal{P}| = t$. This leads to suboptimal results as the pairings are not exact and, at iteration k, changing atom $\boldsymbol{d} \in \mathcal{P}_k$ does not necessarily decrease the error as it might affect representations where $\boldsymbol{d}$ pairs with atoms not in $\mathcal{P}_k$.

7.6.3 *Tensor Singular Valued Decomposition*

Let us shift our focus towards an alternative SVD-based tensor decomposition termed t-SVD. At its core stands the t product between tensors $\boldsymbol{A} \in \mathbb{R}^{n_1 \times d \times n_3}$ and $\boldsymbol{B} \in \mathbb{R}^{d \times n_2 \times n_3}$

$$\boldsymbol{C} = \boldsymbol{A} * \boldsymbol{B}, \tag{7.90}$$

producing $\boldsymbol{C} \in \mathbb{R}^{n_1 \times n_2 \times n_3}$ that can be implemented through the mode-2 matricization

$$\boldsymbol{C}_{(2)} = (\boldsymbol{A} * \boldsymbol{B})_{(2)} = \boldsymbol{B}_{(2)} \left[\boldsymbol{A}_{(2)} \boldsymbol{P} \; \boldsymbol{A}_{(2)} \boldsymbol{P}^2 \; \boldsymbol{A}_{(2)} \boldsymbol{P}^3 \cdots \boldsymbol{A}_{(2)} \boldsymbol{P}^{n_3} \right], \tag{7.91}$$

where $\boldsymbol{P} \in \mathbb{R}^{n_1 n_3 \times n_1 n_3}$ is the d-block circular permutation matrix

$$\boldsymbol{P} = \begin{bmatrix} \boldsymbol{O} & \boldsymbol{I}_{n_1 n_3 - d} \\ \boldsymbol{I}_d & \boldsymbol{O} \end{bmatrix} \tag{7.92}$$

whose operation on blocks is identical to the one (7.4) applies on matrix elements. A simple example will clear things up. If $\boldsymbol{A}$ and $\boldsymbol{B}$ are appropriately dimensioned tensors, each having three frontal slices $\boldsymbol{A}_{\{1,2,3\}}$ and $\boldsymbol{B}_{\{1,2,3\}}$, then

$$\boldsymbol{C}_{(2)}^T = (\boldsymbol{A} * \boldsymbol{B})_{(2)}^T = \begin{bmatrix} \boldsymbol{A}_1 & \boldsymbol{A}_3 & \boldsymbol{A}_2 \\ \boldsymbol{A}_2 & \boldsymbol{A}_1 & \boldsymbol{A}_3 \\ \boldsymbol{A}_3 & \boldsymbol{A}_2 & \boldsymbol{A}_1 \end{bmatrix} \begin{bmatrix} \boldsymbol{B}_1 \\ \boldsymbol{B}_2 \\ \boldsymbol{B}_3 \end{bmatrix}, \tag{7.93}$$

Algorithm 7.9: t-SVD

Data: tensor $\boldsymbol{T} \in \mathbb{R}^{n_1 \times n_2 \times n_3}$
Result: $\boldsymbol{U} \in \mathbb{R}^{n_1 \times n_1 \times n_3}$, $\boldsymbol{V} \in \mathbb{R}^{n_2 \times n_2 \times n_3}$, $\boldsymbol{S} \in \mathbb{R}^{n_1 \times n_2 \times n_3}$

```
1 T̂ = FFT(T)
2 for i = 1 : n3 do
3   [Û_i, Ŝ_i, V̂_i] = SVD(T̂_i)
4 U = IFFT(Û), S = IFFT(Ŝ), V = IFFT(V̂)
```

where we transpose so that the mode-2 operation stores the tubes as rows instead of columns. Given that this is basically a convolution, t-product is efficiently calculated similarly to (7.25)

$$\boldsymbol{C}_k = \text{IFFT}(\hat{\boldsymbol{A}}_k \hat{\boldsymbol{B}}_k), \tag{7.94}$$

where the hat denotes the tensor FFT along the third dimension.

Before moving forwards, we need to redefine a couple of matrix notions in tensor speak. Transposing of $\boldsymbol{T} \in \mathbb{R}^{n_1 \times n_2 \times n_3}$ is done by first transposing each frontal slice and then reversing the order of the slices from 2 to n_3. The tensor $\boldsymbol{I} \in \mathbb{R}^{n \times n \times n_3}$ that has its first frontal slice $\boldsymbol{I}_1$ equal to the identity matrix $\boldsymbol{I}_n$ and the rest equal to $\boldsymbol{O}_n$ is called the identity tensor. A tensor $\boldsymbol{U} \in \mathbb{R}^{n \times n \times n_3}$ is orthogonal if

$$\boldsymbol{U} * \boldsymbol{U}^T = \boldsymbol{U}^T * \boldsymbol{U} = \boldsymbol{I}, \tag{7.95}$$

where $*$ is the t-product. A tensor $\boldsymbol{D} \in \mathbb{R}^{n \times n \times n_3}$ is called f-diagonal if each frontal slice is diagonal.

With that, the tensor Single Value Decomposition (t-SVD) of $\boldsymbol{T} \in \mathbb{R}^{n_1 \times n_2 \times n_3}$ is

$$\boldsymbol{T} = \boldsymbol{U} * \boldsymbol{S} * \boldsymbol{V}^T, \tag{7.96}$$

where $\boldsymbol{U} \in \mathbb{R}^{n_1 \times n_1 \times n_3}$ and $\boldsymbol{V} \in \mathbb{R}^{n_2 \times n_2 \times n_3}$ are orthogonal tensors and $\boldsymbol{S} \in \mathbb{R}^{n_1 \times n_2 \times n_3}$ is f-diagonal. The efficient implementation of the t-SVD decomposition is described in Algorithm 7.9. Following (7.94), we take the Fourier transform on the third dimension of the input tensor in step 1 and then proceed to step 2 where we compute the singular value decomposition of each frontal slice from the result $\hat{\boldsymbol{T}}$ (step 3). At the end we compute the inverse transform to obtain the ingredients for (7.96). Step 4 is often skipped or delayed as much as possible if t-SVD is part of a larger algorithm where further t-product operations await.

The optimal approximation of t-SVD theorem [108] states that

$$\boldsymbol{T}^{(k)} = \sum_{i=1}^{k} \boldsymbol{U}_{:,i,:} * \boldsymbol{S}_{i,i,:} * \boldsymbol{V}^T_{:,i,:}, \tag{7.97}$$

with $k < \min(n_1, n_2)$, is the optimal solution of

$$\boldsymbol{T}^{(k)} = \arg\min_{A \in \mathcal{A}} \|\boldsymbol{T} - \boldsymbol{A}\|_F \tag{7.98}$$

where $\mathcal{A}$ is the set of all k-block t-products between two tensors.

7.6.4 t-SVD Dictionary Update

K-TSVD [219] uses t-product operations to build the dictionary and its sparse representation. In the following we will use $\bar{\boldsymbol{T}}_k$ to denote lateral slice k of tensor $\boldsymbol{T}$. This notation will come in handy because the input data is now organized as lateral slices $\bar{\boldsymbol{Y}}_k$ that are represented by the "t-linear" combination of 2D dictionary atoms $\bar{\boldsymbol{D}}_i$ and their associated tube coefficients $\bar{\boldsymbol{x}}_i$

$$\bar{\boldsymbol{Y}}_k = \bar{\boldsymbol{D}}_{\mathcal{I}} * \bar{\boldsymbol{X}}_{k_{\mathcal{I}}} = \sum_{i=1}^{s} \bar{\boldsymbol{D}}_i * \bar{\boldsymbol{x}}_{k_i}. \tag{7.99}$$

In Fig. 7.4 we can see that this process is similar to standard sparse representation (1.2) with data extruded in the third dimension. Indeed, we can adapt the optimization problem (1.6), transforming it into

$$\begin{aligned} \min_{\bar{\boldsymbol{X}}_k} \quad & \left\|\bar{\boldsymbol{Y}}_k - \bar{\boldsymbol{D}}\bar{\boldsymbol{X}}_k\right\|_F^2 \\ \text{s.t.} \quad & \|\bar{\boldsymbol{X}}_k\|_{\text{TS}} \end{aligned} \tag{7.100}$$

where $\|.\|_{\text{TS}}$ counts the number of nonzero tubes in $\bar{\boldsymbol{X}}_k$ (see also Fig. 7.5).

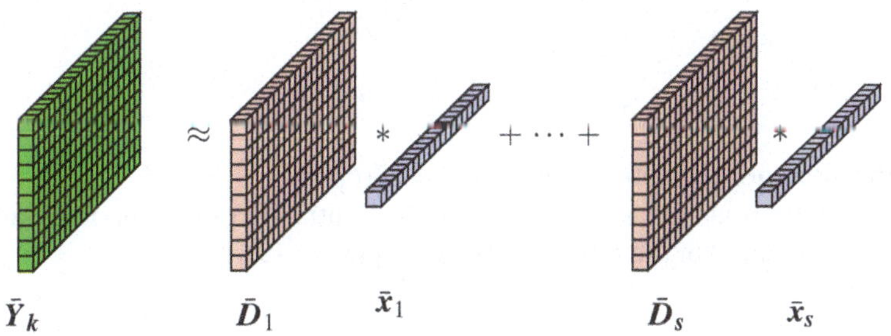

Fig. 7.4 The lateral slide $\bar{\boldsymbol{Y}}_k$ is approximated through the t-linear combination of s 2D atoms whose coefficients are the $\bar{\boldsymbol{x}}_i$ nonzero tubes of the sparse representation $\bar{\boldsymbol{X}}_k$

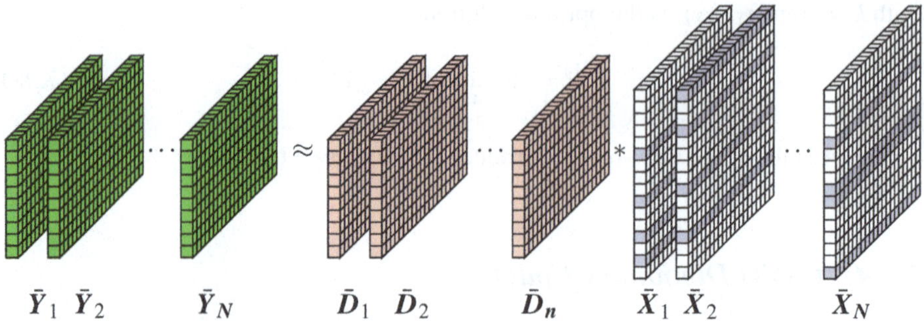

Fig. 7.5 Dictionary learning using the t-product. The nonzero tubes of $\boldsymbol{X}$ are marked blue

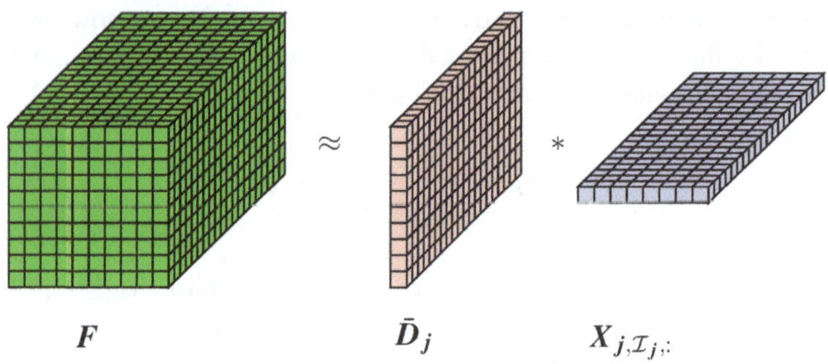

Fig. 7.6 Atom update: $\boldsymbol{F}_{\mathcal{I}_j}$ and $\boldsymbol{X}_{\mathcal{I}_j}$ are the signals and representation tubes using atom j

When learning a dictionary $\boldsymbol{D}$, consisting of n lateral atoms, we require a large training set $\boldsymbol{Y}$ built with N lateral 2D signals. As depicted in Fig. 7.5, the resulting sparse representations $\boldsymbol{X}$ have at most s nonzero tubes each and the standard DL problem (2.2) is rewritten as

$$\begin{aligned} \min_{\boldsymbol{D},\boldsymbol{X}} \quad & \|\boldsymbol{Y} - \boldsymbol{D} * \boldsymbol{X}\|_F^2 \\ \text{s.t.} \quad & \|\boldsymbol{X}_\ell\|_{\text{TS}} \leq s, \ \ell = 1 : N \end{aligned} \tag{7.101}$$

where the atom normalization constraint was dropped.

Figure 7.6 presents the adaptation of the sequential dictionary update stage to the t-product approach. On the left side, the fixed part (3.19) is

$$\boldsymbol{F} = \left[\boldsymbol{Y} - \sum_{i \neq j} \bar{\boldsymbol{D}}_i * \bar{\boldsymbol{X}}_i \right]_{\mathcal{I}_j}, \tag{7.102}$$

Algorithm 7.10: K-TSVD Dictionary Update

Data: signals set $\boldsymbol{Y} \in \mathbb{R}^{m \times N \times m}$
current dictionary $\boldsymbol{D} \in \mathbb{R}^{m \times n \times m}$
representation matrix $\boldsymbol{X} \in \mathbb{R}^{n \times N \times m}$
Result: updated dictionary $\boldsymbol{D}$

1 Compute error $\boldsymbol{E} = \boldsymbol{Y} - \boldsymbol{D} * \boldsymbol{X}$
2 **for** $j = 1 : n$ **do**
3 Modify error: $\boldsymbol{F} = \boldsymbol{E}_{\mathcal{I}_j} + \bar{\boldsymbol{D}}_j * \boldsymbol{X}_{j,\mathcal{I}_j,:}$
4 Compute first order t-SVD approximation $\boldsymbol{F}^{(1)} = \boldsymbol{U}_{:,1,:} * \boldsymbol{S}_{1,1,:} * \boldsymbol{V}^T_{:,1,:}$
5 Update atom and representation: $\bar{\boldsymbol{D}}_j = \boldsymbol{U}_{:,1,:}$, $\boldsymbol{X}_{j,\mathcal{I}_j,} = \boldsymbol{S}_{1,1,:} * \boldsymbol{V}^T_{:,1,:}$
6 Recompute error: $\boldsymbol{E}_{\mathcal{I}_j} = \boldsymbol{F} - \bar{\boldsymbol{D}}_j * \boldsymbol{X}_{j,\mathcal{I}_j,:}$

and the atom-update problem (3.20) becomes

$$\min_{\bar{\boldsymbol{D}}_j} \quad \left\| \boldsymbol{F} - \bar{\boldsymbol{D}}_j * \boldsymbol{X}_{j,\mathcal{I}_j,:} \right\|_F^2, \tag{7.103}$$

where $\boldsymbol{X}_{j,\mathcal{I}_j,:}$ is the horizontal slice, corresponding to atom j, restricted to the set of nonzero tubes $\mathcal{I}_j$.

In the sparse coding step, K-TSVD solves (7.100) by moving the sparsity constraint in the objective function

$$\min_{\bar{\boldsymbol{X}}_k} \quad \|\boldsymbol{Y} - \boldsymbol{D} * \boldsymbol{X}\|_F^2 + \lambda \, \|\boldsymbol{X}\|_{1,1,2} \tag{7.104}$$

and switching to a convex norm

$$\|\boldsymbol{T}\|_{1,1,2} = \sum_{i=1}^{n_1} \sum_{j=1}^{n_2} \|\boldsymbol{T}_{i,j,:}\|_F \tag{7.105}$$

such that the problem can be solved via ADMM. By aggregating the l_2 norms of all nonzero tubes, the norm reformulation produces a globally sparse $\boldsymbol{X}$ instead of the familiar set of N individually s-sparse representations. This is an extension of the approaches based on the ℓ_1 penalized objective (5.4).

The dictionary update stage solves (7.103) by computing the first order t-SVD approximation of $\boldsymbol{F}$ for each atom j. Algorithm 7.10 summarizes the necessary operations. Step 1 starts by computing the initial approximation error to be used by each atom (step 3) to compute the fixed part. Next, the first order t-SVD decomposition of $\boldsymbol{F}$ is computed (step 4) and its factors are used to update the atom and the associated representations in step 5. This process is repeated for all atoms (step 2). At first sight, the recipe follows K-SVD almost to the letter, but the core t-product operation is very different from standard matrix multiplication.

7.7 Composite Dictionaries

Composite dictionaries [201] are partly designed from training signals, partly from the test signals. The dictionary has the form

$$D = [D_e \; D_i], \tag{7.106}$$

where the external dictionary $D_e \in \mathbb{R}^{m \times n_e}$ is designed beforehand, using a training data set, while the internal dictionary $D_i \in \mathbb{R}^{m \times n_i}$ is designed during the application, using the actual data that are processed. A typical case is that of image denoising, where D_e is designed using clean available images (external examples), while D_i is designed using the noisy images that need denoising (internal examples). Any DL algorithm can be used, but we would normally use a simpler algorithm for the internal dictionary in order to reduce processing time. In any case, the composite dictionary approach is prone to take significantly more time than the usual pre-learned dictionary approach.

Of course, it is natural to take $n_i < n_e$, not only for complexity reasons (D_i must be re-learned repeatedly), but also to give more weight to the clean examples.

It was found that composite dictionaries can give better denoising results, a reason being their ability to take advantage of inner image similarities, that may be better than similarities between different images.

7.7.1 Convex Approach

The original composite framework [201] tackles the sparse representation problem via the convex problem

$$\min_{x,\mu^e,\mu^i} \left\| y - \sum_{i=1}^{n_e} d_i^e \mu_i^e x_i^e - \sum_{j=1}^{n_i} d_j^i \mu_j^i x_j^i \right\|_F^2 + \lambda^e \left\| x^e \right\|_1 + \lambda^i \left\| x^i \right\|_1 \tag{7.107}$$

which is a generalization of the standard basis pursuit problem (1.9), with sparsity obtained through the familiar l_1 norm penalty. The vectors x^e and x^i contain the representation coefficients corresponding to the external and, respectively, internal dictionary atoms. The innovation is the introduction of the weights μ_i^e and μ_j^i, with the role of encouraging the presence of some atoms in the representation, in particular those that are closer to the signal y. If a weight μ_i^e has a large value, then the coefficient x_i^e can be small and thus has a small contribution to $\|\boldsymbol{x}^e\|_1$; the atom $\boldsymbol{d}_i^e$ is encouraged to participate to the representation. On the contrary, a small weight would force a large coefficient, thus a large l_1 norm of the representation vector, which would substantially increase the value of the objective; the atom is rather left out of the representation. Note that (7.107) is the basis pursuit problem (1.9) when $\mu_i^e = \mu_j^i = 1$, for all i, j.

The prerequisites for solving (7.107) are the two dictionaries $\boldsymbol{D}_e$ and $\boldsymbol{D}_i$, learned via K-SVD in [201]. Next comes the choice of the weights vectors $\boldsymbol{\mu}^e$ and $\boldsymbol{\mu}^i$ detailed below. With that, everything but the sparse representations is fixed and we use coordinate descent (or another simple method for convex optimization) to obtain $\boldsymbol{x}^e$ and $\boldsymbol{x}^i$.

The weights are chosen using similarity measures between the signal and the atoms. More precisely, we take

$$\mu_i^e = \kappa(\boldsymbol{d}_i^e, \boldsymbol{y}),\ \mu_j^i = \kappa(\boldsymbol{d}_j^e, \boldsymbol{y}),$$

where $\kappa(\cdot)$ is a function depending on one or more parameters that can give specific roles to the external and internal representations. A simple choice is to assume a normal distribution of the signals around the atoms and thus use the static RBF kernel

$$\kappa(\boldsymbol{d}, \boldsymbol{y}, \omega) = \exp\left(-\omega \|\boldsymbol{y} - \boldsymbol{d}\|^2\right) \tag{7.108}$$

as a similarity measure. We encourage choosing only highly correlated atoms from the internal dictionary via large ω^i values. For the general external dictionary, which is richer in atoms that may be similar to a signal, we become more tolerant and take a lower ω^e.

In practice, the signal distribution often takes the form of an ellipsoid whose shape can be learned via the Mahalanobis distance

$$\kappa(\boldsymbol{d}, \boldsymbol{y}, \boldsymbol{\Omega}) = \exp\left(-(\boldsymbol{y} - \boldsymbol{d})^T \boldsymbol{\Omega}(\boldsymbol{y} - \boldsymbol{d})\right) \tag{7.109}$$

where the $\boldsymbol{\Omega}$ similarity matrix is ideally the inverse of the correlation matrix between $\boldsymbol{d}$ and $\boldsymbol{y}$. The similarity matrices can be learned via (7.107); for example, $\boldsymbol{\Omega}^e$ is optimized via coordinate descent after fixing $\boldsymbol{x}^e$, $\boldsymbol{x}^i$, and $\boldsymbol{\Omega}^i$. The same process is applied for $\boldsymbol{\Omega}^i$.

We gathered in Algorithm 7.11 the necessary operations needed to perform signal denoising. Steps 1 and 2 train the external and internal dictionaries using K-SVD. We picked this learning method to match the choice in [201], but any other DL alternative would do. Step 3 iterates all noisy patches in $\boldsymbol{Y}_i$ in order to perform sparse representation and denoising. A few choices for the parameters of step 4 were described earlier. Once $\boldsymbol{\mu}^i$ and $\boldsymbol{\mu}^e$ are fixed, we can perform sparse representation by solving the convex problem (7.107) via Coordinate Descent. Using the composite dictionary (step 6), the representation $\boldsymbol{D}\boldsymbol{x}_k$ produces the denoised signal $\boldsymbol{y}_k^c$ in step

Algorithm 7.11: Denoising via Composite Convex DL

Data: external signals set $\boldsymbol{Y}_e \in \mathbb{R}^{m \times N_e}$
noisy signals set $\boldsymbol{Y}_i \in \mathbb{R}^{m \times N_i}$
sparsity regularization parameters λ^i and λ^e
Result: denoised signals $\boldsymbol{Y}^c \in \mathbb{R}^{m \times N_i}$

```
Train global dictionary: D_e = K-SVD(Y_e)
Train dictionary with noisy set: D_i = K-SVD(Y_i)
for k = 1 : N_i do
    Choose or learn weights μ^e and μ^i
    Represent: solve (7.107) via Coordinate Descent to obtain x_k
    Compose: D = [D_e D_i]
    Denoise: y_k^c = D x_k
```

7. As stated in the introduction of this section and also in Sect. 2.4.1, this is based on the assumption that the information lost through this approximation is the actual noise.

7.7.2 *Composite Dictionaries with Orthogonal Blocks*

Shifting towards l_0 solutions, we explored [96] dictionary composition in the context of SBO from Algorithm 7.5. During training, SBO grows the dictionary size by adding new orthogonal blocks. This fits naturally in the current framework. A first idea is to take an SBO trained dictionary $\boldsymbol{D}_e$ and combine it with a new orthogonal basis $\boldsymbol{D}_i$ trained on the current noisy image patches $\boldsymbol{Y}$.

We use 1ONB to train the internal dictionary as described in step 1 from Algorithm 7.12. The twist comes during representation where we use (7.20) to select the basis $\boldsymbol{Q}_b$ from $\boldsymbol{D}_e$ that best fits the current patch $\boldsymbol{y}_k$ (step 3). The composite dictionary (7.106) is built by juxtaposing this block with the internal block $\boldsymbol{D}_i$ in step 4. Now, with the composite dictionary $\boldsymbol{D}$, we employ OMP, preferably called like in (2.19), to obtain the final sparse patch representation $\boldsymbol{x}_k$ (step 5) thus producing the denoised signal $\boldsymbol{y}_k^c$ in step 6.

The advantage of this method over Algorithm 7.11 is speed. Experiments also showed an improvement in representation quality only with loosened sparsity constraints ($s > \sqrt{m}$). To mediate this disadvantage, while still being speed conscious, we can expand the training of the internal dictionary from step 1 to include L_i orthogonal bases, leading to another smaller SBO dictionary as described in Algorithm 7.13. During representation, this approach requires extra operations for selecting the fittest orthogonal basis from the internal dictionary in step 4. Thus the final dictionary from step 5 consists of two orthogonal bases, $\boldsymbol{Q}_i$ from the internal

Algorithm 7.12: SBO-C1

Data: signals set $\boldsymbol{Y} \in \mathbb{R}^{m \times N}$
external SBO dictionary $\boldsymbol{D}_e \in \mathbb{R}^{m \times L_e m}$
Result: denoised signals $\boldsymbol{Y}^c \in \mathbb{R}^{m \times N}$

```
1 Train orthogonal block with noisy set: D_i = 1ONB(Y)
2 for k = 1 : N do
3     Choose block Q_b of D_e: b = arg max_{j=1:L_e}(E_{x_k^j}), as described around (7.20)
4     Compose: D = [Q_b D_i]
5     Represent: x_k = OMP(D, y_k)
6     Denoise: y_k^c = D x_k
```

Algorithm 7.13: SBO-C

Data: signals set $\boldsymbol{Y} \in \mathbb{R}^{m \times N}$
external SBO dictionary $\boldsymbol{D}_e \in \mathbb{R}^{m \times L_e m}$
number of internal dictionary orthogonal blocks $L_i < L_e$
Result: denoised signals $\boldsymbol{Y}^c \in \mathbb{R}^{m \times N}$

```
1 Train L_i orthogonal blocks with noisy set: D_i = SBO(Y, L_i)
2 for k = 1 : N do
3     Choose block Q_e of D_e: e = arg max_{j=1:L_e}(E_{x_k^j})
4     Choose block Q_i of D_i: i = arg max_{j=1:L_i}(E_{x_k^j})
5     Compose: D = [Q_e Q_i]
6     Represent: x_k = OMP(D, y_k)
7     Denoise: y_k^c = D x_k
```

dictionary and $\boldsymbol{Q}_e$ from the external dictionary. The number L_i of internal bases has to be small not only for avoiding the extra complexity but also for minimizing the risk of modeling the actual noise in our internal dictionary.

Finally, our third proposal is a hybrid: train an external unstructured dictionary with a slower but more accurate DL algorithm, and then train a fast specialized orthogonal basis via SBO for the internal samples. More specifically, the input dictionary $\boldsymbol{D}_e$ in Algorithm 7.13 is obtained in standard way via AK-SVD, thus canceling step 3 because $\boldsymbol{Q}_e$ is replaced by $\boldsymbol{D}_e$. The rest of the algorithm remains unchanged. We denote this hybrid AK-S. Note that, while the off-line part of AK-S is slower, the online part involves SBO like in SBO-C, thus has the same complexity.

In denoising experiments reported in [96], we found that AK-S was the top performer (in terms of denoising quality), beating also unstructured composite dictionaries, while SBO-C closely followed behind.

Problems

P 7.1 The ratio $\rho(\boldsymbol{v}) = \|\boldsymbol{v}\|_1 / \|\boldsymbol{v}\|_2$ is used in Sect. 7.2.2 as a sparsity measure for a vector $\boldsymbol{v} \in \mathbb{R}^m$. Show that $\rho(\boldsymbol{v}) \geq 1$. What vectors satisfy $\rho(\boldsymbol{v}) = 1$? ("Draw" the unit spheres in the 1- and 2-norms and notice that the former is contained in the latter. What are the common points?)

Consider vectors $\boldsymbol{u}, \boldsymbol{v} \in R^m$ such that $\|\boldsymbol{u}\|_0 = s$, $\|\boldsymbol{v}\|_0 > s$, and $\boldsymbol{u}$ and $\boldsymbol{v}$ have $s-1$ common nonzero elements; show that $\rho(\boldsymbol{u}) \leq \rho(\boldsymbol{v})$.

Consider now vectors $\boldsymbol{x}, \boldsymbol{y} \in R^m$. Is it possible that $\rho(\boldsymbol{x}) < \rho(\boldsymbol{y})$ but $\|\boldsymbol{x}\|_0 > \|\boldsymbol{y}\|_0$?

P 7.2 Consider the dictionary update for an atom $\boldsymbol{d}$ of $\boldsymbol{D}_2$ in the multiscale dictionary (7.9). Show that, adapting the notation from Proposition 3.4, the optimization problem has the form

$$\begin{aligned} \min_{\boldsymbol{d}} \quad & \left\|\boldsymbol{F}_u - \boldsymbol{d}\boldsymbol{x}_u^T\right\|_F^2 + \left\|\boldsymbol{F}_l - \boldsymbol{d}\boldsymbol{x}_l^T\right\|_F^2 \\ \text{s.t.} \quad & \|\boldsymbol{d}\| = 1 \end{aligned}$$

where the indices u and l correspond to the upper and lower half of the signals, respectively. Show that the solution is

$$\boldsymbol{d} = \boldsymbol{F}_u\boldsymbol{x}_u + \boldsymbol{F}_l\boldsymbol{x}_l$$

followed by normalization.

P 7.3 Given a signal $\boldsymbol{y} \in \mathbb{R}^m$ and an orthogonal dictionary $\boldsymbol{Q} \in \mathbb{R}^{m\times m}$, show that the optimal representation $\boldsymbol{x}$ with sparsity level s that minimizes $\|\boldsymbol{y} - \boldsymbol{D}\boldsymbol{x}\|$ is obtained by setting to zero all but the s largest (in absolute value) entries of $\boldsymbol{Q}^T\boldsymbol{y}$.

P 7.4 Show that the solution of the sparse coding problem (7.22) with ℓ_0 penalty and orthogonal dictionary is indeed given by the hard thresholding operation (7.23).

P 7.5 Adapt the OMP algorithm to the shift invariant dictionary structure (7.26). Show that an efficient implementation has a complexity no more than $O(\log m)$ larger than that of standard OMP with an $m \times n$ dictionary.

P 7.6 Prove the relationship between the Khatri–Rao product and the Hadamard product expressed in (7.63).

P 7.7 Show that $\|\boldsymbol{A} - \boldsymbol{B} \otimes \boldsymbol{C}\|_F^2 = \left\|\mathcal{R}(\boldsymbol{A}) - \text{vec}(\boldsymbol{B})\text{vec}(\boldsymbol{C})^T\right\|_F^2$ for $\boldsymbol{A} \in \mathbb{R}^{n_1n_2\times m_1m_2}$, $\boldsymbol{B} \in \mathbb{R}^{n_1\times m_1}$ and $\boldsymbol{C} \in \mathbb{R}^{n_2\times m_2}$, where the operator $\mathcal{R}$ is defined in (7.85).

P 7.8 Write a 2D OMP version for the dictionary (7.84), which is a sum of Kronecker products, and analyze its complexity. When is it still more convenient than standard OMP?

Chapter 8
Classification

Abstract Classification is an application where sparse representation intuitively may have a great contribution, due to its ability to grasp the essentials of a signal. We present two large categories of methods. In the first, a dictionary is learned from the training signals defining each class. Classification is based on the residuals of the representations on each individual dictionary or, more often, on their union. An important DL goal modification is to encourage not only each dictionary to well represent the signals from its own class, but also to badly represent the signals from other classes, thus gaining in discriminative power. The second category is that where a single dictionary is used for representation and a classifier matrix is also trained to extract class information from the sparse representation matrix. The two most prominent methods are discriminative DL and label consistent DL. Other methods that fall somewhat in between the categories are also presented. We apply the methods to fault detection in a water network where a small number of sensors can be used to determine the location of faults.

8.1 Classification Problem

The methods presented in this chapter aim to solve the standard *supervised classification* problem. Given a test signal $\boldsymbol{y} \in \mathbb{R}^m$, we need to decide to which of the classes $\mathcal{C}_i$, $i = 1 : c$, it belongs. The classes are typically defined through matrices $\boldsymbol{Y}_i \in \mathbb{R}^{m \times N_i}$ of training signals, one for each class. We continue to denote $N = \sum_{i=1}^{c} N_i$ the total number of training signals. We associate a label matrix $\boldsymbol{H} \in \mathbb{R}^{c \times N}$ with the classification problem, containing the labels of the classes; more precisely, if the signal $\boldsymbol{y}_j$ belongs to class $\mathcal{C}_i$, then the column $\boldsymbol{h}_j$ is the unit vector with 1 on position i and zeros elsewhere. For example, if there are three classes and six training vectors, two in each class, the label matrix

$$\boldsymbol{H} = \begin{bmatrix} 1\,1\,0\,0\,0\,0 \\ 0\,0\,1\,1\,0\,0 \\ 0\,0\,0\,0\,1\,1 \end{bmatrix} \tag{8.1}$$

B. Dumitrescu, P. Irofti, *Dictionary Learning Algorithms and Applications*,
https://doi.org/10.1007/978-3-319-78674-2_8

says that the first two signals belong to $\mathcal{C}_1$, the next two to $\mathcal{C}_2$ and the last two to $\mathcal{C}_3$.

In practical applications we seldom work with raw signals, due to their large size and low information density. They are replaced with vectors of features computed with some meaningful processing of the signals. For example, a segment of a speech signal could be characterized by (among others) its energy, maximum amplitude, zero crossing rate, fundamental frequency, and its power in some relevant frequency bands (e.g., organized following the perception of the human auditory system, with bandwidths that grow logarithmically), instead of the raw speech samples. For the main purpose of this chapter, which is classification, we ignore such preprocessing and simply denote $\boldsymbol{y}$ the vector to be classified.

Some of the methods can be extended to *unsupervised* classification, in which we are given the training samples $\boldsymbol{Y} \in \mathbb{R}^{m \times N}$ but no information regarding the classes to which they belong. So, the first purpose is to determine the classes $\mathcal{C}_i$, $i = 1 : c$, where the number c of classes is given or should be found as well. After finding the classes, the problem of classifying new signals is the same as in the supervised case. An even more complicated variation, which we do not treat here, is the possibility to change the classes as new signals arrive.

Our aim in this chapter is to present classification methods that use sparse representations and dictionary learning. Their underlying motivation is again the ability of sparse representations to catch the specific of a signal. We will cover several methods, but we start with a simple example that suggests the principle of most methods. Since the topic of classification is vast and well covered, we do not discuss methods unrelated to sparse representations.

Example 8.1 Figure 8.1 presents a bird's-eye view of the DL classification process. Starting from the training signals $\boldsymbol{Y}$ that are split into $c = 3$ distinct classes of $N_i = 7$ signals each ($i = 1 : 3$), the training process produces a dictionary $\boldsymbol{D}$ whose atoms are grouped in three sets of four atoms each. This is the simple situation where a group consists of atoms specialized in representing signals from a single class. Ideally, after the learning process is complete, when signals from the first class are represented with the dictionary $\boldsymbol{D}$, only the atoms from the first group are used in their sparse representation, and so on for the rest of the classes.

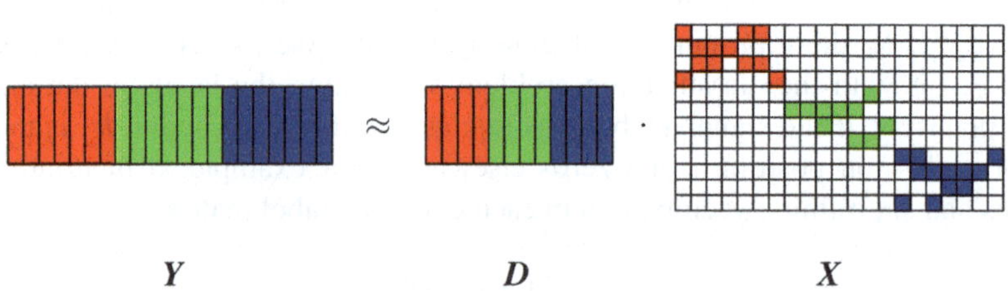

Fig. 8.1 Ideal dictionary driven classification. The unclassified signals $\boldsymbol{Y}$ belong to either of the three classes: red, green, blue. Dictionary atoms are specialized in representing a single distinctive class

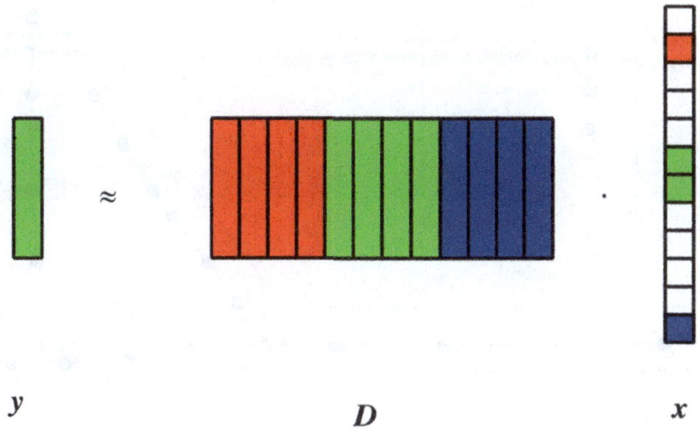

Fig. 8.2 Common situation when classifying a signal $\boldsymbol{y}$ using learned dictionary $\boldsymbol{D}$. The signal might belong to one of three classes: red, green, blue. Dictionary atoms are specialized in representing a single distinctive class. Classification performed by inspecting the nonzero coefficients of $\boldsymbol{x}$ is not necessarily obvious

If dictionary $\boldsymbol{D}$ is structured as a set of smaller class-specific dictionaries corresponding to the red, green, and blue parts, the learning process specializes each small dictionary to be very good at representing its assigned class. When $\boldsymbol{D}$ is viewed as a whole, and thus unstructured, the learning process may also instill discriminative properties to its atoms, such that each is very good at representing only one of the three classes.

In practice a signal representation often includes atoms from multiple classes, as described in Fig. 8.2; thus, classification based on the indices of the chosen atoms becomes difficult. A possible solution when $\boldsymbol{D}$ is composed of smaller class-specific dictionaries is to perform classification by looking at which of the class dictionaries provides the smallest representation error. The unstructured case is a bit more complex and will be discussed separately in the following sections, but essentially we try to determine which atoms have a more significant contribution to the representation. ■

8.2 Water Networks Application

In this section we present a real-life classification problem, where the objective is to detect and isolate faults that take place in a water distribution network. The mathematical model of such systems is laborious and consists of nonlinear equations that are intractable in practice. In [102] we proposed a fast alternative that uses DL to train a discriminative dictionary on past fault data such that new incoming data can be identified as faults and, if so, localized through classification.

A water network can be represented as a graph whose edges represent the pipes and the nodes are either water tanks or junctions that distribute the current flow to users and connecting pipes. Figure 8.3 depicts the small Hanoi water network, where we have a single water tank supplying water to 31 junction nodes. This is a weighted graph where the edge weight represents the pipe length in meters.

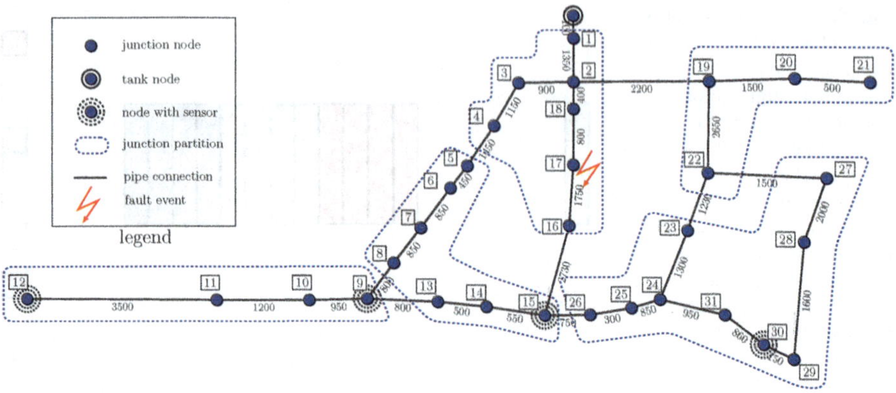

Fig. 8.3 The small Hanoi water network

One of the main challenges with existing water networks is fault detection and isolation (FDI). For example, the leak caused by a broken pipe near junction 17 in Fig. 8.3 might be indistinguishable from user demand, so the first task is fault detection: knowing that something bad happened. Because water networks span across large areas of terrain with varied landforms, next we need to perform fault isolation: knowing the whereabouts of the leak.

The amount of water waste depends on how fast a fault event is detected and isolated such that repair teams are dispatched right away and need to cover as little ground as possible.

We observe the network behavior via water pressure values recorded at junction nodes. It was shown [28] that pressure dynamics are described through the nonlinear combination of local and neighboring influences that are hard to apply in practice.

A simpler way of investigating pressure changes is to compare the current values to a known nominal behavior. The difference between the two values is called a residual. For example, the nominal behavior can represent pressure measurements taken during the night when user demand is negligible. When a fault event takes place the variations from the nominal pressure are large enough to be observed by the nearest junction and even by its neighbors. Thus, such measurements would be enough to provide us with both detection and isolation; the problem is that most real-world water networks are vast, with usually thousands of junctions and pipes, and this approach requires placing an equal amount of water pressure sensors. This is impractical because sensors are expensive and hard to install. Indeed, existing networks have only a few sensors compared to the total number of junctions.

Residual training data is often built from (simulated) sensor measurements corresponding to fault events that took place in each junction node with multiple magnitudes. For a network with J junctions and m sensors where M fault magnitudes were tested for each junction, the training matrix $\boldsymbol{Y} \in \mathbb{R}^{m \times N}$ has $N = JM$ data items.

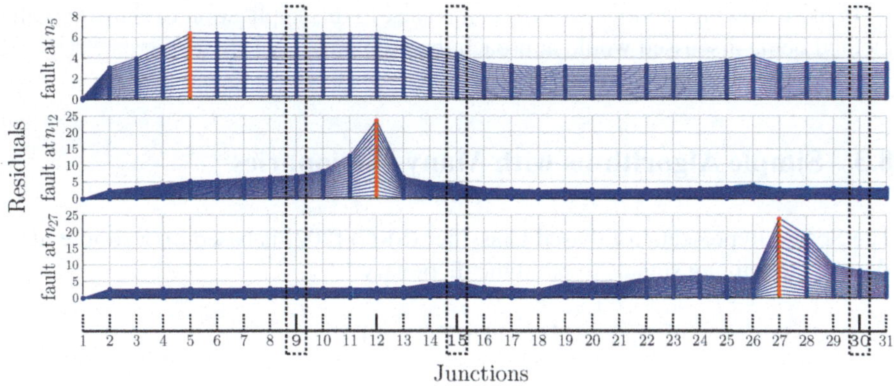

Fig. 8.4 Residual signals for three fault scenarios in the Hanoi water network. The junctions with leakages are shown in red

The Hanoi network from Fig. 8.3 has sensors in junctions 9, 12, 15, and 30. Let us assume we have access to residual pressure data recorded by these sensors as described above. Based on these residuals, our goal is to train a dictionary $\boldsymbol{D}$ with whom to perform classification of new sensor data. If the sparse representation of the new residual from the four sensors is not null, then a fault took place and its location can be retrieved as suggested in Example 8.1 and later explained in detail. In Fig. 8.4 we illustrate the residuals for each of the 31 junctions for three fault scenarios: leakage in nodes 5, 12, and 27. Each blue line represents a different fault magnitude. The magnitude of a leak is measured in the amount of liters wasted per minute. As expected, residual values increase proportional to the fault magnitude and peak in the junction directly affected by the fault. Only the data inside the dotted rectangles are available for classification.

A simple choice for classification classes is to have each class represent a single junction node. This is ideal if we have a very small network with sufficient sensors because the FDI process is fast and exact. Unfortunately, this is rarely the case even for the small Hanoi network. Consider the available sensors and the fault in node 17. It is very likely that, due to this particular scenario, the sensor measurements for a fault in node 17 be very similar to one in node 18 which could lead to misclassification. Of course, in reality we have to deal with much larger networks with insufficient sensors where such errors will often occur if each junction node has its own class. In practice it turns out that such precision is not required and it is good enough if we can isolate the fault within a given neighborhood. Thus, we partition the network in groups of neighboring junctions where each class represents a partition. Figure 8.3 shows (with blue dotted lines) a split into five partitions of the Hanoi network. Using this partitioning scheme, if the fault event in junction 17 is processed and properly classified, then we inform the repair team that the leak took place somewhere between junctions 1, 2, 3, 4, 16, 17, and 18.

In the following sections we will come back to this application to show how the method being discussed would approach the problem.

8.3 Simple Algorithms with Many Dictionaries

The Sparse Representation-based Classifier (SRC) [207] is based on a simple idea. We use the training signals to build the dictionary

$$D = [Y_1 \; Y_2 \; \dots \; Y_c] \tag{8.2}$$

with $n = N$ atoms; we normalize the atoms (but still denote D the dictionary). Thus, if we were to adapt Fig. 8.1, the signals and the dictionary would be of equal size and the sparse representations matrix X would become a square diagonal matrix whose entries are the signal norms. There is no dictionary learning in SRC.

The signal y is represented with a sparse representation x, by solving (1.6) with OMP or other algorithm mentioned in Chap. 1. We can assume that, if the signal belongs to a class, then the dictionary atoms from that class are more likely to be present in the representation. This phenomenon is depicted in Fig. 8.2. To quantify this intuitive heuristic, let $\delta_i(x)$ be the vector containing only the nonzero entries of x corresponding to the atoms in class i. We can compute the representation error for each such partial vector,

$$e_i(y) = \|y - D\delta_i(x)\|. \tag{8.3}$$

The SRC method assigns the signal y to the class with minimum error. Obviously, a training signal has zero error, since its representation consists of a single atom (the normalized signal); hence, the method is consistent, in the sense that the training signals are correctly classified.

An alternative to the classification based on the error (8.3) is to represent the signal y with each of the dictionaries

$$y \approx D_i x_i, \;\; \|x_i\|_0 \le s, i = 1 : c \tag{8.4}$$

where s is some imposed sparsity. Then, the signal is assigned to the class with smallest representation error $\|y - D_i x_i\|$.

Example 8.2 For water networks the SRC dictionary D is built from all of the available sensor measurements. Even though the number of atoms per class is large, it is likely to encounter misclassification due to the fact that atoms across different classes might be almost linearly dependent because of the way the sensors are placed (as discussed in Sect. 8.2). If two similar atoms exist, one from the true class and the other from a random class, what would prevent OMP from picking the wrong one? ■

There are several possible variations on the SRC theme, starting from possible drawbacks of the method. One of them is the size of the dictionary, which may be very large, depending on the number of available training signals; of course, we could use only some of the signals, but wasting some information by doing so. Also, the number of atoms may be different from class to class, which is right if the training samples are obtained according to the probability of each class, but may be wrong otherwise. These drawbacks can be alleviated by introducing DL in the equation, thus replacing the raw signals used by SRC with trained atoms.

The methods proposed in [179, 213] (for texture classification and face recognition, respectively) train distinct dictionaries $\boldsymbol{D}_i$ of size $n_i < N_i$ from the signals $\boldsymbol{Y}_i$; in principle, any of the DL algorithms presented in Chap. 3 can be used. Then, the dictionary (8.2) is replaced with

$$\boldsymbol{D} = [\boldsymbol{D}_1\ \boldsymbol{D}_2\ \dots\ \boldsymbol{D}_c],\ \boldsymbol{D}_i \in \mathbb{R}^{m \times n_i}. \tag{8.5}$$

Figure 8.1 accurately describes this method with the observation that the red, green, and blue atom groups are in fact the separately trained dictionaries $\boldsymbol{D}_1$, $\boldsymbol{D}_2$, and $\boldsymbol{D}_3$. The advantage is a smaller (note that now the dictionary has $n = \sum_{i=1}^{c} n_i$ atoms) and possibly better dictionary; we can tune the size of each dictionary $\boldsymbol{D}_i$ according to a priori information on the respective class. However, unlike SRC, there is no guarantee that a training signal will be correctly classified. A similar approach using orthogonal dictionary blocks for each class was studied in [171] and discussed in Sect. 7.3.3.

Example 8.3 For the water network case, this involves training dictionaries $\boldsymbol{D}_i$, with $i \in J$, on the residual data $\boldsymbol{Y}_i \in \mathbb{R}^{m \times M}$ specific to each node. This is an improvement from standard SRC, but, while it minimizes the risk, it still does not alleviate the problems earlier described as we will see in the following sections. ■

Another drawback, that applies also to (8.5), is the possible existence of very similar atoms in several dictionaries $\boldsymbol{D}_i$, due to common properties of several classes. In this case, a sparse representation may almost arbitrarily choose one of these atoms, whose contribution to classification becomes irrelevant. It would be useful to have dictionaries whose atoms are clearly distinct from all other atoms of other dictionaries. The solution proposed in [149] adds a coherence penalty to the representation error. The dictionary (8.5) is optimized using the objective

$$\sum_{i=1}^{c} \|\boldsymbol{Y}_i - \boldsymbol{D}_i \boldsymbol{X}_i\|_F^2 + \gamma \sum_{i=1}^{c} \sum_{j \neq i} \left\| \boldsymbol{D}_i^T \boldsymbol{D}_j \right\|_F^2. \tag{8.6}$$

The first sum is simply the error $\|\boldsymbol{Y} - \boldsymbol{D}\boldsymbol{X}\|_F^2$. The terms $\left\| \boldsymbol{D}_i^T \boldsymbol{D}_j \right\|_F^2$ measure the overall distance between the atoms from dictionaries allocated to two classes; they are similar to the coherence measure in (4.26), with the difference that intra-class coherence is not optimized. Several algorithms can be imagined for solving (8.6), including adaptations of those from Sect. 4.4, taking into account that a signal is represented with only atoms from the same dictionary $\boldsymbol{D}_i$. In [149] the objective also contains an ℓ_1 penalty for enforcing sparsity, instead of having explicit constraints on the sparsity level.

Remark 8.4 This type of approach can be adapted to unsupervised learning with given number c of classes. We start with a given set of dictionaries, used for representing the signals and thus classifying them based on (8.3) or (8.4); this is the assignment step. Then, the dictionaries are updated by decreasing (8.6) based on the current classes. The assignment and update steps can be repeated until the classes stay unchanged. This procedure is very similar with the K-means clustering algorithm. Like in many alternate optimization algorithms, the initialization is important. The number of classes can also be discovered by starting with a single dictionary and then splitting one class (for example, the most populated) into two new classes. ■

Remark 8.5 The SRC method and its variants can be used to solve a related problem, that of source separation. For simplicity, we discuss only the case of $c = 2$ sources. Assume that the signal $\boldsymbol{y}$ is obtained by adding two signals coming from different sources, meaning that $\boldsymbol{y} = \boldsymbol{y}_1 + \boldsymbol{y}_2$. Separation means recovering $\boldsymbol{y}_1$ and $\boldsymbol{y}_2$ from $\boldsymbol{y}$ and is in principle impossible, unless we possess some additional information. If training signals are available from the two sources, we can learn a dictionary for each source, such that $\boldsymbol{y}_i \approx \boldsymbol{D}_i\boldsymbol{x}_i$, $i = 1 : 2$ (or simply form $\boldsymbol{D}_i$ with the available signals as atoms). Building $\boldsymbol{D} = [\boldsymbol{D}_1\ \boldsymbol{D}_2]$, we compute the sparse representation

$$\boldsymbol{y} \approx \boldsymbol{D}\boldsymbol{x} = \boldsymbol{D}_1\boldsymbol{x}_1 + \boldsymbol{D}_2\boldsymbol{x}_2 = \boldsymbol{y}_1 + \boldsymbol{y}_2,$$

thus retrieving the two components $\boldsymbol{y}_1$, $\boldsymbol{y}_2$ by simply splitting the sparse representation into the parts corresponding to the two (sub)dictionaries. ■

8.4 Discriminative DL with Many Dictionaries

The above classification methods encourage good representation properties obtained by learning the dictionaries, but do not train explicitly discriminative properties. Only the method based on (8.6) attempts to enhance the differences between dictionaries, but this is only indirectly related to classification performance. In essence, good discrimination means that the dictionary associated with a class represents well the signals from that class, but poorly the signals from all the other classes. We are willing to trade some of the own-class representation power for getting much worse other-class representations.

8.4.1 DL with Discriminative Penalty Functions

One of the first discriminative DL algorithms was presented in [126]. It uses a collection of dictionaries (8.5). We denote

$$e_{j,\ell} = \|\boldsymbol{y}_j - \boldsymbol{D}_\ell \boldsymbol{x}_{\ell j}\|^2 \tag{8.7}$$

the error obtained when representing the signal $\boldsymbol{y}_j$ with the dictionary $\boldsymbol{D}_\ell$. We denote $\mathcal{C}(j)$ the class to which $\boldsymbol{y}_j$ belongs. Good discrimination means small values $e_{j,\mathcal{C}(j)}$, when representing signals within the own class, and large values $e_{j,\ell}$, $\ell \neq \mathcal{C}(j)$, when representing signals from the other classes. Like in (8.6), we achieve this goal by penalizing the representation error, this time having the representation errors as arguments:

$$\min_{\boldsymbol{D}_i, i=1:c} \sum_{i=1}^{c} \|\boldsymbol{Y}_i - \boldsymbol{D}_i \boldsymbol{X}_{ii}\|_F^2 + \alpha \sum_{j=1}^{N} g(e_{j,1}, \ldots, e_{j,c}). \tag{8.8}$$

The parameter $\alpha > 0$ allows the trade-off between representation and discrimination. The function g should have small values when the signal is correctly classified, i.e., when $e_{j,\mathcal{C}(j)} \leq e_{j,\ell}$, $\ell = 1 : c$, and large values when the signal is misclassified, case in which an index $\ell \neq \mathcal{C}(j)$ exists such that $e_{j,\ell} < e_{j,\mathcal{C}(j)}$. This is achieved in [126] via a softmax discriminative function

$$g(e_{j,1}, \ldots, e_{j,c}) = \log \left(\sum_{\ell=1}^{c} \mathrm{e}^{-\lambda(e_{j,\ell} - e_{j,\mathcal{C}(j)})} \right), \tag{8.9}$$

where λ is a positive constant. Indeed, if the signal is correctly classified, then all the exponents are nonnegative (one of them being exactly zero) and so the function has a positive value near zero. If the case of a misclassification, at least an exponent becomes positive and thus the value of the function grows.

The optimization of (8.8) can be done along the lines of the basic algorithm. The sparse coding step, in which each signal is represented with each dictionary, is the standard one. In the dictionary update step, all atoms of (8.5) are successively optimized. The key to obtaining efficient algorithm is to replace the softmax function (8.9) with a local linearization. Thus, the optimization objective becomes quadratic and a K-SVD-like algorithm can be applied. The initialization is made by training each dictionary with the signal from its class, i.e. based only on reconstruction properties.

The parameters α and λ play an important role in the success of the algorithm and it is clearly better to vary them in the DL iterations. For example, it was proved advantageous to start with low values of λ and α, such that good representation is favored (when one of the parameters is zero, then the discriminative penalty in (8.8) is a constant) and then increase them to enhance the discrimination.

A somewhat related idea was presented in [214] and is based on Fisher discrimination. Like in (8.8), a representation and a discrimination term are combined, using a dictionary (8.5). The representation term contains also elements of discrimination and, for class $\mathcal{C}_i$, has the form

$$f_i(\boldsymbol{Y}_i, \boldsymbol{D}, \boldsymbol{X}_i) = \|\boldsymbol{Y}_i - \boldsymbol{D}\boldsymbol{X}_i\|_F^2 + \|\boldsymbol{Y}_i - \boldsymbol{D}_i\boldsymbol{X}_{ii}\|_F^2 + \sum_{\ell=1,\ell\neq i}^{c} \|\boldsymbol{D}_\ell \boldsymbol{X}_{\ell i}\|_F^2. \tag{8.10}$$

The first term imposes good representation of the signals from $\mathcal{C}_i$ using the full dictionary. The second term requires that the same signals are also well represented using only the dictionary associated with their class. The third term is meant to explicitly reduce the power of the coefficients from $\boldsymbol{X}$ that correspond to the dictionaries of other classes.

The discriminative term is built via the within-class scatter of the representation coefficients $\boldsymbol{X}$, defined by

$$S_W(\boldsymbol{X}) = \sum_{i=1}^{c} \sum_{\boldsymbol{x}_j \in \mathcal{C}_i} (\boldsymbol{x}_j - \boldsymbol{m}_i)(\boldsymbol{x}_j - \boldsymbol{m}_i)^T, \tag{8.11}$$

and the between-class scatter

$$S_B(\boldsymbol{X}) = \sum_{i=1}^{c} N_i(\boldsymbol{m}_i - \boldsymbol{m})(\boldsymbol{m}_i - \boldsymbol{m})^T. \tag{8.12}$$

In these definitions, $\boldsymbol{m}_i$ is the average of the representations from $\boldsymbol{X}_i$ and $\boldsymbol{m}$ that of representations from $\boldsymbol{X}$ (remind that N_i is the number of signals from class $\mathcal{C}_i$). For better discrimination, the within-class scatter should be small and the between-class scatter should be large. The discriminative term is

$$g(\boldsymbol{X}) = \operatorname{tr}(S_W(\boldsymbol{X})) - \operatorname{tr}(S_B(\boldsymbol{X})) + \eta\|\boldsymbol{X}\|_F^2. \tag{8.13}$$

The last term is added for regularization purposes and makes the function convex if η is large enough.

Finally, the optimization problem is

$$\min_{\boldsymbol{D},\boldsymbol{X}} \sum_{i=1}^{c} f_i(\boldsymbol{Y}_i, \boldsymbol{D}, \boldsymbol{X}_i) + g(\boldsymbol{X}) + \lambda\|\boldsymbol{X}\|_{1,1}, \tag{8.14}$$

the last term inducing sparsity. The objective is convex when either $\boldsymbol{D}$ or $\boldsymbol{X}$ is fixed, so the standard alternating two-step approach can be adapted. When $\boldsymbol{D}$ is fixed, each representation $\boldsymbol{X}_i$ is improved while keeping the others fixed. The dictionaries are updated atom by atom by solving a quadratic problem.

The classification is made by combining the representation error and the distance of the representation coefficients to the averages $\boldsymbol{m}_i$ of the classes.

Other discrimination solutions were proposed in [223]. First of all, the metric of the error (8.7) is changed by introducing a projection matrix $\boldsymbol{P} \in \mathbb{R}^{\tilde{m} \times m}$, with $\tilde{m} \leq m$, with the purpose of enhancing the discriminative properties. The error becomes

$$e_{j,\ell} = \|\boldsymbol{P}(\boldsymbol{y}_j - \boldsymbol{D}_\ell \boldsymbol{x}_{\ell j})\|^2. \tag{8.15}$$

It is hoped that the projection, which is a variable matrix to be learned from the data, can move the error to a subspace where sparse representations can achieve better classification properties. For better numerical behavior, the projection matrix has orthogonal rows, meaning that $\boldsymbol{P}\boldsymbol{P}^T = \boldsymbol{I}$.

The second contribution regards the discrimination function, which is an alternative to (8.9) or (8.13). Assuming that the learning process is done signal by signal (like in online algorithms), the discrimination function is

$$g(e_{j,1}, \ldots, e_{j,c}) = \frac{1}{1 + \exp\left(\beta(1 - \frac{e_{j,C(j)}}{\min_{i \neq j} e_{j,i}})\right)}, \tag{8.16}$$

where $\beta > 0$ is a parameter that dictates the sharpness of the transition from 0 to 1 of the sigmoid function (8.16). Note that if $e_{j,C(j)} < \min_{i \neq j} e_{j,i}$, which means that the signal is correctly classified, then the function value is less than 0.5; misclassification produces a value larger than 0.5.

The standard alternating optimization strategy can be used, based on gradient steps. When optimizing $\boldsymbol{D}$ or $\boldsymbol{P}$ based on (8.16), the incorrect class that gives the minimum error is fixed by the current values of the variables. Hence, the gradients of the function g with respect to the variables can be easily computed. The gradient formulas can be found in [223]; let us note that the gradient of (8.15) with respect to $\boldsymbol{P}$ is

$$2\boldsymbol{P}(\boldsymbol{y}_j - \boldsymbol{D}_\ell \boldsymbol{x}_{\ell j})(\boldsymbol{y}_j - \boldsymbol{D}_\ell \boldsymbol{x}_{\ell j})^T.$$

Once $\boldsymbol{P}$ is updated with a gradient step, its rows must be orthonormalized, which can be done through a simple QR factorization. Although there are no convergence guarantees, small enough gradient steps lead to a decrease of the optimization objective.

8.4.2 Adding a Shared Dictionary

In order to alleviate the problem of similar atoms among different class dictionaries $\boldsymbol{D}_i$, the Fisher discrimination (8.10) is extended in [196] to include a shared dictionary $\boldsymbol{D}_0$. This special dictionary avoids atoms that are class specific, but also those that model existing noise, by maintaining a low-rank and forcing similar representation coefficients across classes.

The new dictionary becomes

$$\bar{\boldsymbol{D}} = \begin{bmatrix} \boldsymbol{D} & \boldsymbol{D}_0 \end{bmatrix}, \tag{8.17}$$

which extends the sparse representations

$$\bar{\boldsymbol{X}}_i = \begin{bmatrix} \boldsymbol{X}_{ii} \\ \boldsymbol{X}_{i0} \end{bmatrix} \tag{8.18}$$

to accommodate the coefficients of the shared dictionary atoms. These extensions are also depicted in Fig. 8.5. The yellow atoms make up the shared dictionary $\boldsymbol{D}_0$ that is used across classes for representation. The adjustments necessary to change the representation term (8.10) into

$$\begin{aligned}\bar{f}_i(\boldsymbol{Y}_i, \bar{\boldsymbol{D}}, \bar{\boldsymbol{X}}_i) &= \|\boldsymbol{Y}_i - \bar{\boldsymbol{D}}\bar{\boldsymbol{X}}_i\|_F^2 + \|\boldsymbol{Y}_i - \boldsymbol{D}_i\boldsymbol{X}_{ii} - \boldsymbol{D}_0\boldsymbol{X}_{i0}\|_F^2 + \sum_{\ell=1,\ell\neq i}^{c} \|\boldsymbol{D}_\ell\boldsymbol{X}_{\ell i}\|_F^2 \\ &= f_i(\boldsymbol{Y}_i - \boldsymbol{D}_0\boldsymbol{X}_{i0}, \boldsymbol{D}, \boldsymbol{X}_i) = f_i(\bar{\boldsymbol{Y}}_i, \boldsymbol{D}, \boldsymbol{X}_i)\end{aligned} \tag{8.19}$$

and discrimination term (8.13) into

$$\bar{g}(\boldsymbol{X}) = g(\boldsymbol{X}) + \|\boldsymbol{X}_0 - \boldsymbol{M}_0\|_F^2 \tag{8.20}$$

are minimal. The extra term in (8.20) ensures that the new shared coefficients have similar magnitude across training signals from all classes; the matrix $\boldsymbol{M}_0$ has identical columns, equal to the average of the columns of $\boldsymbol{X}_0$. Finally, the new optimization problem is

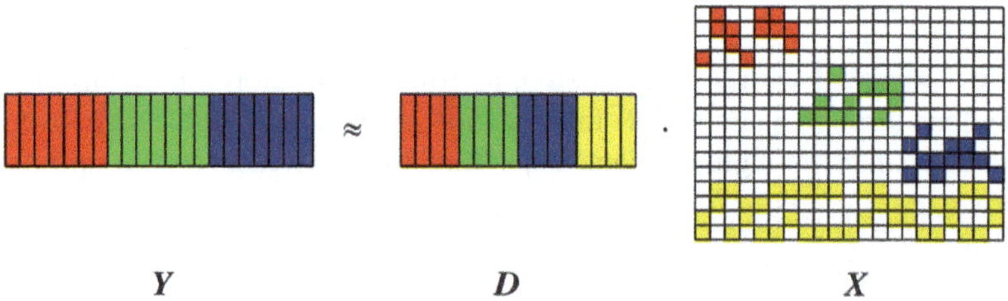

Fig. 8.5 Ideal dictionary driven classification with common dictionary. The unclassified signals $\boldsymbol{Y}$ belong to either of the three classes: red, green, blue. Dictionary atoms are specialized in representing the distinctive features of a single class, while the features common to all classes are left to the common yellow atoms

$$\min_{\bar{D},\bar{X}} \sum_{i=1}^{c} f_i(\bar{Y}_i, D, X_i) + \bar{g}(\bar{X}) + \lambda \left\|\bar{X}\right\|_{1,1} + \eta \left\|D_0\right\|_*, \tag{8.21}$$

the last term forcing a low-rank shared dictionary by penalizing the objective via the convex nuclear norm.

DL-COPAR [112] takes things a step further by being more aggressive in ensuring good representation with just the class specific and shared dictionary by explicitly forcing the coefficients outside these two dictionaries to be very small and penalizing inter-class atom coherence. Preserving earlier notations of D and X extensions, the representation term of the objective is similar to (8.19) and has the form

$$f_i(Y_i, \bar{D}, \bar{X}_i) = \left\|Y_i - \bar{D}\bar{X}_i\right\|_F^2 + \left\|Y_i - D_i X_{ii} - D_0 X_{i0}\right\|_F^2 + \left\|\tilde{X}_i\right\|_F^2, \tag{8.22}$$

where the third term of (8.19) is replaced with a firm penalty on the coefficients of the current class representations that lie outside the entries corresponding to either the class dictionary D_i or the shared dictionary D_0. We denote this set with $\tilde{X}_i$ and ideally expect $\left\|\tilde{X}_i\right\|_F = 0$. Coherence penalization is introduced in the overall optimization problem

$$\min_{\bar{D},\bar{X}} \sum_{i=1}^{c} f_i(Y_i, \bar{D}, \bar{X}_i) + \lambda \left\|\bar{X}\right\|_{1,1} + \eta \sum_{i=0}^{c} \sum_{j\neq i} \left\|D_i^T D_j\right\|_F^2 \tag{8.23}$$

which ensures that the inclusion of atoms representing common patterns arising in several classes is avoided. Unlike [149], this includes the shared dictionary, which makes sense because a strong correlation between an atom from D_0 with any atom amongst the class specific dictionaries leads to lower discrimination. Given its convex form, DL-COPAR solves the representation problem via lasso methods and performs dictionary refinement by updating one atom at time, keeping everything else fixed, and setting the resulting gradient to zero. Normalization is performed in post-processing and the affected representations are multiplied with the updated atom norm.

Example 8.6 When applying these methods to water network FDI, we initialize each class specific dictionary D_i by performing a short training round on the class specific residual measurements (just like in the former example) and then proceed to perform dictionary learning by taking into account the whole dictionary D following one of the Fisher variants. This last part handles the problem of similar atoms sneaking in across multiple classes. If using a shared dictionary, we initialize it by training on the entire sensor data and then proceed by solving the optimization problem (8.23). When performing FDI, and thus classification of a new residual signal, we operate like in SRC or related methods, but we ignore the sparse coefficients corresponding to atoms from D_0. ■

8.5 Discriminative DL with a Single Dictionary

Unlike the previous methods, which train a dictionary for each class, the next methods operate with a single dictionary for representation, helped by other classifying tools. The atoms are no longer explicitly specialized for one class, but they can specialize during the DL process. However, they are not used directly for classification, which is made via a specially designed classifier matrix.

8.5.1 Classification Using the Representation Matrix

The probably simplest idea is to completely separate the DL process and the classification. So, the unique dictionary $\boldsymbol{D}$ is learned using all the available signals gathered in the matrix $\boldsymbol{Y}$ from (8.2) (which we only rename here). This is a standard DL problem solved with the preferred DL algorithm, whose outcome is the dictionary $\boldsymbol{D}$ and the representation matrix $\boldsymbol{X}$. The classification task relies on the assumption that the representation matrix should contain all the information that is necessary. However, we cannot use it directly, like in the previous methods, since we have a single dictionary and we do not know which atoms are more likely to be representative for a certain class; in fact, we even do not know if the atoms are specialized or not.

So, we need to build a transformation of the representation matrix $\boldsymbol{X}$ that leads to classification information. Remind that $\boldsymbol{H}$ is the label matrix, as described around (8.1). Ideally, we should obtain a transformation $\boldsymbol{W} \in \mathbb{R}^{c \times n}$, called also classifier matrix, such that $\boldsymbol{W}\boldsymbol{X} = \boldsymbol{H}$, thus directly transforming representations into labels; classification would become obvious. However, perfect equality is impossible and thus the classifier should be taken as the solution of the regularized least-squares problem

$$\min_{\boldsymbol{W}} \|\boldsymbol{H} - \boldsymbol{W}\boldsymbol{X}\|_F^2 + \gamma \|\boldsymbol{W}\|_F^2, \tag{8.24}$$

where γ is a small positive constant with the role of making the problem better conditioned. The solution of (8.24) is

$$\boldsymbol{W} = \boldsymbol{H}\boldsymbol{X}^T(\boldsymbol{X}\boldsymbol{X}^T + \gamma \boldsymbol{I})^{-1}. \tag{8.25}$$

Once this matrix is available, we classify a test signal $\boldsymbol{y}$ following the two steps of the learning algorithm. Firstly, the signal is represented using the trained dictionary as $\boldsymbol{y} \approx \boldsymbol{D}\boldsymbol{x}$, with sparse $\boldsymbol{x}$. Secondly, the representation $\boldsymbol{x}$ is used for classification, by computing $\boldsymbol{h} = \boldsymbol{W}\boldsymbol{x}$. Ideally, this should be a unit vector, with 1 in the position of the class to which the signal belongs. The best approximation of $\boldsymbol{h}$ with a unit vector is to select the class index of the largest element, $i = \arg\max_{j=1:c} h_j$.

8.5.2 Discriminative DL

A possible drawback of the above method is the separation between training the dictionary $\boldsymbol{D}$, and thus obtaining the representations of the training signals, and computing the classifying matrix $\boldsymbol{W}$.

Discriminative DL solves this issue by jointly designing $\boldsymbol{W}$ and $\boldsymbol{D}$. The method was proposed in [145, 220] and the name "discriminative DL" is often associated with it despite the existence of other discriminative DL techniques, some of them already presented above. The optimization problem is

$$\min_{\boldsymbol{D},\boldsymbol{W},\boldsymbol{X}} \|\boldsymbol{Y} - \boldsymbol{D}\boldsymbol{X}\|_F^2 + \alpha\|\boldsymbol{H} - \boldsymbol{W}\boldsymbol{X}\|_F^2 \tag{8.26}$$

The first term of (8.26) is the usual sparse approximation error. The second is meant to train the classifier $\boldsymbol{W}$ that is used directly for classification, like in (8.24). The connection between the representation and classification terms is made by the (unique) representation matrix $\boldsymbol{X}$. A similar problem is (5.38), that was discussed in Sect. 5.6 in relation with task-driven DL.

We can see (8.26) as the simultaneous optimization of two dictionaries, $\boldsymbol{D}$ and $\boldsymbol{W}$, that are used to represent two training signal sets, $\boldsymbol{Y}$ and $\boldsymbol{H}$, using the same sparse representation matrix $\boldsymbol{X}$. With this interpretation in mind, the problem (8.26) can be formulated equivalently as

$$\min_{\boldsymbol{D},\boldsymbol{W},\boldsymbol{X}} \left\| \begin{bmatrix} \boldsymbol{Y} \\ \sqrt{\alpha}\boldsymbol{H} \end{bmatrix} - \begin{bmatrix} \boldsymbol{D} \\ \sqrt{\alpha}\boldsymbol{W} \end{bmatrix} \boldsymbol{X} \right\|_F^2 . \tag{8.27}$$

So, it has the same form as the basic DL problem (2.2): the matrices $\boldsymbol{H}$ and $\boldsymbol{W}$ can also be seen as extensions of the signal matrix and the dictionary, respectively. Hence, the problem (8.27) can be solved with any DL algorithm. In particular, K-SVD is used in [220]; in [145] the algorithm is different, as it loops through separate optimizations of $\boldsymbol{D}$ and $\boldsymbol{W}$, the latter one via (8.25).

For initializing $\boldsymbol{D}$, we focus only on the representation error and hence run a few iterations of an algorithm for the standard DL problem. The DL algorithm produces also a representation matrix $\boldsymbol{X}$ which is inserted in (8.25) to produce an initial classifier matrix $\boldsymbol{W}$.

After solving (8.27) the resulting $\boldsymbol{D}$ and $\boldsymbol{W}$ are not yet ready to use, due to the lack of normalization. If a standard DL algorithm, like K-SVD, has been used, the atoms of the extended dictionary that stacks $\boldsymbol{D}$ and $\boldsymbol{W}$ are normalized. To normalize the dictionary $\boldsymbol{D}$, we replace $\boldsymbol{d}_j$ with $\boldsymbol{d}_j/\|\boldsymbol{d}_j\|$, as usual. Denoting $\boldsymbol{\Delta} = \text{diag}(\ldots, \|\boldsymbol{d}_j\|, \ldots)$, the normalized dictionary is $\boldsymbol{D}\boldsymbol{\Delta}^{-1}$, and the new representations are $\boldsymbol{\Delta}\boldsymbol{X}$, as we want the product $\boldsymbol{D}\boldsymbol{X}$ to stay constant. Accordingly, the new classifying matrix is $\boldsymbol{W}\boldsymbol{\Delta}^{-1}$, such that $\boldsymbol{W}\boldsymbol{X}$ stays constant. So, the classifying vectors normalization is made by replacing $\boldsymbol{w}_j$ with $\boldsymbol{w}_j/\|\boldsymbol{d}_j\|$.

Once the matrices $\boldsymbol{D}$ and $\boldsymbol{W}$ are available, the classification of a new signal $\boldsymbol{y}$ is performed like in the previous method, by looking at the largest element of $\boldsymbol{W}\boldsymbol{x}$, where $\boldsymbol{x}$ is the sparse representation of $\boldsymbol{y}$ via $\boldsymbol{D}$.

8.5.3 Label Consistent DL

Label consistent DL [105] makes one step forward and tries to improve the classification performance by implicitly forcing the unique dictionary to have a structure like (8.5) (with possible permutations), i.e., one in which the atoms are specialized in representing the signals from a single class. The optimization problem is

$$\min_{\boldsymbol{D},\boldsymbol{W},\boldsymbol{A},\boldsymbol{X}} \|\boldsymbol{Y} - \boldsymbol{D}\boldsymbol{X}\|_F^2 + \alpha\|\boldsymbol{H} - \boldsymbol{W}\boldsymbol{X}\|_F^2 + \beta\|\boldsymbol{Q} - \boldsymbol{A}\boldsymbol{X}\|_F^2 \tag{8.28}$$

The first two terms are identical with those in (8.26) and their role is clear. The third term is meant to add more discriminative power. The matrix $\boldsymbol{Q}$ has n rows, one for each atom; a column j has values of one in the positions of the atoms associated with the class of $\boldsymbol{y}_j$ and zeros elsewhere. For example, if there are $N = 9$ signals split into $c = 3$ classes, we use a dictionary with $n = 6$ atoms and two atoms are meant to represent each class, the label consistency matrix is

$$\boldsymbol{Q} = \begin{bmatrix} 1&1&1&0&0&0&0&0&0 \\ 1&1&1&0&0&0&0&0&0 \\ 0&0&0&1&1&1&0&0&0 \\ 0&0&0&1&1&1&0&0&0 \\ 0&0&0&0&0&0&1&1&1 \\ 0&0&0&0&0&0&1&1&1 \end{bmatrix}.$$

Approximating $\boldsymbol{Q}$ with a linear transformation of the representation matrix $\boldsymbol{X}$ ensures that only a limited number of atoms combinations can characterize a class and that these combinations are fairly well different from those for other classes.

The problem can be transformed using the same trick as when passing from (8.26) to (8.27), obtaining

$$\min_{\boldsymbol{D},\boldsymbol{W},\boldsymbol{A},\boldsymbol{X}} \left\| \begin{bmatrix} \boldsymbol{Y} \\ \sqrt{\alpha}\boldsymbol{H} \\ \sqrt{\beta}\boldsymbol{Q} \end{bmatrix} - \begin{bmatrix} \boldsymbol{D} \\ \sqrt{\alpha}\boldsymbol{W} \\ \sqrt{\beta}\boldsymbol{A} \end{bmatrix} \boldsymbol{X} \right\|_F^2 \tag{8.29}$$

The problem can be solved with K-SVD or other standard DL algorithm. Once the solution is obtained, normalization is similar to that of discriminative DL. Only the initialization is different. To ensure from the beginning a structure resembling (8.5), small dictionaries $\boldsymbol{D}_i$, $i = 1 : c$, are trained with only the signals from class C_i. The number of atoms for each class is chosen proportionally with the number of training signals (or using other prior information) and the matrix $\boldsymbol{Q}$ is built accordingly. Then, the classifier $\boldsymbol{W}$ is initialized with (8.25) and the matrix $\boldsymbol{A}$ is initialized with a similar formula with $\boldsymbol{Q}$ replacing $\boldsymbol{H}$.

The classification of a new signal is performed like in the other methods from this section. So, the matrix $\boldsymbol{A}$ is not used for classification; its only role is to help structuring the dictionary in the learning process.

Example 8.7 We used EPANET [155] to simulate faults in the Hanoi water network from Fig. 8.3. The software allowed us to place sensors and record pressure measurement in order to form residual training and testing data as described in Sect. 8.2; we generated $N = 496$ signals, with 16 different faults magnitude for each of the 31 junctions. With these data we trained and used the discriminative DL and label consistent DL (LC-DL) methods, both based on AK-SVD, to perform FDI on the Hanoi network.

First, we used AK-SVD to initialize small dictionaries $\boldsymbol{D}_i$ as described after (8.29). Then, we ran discriminative DL and LC-DL on the training data. We allowed a maximum sparsity of $s = 4$ with an even number of atoms n_i distributed for each class. At the end we used the final dictionary $\boldsymbol{D}$ and the associated classifier $\boldsymbol{W}$ on new sensor data. A class was either made up of a single junction, thus leading to 31 classes, or a collection of junctions, as marked in Fig. 8.3, totaling to 5 distinct classes. In the first case we used $n_i = 8$ and in the second case $n_i = 51$ atoms per class, leading to dictionaries of $n = 248$ and, respectively, $n = 255$ atoms. In both scenarios we used the same parametrization, initialization, and sensor placement. We varied the number of available sensors m from 2 to 10 in increments of one. The successful classification rates are plotted in Fig. 8.6.

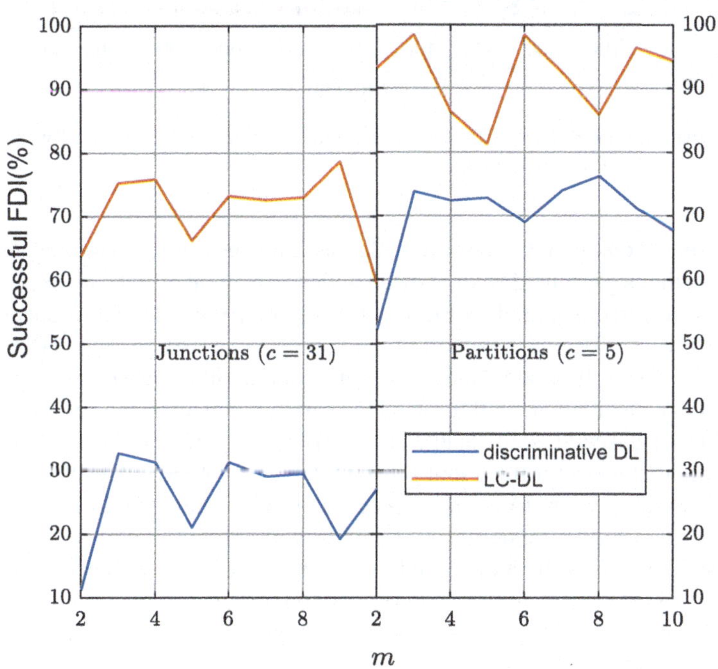

Fig. 8.6 Successful percentage rate of fault detection and isolation when using discriminative DL and LC-KSVD with $m = 2 : 10$ pressure sensors. In the first panel every junction is a classification class, while in the second we partitioned the Hanoi network in five sets

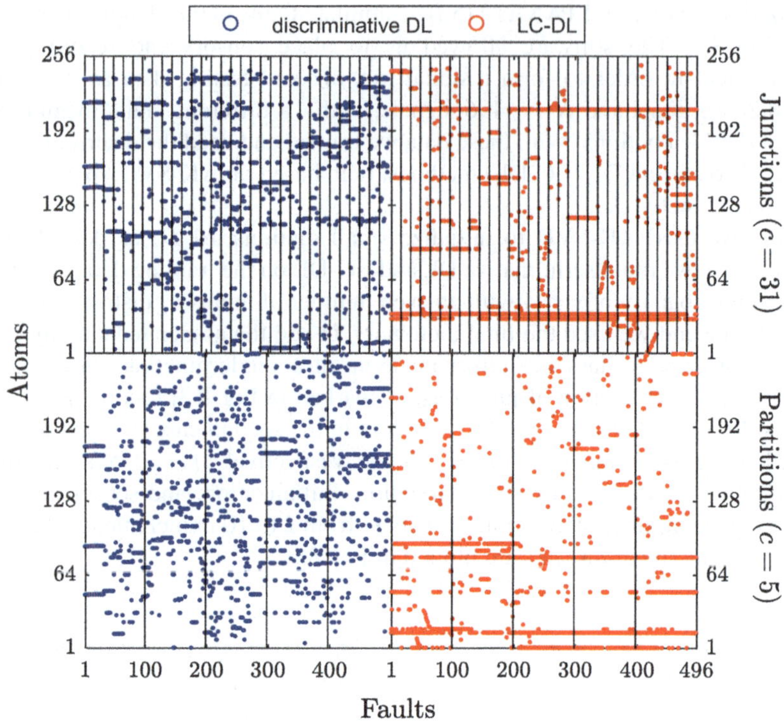

Fig. 8.7 Sparse representation coefficients used for the classification of test data. The vertical lines delimit signals from different classes

In the partitions panel, where few classes are used (thus permitting larger n_i), the two methods present similar results, with discriminative DL always coming in second. As clearly depicted in the junctions panel, the extra atom specialization strategy of LC-DL begins to pay off as the number of classes increases (thus forcing smaller n_i values). It is interesting to observe that a larger number of sensors does not implicitly mean a better FDI rate. This can be blamed on the sensor placement strategy, made based on some heuristics described in [102], but it might also be due to suboptimal choices for the α and β parameters. Following the results in [105], we used $\alpha = 4$ and $\beta = 16$. In our examples we used the same values no matter of the number of classes c or the number of sensors m.

Figure 8.7 presents the atoms used by the sparse representations to represent the testing signals. The existing dictionary was trained on data from $m = 6$ sensors. The top panels depict the results where each junction node is represented by a single class, and the bottom panels present the cases where node partitioning was used. The vertical helper lines delimit signals from different classes. We can observe that on real data the class separation of atoms is far from ideal. Some of the inter class horizontal "lines" can be avoided by enforcing a separate shared dictionary. Still, even in this configuration, LC-DL experiments show good results that are mostly due to the small magnitude of the coefficients corresponding to the common atoms, in contrast with the large magnitude of the coefficients corresponding to class-specific atoms. ■

8.6 Other Developments

There are several other methods that use DL for classification, but cannot be neatly categorized together with the already presented methods. We give here the most important features of a few interesting methods.

Latent dictionary learning [216] attempts a compromise between single dictionary and one-dictionary-per-class approaches. There is a unique dictionary, but each atom can be assigned to one or several classes; the assignment is made during learning. The objective is a variation on (8.6); a simplified form is

$$\sum_{i=1}^{c} \|\boldsymbol{Y}_i - \boldsymbol{D}\mathrm{diag}(\boldsymbol{v}_i)\boldsymbol{X}_i\|_F^2 + \gamma \sum_{i=1}^{c} \sum_{\ell \neq i} \sum_{j=1}^{n} \sum_{k \neq j} v_{ij} \left(\boldsymbol{d}_j^T \boldsymbol{d}_k\right)^2 v_{\ell k}. \tag{8.30}$$

We associate a vector $\boldsymbol{v}_i \in \mathbb{R}^n$ with class C_i, with the role of deciding what atoms are used for that class. Note that the full dictionary $\boldsymbol{D}$ is used, at least potentially, to represent signals from all classes. However, if $v_{ij} = 0$, then the atom $\boldsymbol{d}_j$ is not used for signals from class C_i. If $v_{ij} \neq 0$, then the atom is used; moreover, it can appear in more than a single class, thus also providing a generalization of the shared dictionary presented in Sect. 8.4.2. We can say that the atoms are in a latent state, from where they can become active or not as dictated by the vectors $\boldsymbol{v}_i$. Since these vectors have a clear significance of weights, it is natural to impose nonnegativity: $v_{ij} \geq 0$, for all $i = 1 : c$, $j = 1 : n$. Also, in order to balance atom contributions to different classes, we add to (8.30) the constraint

$$\sum_{j=1}^{n} v_{ij} = 1, \quad i = 1 : c,$$

i.e., we impose equal sums of the vectors (it is not relevant if the above constant is 1 or has another value). The second term in (8.30) encourages incoherence of the active atoms; note that there is no penalty for the atoms associated with the same class.

The optimization process is started with a set of vectors $\boldsymbol{v}_i$, which is kept fixed while the dictionary $\boldsymbol{D}$ is optimized through a typical iterative process. Once the final dictionary is available, the vectors $\boldsymbol{v}_i$ are optimized at their turn, by solving a quadratic programming problem (the objective (8.30) is quadratic in $\boldsymbol{v}_i$ and the constraints are linear).

The classification of a test signal $\boldsymbol{y}$ is done by computing the sparse representations that minimize $\|\boldsymbol{y} - \boldsymbol{D}\text{diag}(\boldsymbol{v}_i)\boldsymbol{x}\|$, $i = 1 : c$ (with the same sparsity level or using the same ℓ_1 penalty) and choosing the class for which the residual is minimum.

Other DL methods, like that from [173], organize the classes following a tree hierarchy, each node of the tree being associated with a dictionary. A root dictionary $\boldsymbol{D}_{0,0}$ is used to represent all signals, the representations being used to discriminate among several groups of classes, each represented by a branch of the tree. The procedure is repeated down to the leaves, each leaf representing a class. Classification at node i at level ℓ in the tree is made by a dictionary $\boldsymbol{D}_{\ell,i}$, composed of two parts: one inherited from the parent, the other specific to the node. For example, at node 2 and level 1, the dictionary is $\boldsymbol{D}_{1,2} = [\boldsymbol{D}_{0,0}\ \boldsymbol{D}^s_{1,2}]$, where $\boldsymbol{D}^s_{1,2}$ is the specific part. For each node i, a classifier matrix $\boldsymbol{W}_i$ is designed to decide on which child the search will be continued; this is similar with the approaches from the previous section, although the design method proposed in [173] is different.

A different kind of hierarchy appears in [123], somewhat related with the multiscale dictionaries (7.9). The objective, inspired by discriminative DL and aimed explicitly at image classification, contains the following terms (besides regularization terms that we ignore here)

$$\|\boldsymbol{Y} - \boldsymbol{D}\boldsymbol{X}\|_F^2 + \lambda_1 \|\mathcal{B}_1(\boldsymbol{X}) - \boldsymbol{D}_1\boldsymbol{X}_1\|_F^2 + \lambda_2 \|\boldsymbol{H} - \boldsymbol{W}\mathcal{B}_2(\boldsymbol{X}_1)\|_F^2 . \tag{8.31}$$

The first term is the usual sparse representation error, where the training matrix $\boldsymbol{Y}$ is made of vectorized image patches; so, the dictionary $\boldsymbol{D}$ has its usual role. The second term is meant to learn a second dictionary $\boldsymbol{D}_1$, with which blocks of several representation vectors are represented. The operators $\mathcal{B}_1$ and $\mathcal{B}_2$ stack groups of columns of their matrix arguments. So, the representations $\boldsymbol{X}_1$ can model relations between adjacent patches of an image. Finally, the third term contains the label and the classifier matrices, like in discriminative DL; the specific here is that the classifier acts on a second-level stack of representations, which can be very large; for example, a column of $\mathcal{B}_2(\boldsymbol{X}_1)$ may correspond to a full image; thus, the decision is taken from a single application of the classifier matrix, not from a combination of the decisions on image patches. The optimization of (8.31) is constrained with the typical sparsity constraints. The classification of a test image, split in vectorized patches that form a matrix $\tilde{\boldsymbol{Y}}$, starts as usual by computing the sparse representations $\tilde{\boldsymbol{X}}$, then by computing the second-level representations by minimizing $\left\|\mathcal{B}_1(\tilde{\boldsymbol{X}}) - \boldsymbol{D}_1\tilde{\boldsymbol{X}}_1\right\|_F$; both problems can be solved with OMP. Then, classification is performed by seeing which element of the vector $\boldsymbol{W}\mathcal{B}_2(\tilde{\boldsymbol{X}}_1)$ has the largest value.

Problems

P 8.1 Adapt the AK-SVD algorithm to the objective (8.6), using the similarity with the AK-SVDc Algorithm 4.2 from Sect. 4.4. In the dictionary update step, when a single atom from a single dictionary is optimized, the objective has a form that is similar to (4.28). Find the expression of the matrix $\bar{\boldsymbol{D}}$.

P 8.2 An alternative to the objective (8.6) (which encourages good separation between atoms from different classes) is

$$\sum_{i=1}^{c} \left(\|\boldsymbol{Y}_i - \boldsymbol{D}_i \boldsymbol{X}_i\|_F^2 + \lambda \|\boldsymbol{X}_i\|_{1,1} \right) + \gamma \sum_{i=1}^{c} \sum_{j \neq i} \left\| \boldsymbol{D}_i^T \boldsymbol{D}_j \right\|_F^2 .$$

Propose an algorithm for minimizing this objective inspired from the APrU Algorithm 5.1.

Extend the algorithm to the objective (8.23).

P 8.3 The discriminative DL problem (8.27) implies a normalization of the solution, as described in Sect. 8.5.2. Propose modifications of the K-SVD and AK-SVD that keep the dictionary $\boldsymbol{D}$ normalized at all times and hence provide directly the correct classification matrix $\boldsymbol{W}$. Will these modified algorithms give the same results as the standard algorithms followed by normalization?

P 8.4 Debate on the usefulness of recomputing the classifier matrix $\boldsymbol{W}$ with (8.25) after solving the discriminative DL (8.26) or label consistent DL (8.28) problems. Note that the approach is still different from that of Sect. 8.5.1. How much would you expect the matrix $\boldsymbol{W}$ to be changed by (8.25)? Will the objective of (8.26) or (8.28) decrease?

P 8.5 We want to add discriminative features to DL with separable dictionaries as formulated in (7.43) in order to obtain a classification method. Inspired by the discriminative DL problem (8.26) and denoting $\boldsymbol{h}_k$ the kth column of the label matrix associated with the classification problem, we can propose the objective

$$\sum_{k=1}^{N} \left(\left\| \boldsymbol{Y}_k - \boldsymbol{D}_1 \boldsymbol{X}_k \boldsymbol{D}_2^T \right\|_F^2 + \alpha \|\boldsymbol{h}_k - \boldsymbol{W} \mathrm{vec}(\boldsymbol{X}_k)\|^2 \right)$$

where the classifier matrix $\boldsymbol{W} \in \mathbb{R}^{c \times n_1 n_2}$ is applied to the vectorized sparse representation. An alternative objective is

$$\sum_{k=1}^{N} \left(\left\| \boldsymbol{Y}_k - \boldsymbol{D}_1 \boldsymbol{X}_k \boldsymbol{D}_2^T \right\|_F^2 + \alpha \|\boldsymbol{h}_k - \boldsymbol{W}_1 \boldsymbol{X}_k \boldsymbol{w}_2\|^2 \right)$$

where $\boldsymbol{W}_1 \in \mathbb{R}^{c \times n_1}$ and $\boldsymbol{w}_2 \in \mathbb{R}^{n_2}$. Debate the possible advantages of each formulation and outline optimization algorithms based on block coordinate descent.

Chapter 9
Kernel Dictionary Learning

Abstract Sparse representations are linear by construction, a fact that can hinder their use in classification problems. Building vectors of characteristics from the signals to be classified can overcome the difficulties and is automated by employing kernels, which are functions that quantify the similarities between two vectors. DL can be extended to kernel form by assuming a specific form of the dictionary. DL algorithms have the usual form, comprising sparse coding and dictionary update. We present the kernel versions of OMP and of the most common update algorithms: MOD, SGK, AK-SVD, and K-SVD. The kernel methods use many operations involving a square kernel matrix whose size is equal to the number of signals; hence, their complexities are significantly higher than those of the standard methods. We present two ideas for reducing the size of the problem, the most prominent being that based on Nyström sampling. Finally, we show how kernel DL can be adapted to classification methods involving sparse representations, in particular SRC and discriminative DL.

9.1 Kernels

A limitation of sparse representations using dictionaries is the inherent linear character of the representations. For example, a classification method that is based on the atoms selected for representation, like most of those presented in the previous chapter, is never able to discern between two collinear vectors $\boldsymbol{y}$ and $\alpha\boldsymbol{y}$, since they are represented with the same atoms and hence considered to belong to the same class although this may not be true.

An extension to nonlinearity can be done by associating to a vector $\boldsymbol{y} \in \mathbb{R}^m$ a feature vector $\boldsymbol{\varphi}(\boldsymbol{y}) \in \mathbb{R}^{\tilde{m}}$, where $\boldsymbol{\varphi} : \mathbb{R}^m \to \mathbb{R}^{\tilde{m}}$ is a nonlinear function that combines the elements of the argument vector (the signal); although $\boldsymbol{\varphi}(\boldsymbol{y})$ essentially contains no more information than $\boldsymbol{y}$, even if $\tilde{m} > m$, it may be more appropriate for classification. In order to obtain good performances it is customary to take $\tilde{m} \gg m$ or even feature vectors of infinite length. The sparse representation is made for the feature vector instead of the original vector; the representation is still linear, but the nonlinearity of the function $\boldsymbol{\varphi}$ favors the use of different atoms in the representation.

B. Dumitrescu, P. Irofti, *Dictionary Learning Algorithms and Applications*,
https://doi.org/10.1007/978-3-319-78674-2_9

Example 9.1 Consider the simple feature function $\boldsymbol{\varphi} : \mathbb{R}^2 \to \mathbb{R}^3$ defined by

$$\boldsymbol{\varphi}(\boldsymbol{y}) = \begin{bmatrix} y_1 \\ y_2 \\ y_1 y_2 \end{bmatrix}.$$

It results that

$$\boldsymbol{\varphi}(\alpha \boldsymbol{y}) = \alpha \begin{bmatrix} y_1 \\ y_2 \\ \alpha y_1 y_2 \end{bmatrix}.$$

The vectors $\boldsymbol{\varphi}(\boldsymbol{y})$ and $\boldsymbol{\varphi}(\alpha \boldsymbol{y})$ are not collinear, which leaves open the possibility to place them in different classes using sparse representations. ■

Example 9.2 Figure 9.1 illustrates some simple classification problems with two classes (shown with different colors, pink and lime) that pose difficulties to methods based on sparse representations. (Warning: these artificial problems can be easily solved with heuristic approaches, they only suggest the nature of the difficulty.) In the upper left figure, the classes are defined by clearly different directions, hence sparse representations should perform well; the algorithms from Sects. 8.3 and 8.4, which are based on the representation error, have indeed no trouble with this situation; if $\boldsymbol{x}$ is the sparse representation of $\boldsymbol{y}$, then $-\boldsymbol{x}$ is the sparse representation of $-\boldsymbol{y}$ and the representation error is the same in both cases; so, $\boldsymbol{y}$ and $-\boldsymbol{y}$ are always assigned to the same class, as they should. However, a linear classifier $\boldsymbol{W}$ like that designed by discriminative DL (8.27) has contradictory objectives in the two cases; both $\boldsymbol{W}\boldsymbol{x}$ and $-\boldsymbol{W}\boldsymbol{x}$ should be good approximations of a unit vector (e.g., $\boldsymbol{e}_1$ for the first class), which is impossible. In the lower left figure, directions no longer discriminate between classes; now $\boldsymbol{y}$ and $-\boldsymbol{y}$ belong to different classes. The situation is opposite to the previous one: discriminative DL can succeed (in problems with two classes), while error based methods fail. The right figures illustrate cases where all sparse representation methods fail, since some or all directions define both classes. ■

The feature function $\boldsymbol{\varphi}$ could be tailored for each application, but the design effort, which may include a lot of guessing, would be huge. A general approach for creating feature vectors is clearly preferable. Such an approach is based on kernels. Here, the notion of kernel is that used in machine learning and classification, which is different from that used in Sect. 6.6 (taken from statistics); in fact, it can be seen as a generalization. We present here only basic information on kernels and their use; for a more detailed reading, see [94]. A kernel is a function $\kappa : \mathbb{R}^m \times \mathbb{R}^m \to \mathbb{R}_+$, the value $\kappa(\boldsymbol{y}, \boldsymbol{z})$ measuring the similarity between the vectors $\boldsymbol{y}$ and $\boldsymbol{z}$; with classification in mind, one can imagine that $\boldsymbol{y}$ is an already classified signal and $\boldsymbol{z}$ is a signal yet unclassified; the values $\kappa(\boldsymbol{y}, \boldsymbol{z})$ computed over all classified signals can

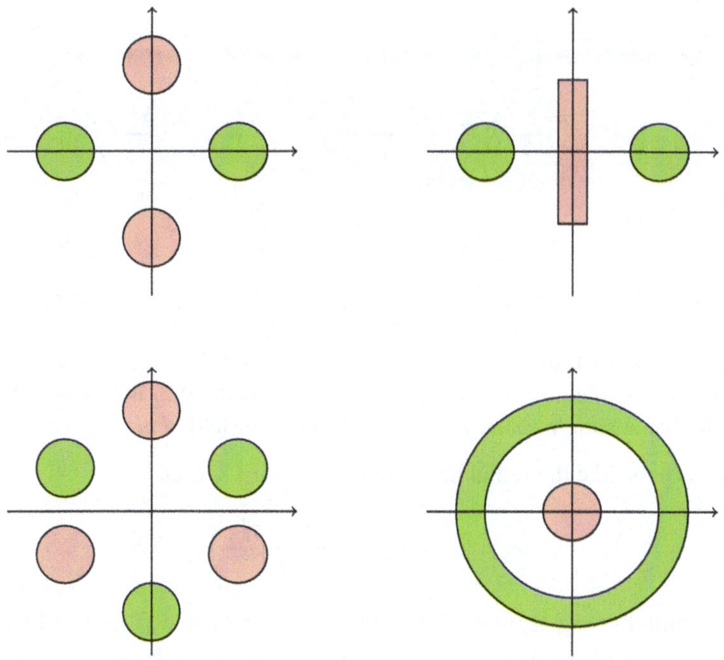

Fig. 9.1 Simple classification examples with two classes

be used for deciding the class to which z belongs. One can define kernels just as a function of $\|y - z\|$, thus appealing to definitions similar to (6.11).

Besides generality, the most important reason for using kernels is complexity. Working directly with feature vectors is possible only if these vectors are short. To cope with the complexity induced by large feature vectors if they were used as such, one employs the *kernel trick*. The trick consists of using Mercer kernels, i.e. kernels that can be expressed as a scalar product of feature vectors:

$$\kappa(\boldsymbol{y}, \boldsymbol{z}) = \boldsymbol{\varphi}(\boldsymbol{y})^T \boldsymbol{\varphi}(\boldsymbol{z}). \tag{9.1}$$

Instead of explicitly computing the scalar product of huge feature vectors, the kernel value depends only on the signals $\boldsymbol{y}$ and $\boldsymbol{z}$.

In practice, one starts directly with kernels that fulfill certain conditions and so can be expressed as in (9.1), without necessarily looking at the form of the feature vector $\boldsymbol{\varphi}(\boldsymbol{y})$. In some cases this is relatively easy.

Example 9.3 The *polynomial kernel* has the form

$$\kappa(\boldsymbol{y}, \boldsymbol{z}) = (\boldsymbol{y}^T \boldsymbol{z} + c)^d \tag{9.2}$$

and small values of d are usually employed. For $d = 2$ one can write

$$
\begin{aligned}
(\boldsymbol{y}^T\boldsymbol{z}+c)^2 &= \left(\sum_{i=1}^{m} y_i z_i + c\right)^2 \\
&= \sum_i y_i^2 z_i^2 + 2\sum_i \sum_{j>i} y_i y_j z_i z_j + 2c\sum_i y_i z_i + c^2 \\
&= \boldsymbol{\varphi}(\boldsymbol{y})^T \boldsymbol{\varphi}(\boldsymbol{z}),
\end{aligned}
$$

where

$$
\boldsymbol{\varphi}(\boldsymbol{y}) = [y_1^2 \ \dots \ y_m^2 \ \sqrt{2}y_1 y_2 \ \dots \ \sqrt{2}y_{m-1}y_m \ \sqrt{2c}y_1 \ \dots \ \sqrt{2c}y_m \ c]^T .
$$

The length of the feature vector is $\tilde{m} = m(m-1)/2 + 2m + 1$, so indeed $\tilde{m} \gg m$. However, the computation of the kernel value through (9.2) needs only few more operations than a simple scalar product for vectors of length m. ■

Other kernels, like the radial basis function (RBF) kernel

$$
\kappa(\boldsymbol{y}, \boldsymbol{z}) = \exp\left(-\frac{\|\boldsymbol{y}-\boldsymbol{z}\|^2}{2\sigma^2}\right), \tag{9.3}
$$

have an infinitely long feature vector. (The same kernel was named Gaussian in Sect. 6.6.) For most practical kernels, the computation of the kernel value has an $O(m)$ complexity, as the above examples suggest.

To adapt DL to the kernel idea, it was proposed in [140] to solve the problem

$$
\begin{aligned}
\min_{\boldsymbol{D},\boldsymbol{X}} \quad & \|\boldsymbol{\varphi}(\boldsymbol{Y}) - \boldsymbol{D}\boldsymbol{X}\|_F^2 \\
\text{s.t.} \quad & \|\boldsymbol{x}_\ell\|_0 \le s, \ \ell = 1:N \\
& \|\boldsymbol{d}_j\| = 1, \ j = 1:n
\end{aligned} \tag{9.4}
$$

Naturally, by $\boldsymbol{\varphi}(\boldsymbol{Y})$ we understand the application of the feature function to each column of $\boldsymbol{Y}$, i.e., to each signal. Note that now the dictionary $\boldsymbol{D}$ has size $\tilde{m} \times n$, hence it is much larger than in (2.2). Solving (9.4) as it stands is certainly difficult, so the problem must be expressed using kernels.

9.2 Dictionary Form

A first result states that the dictionary has a particular form in the kernel DL problem.

Proposition 9.4 ([140]) *The optimal solution of (9.4) has the form*

$$
\boldsymbol{D}^* = \boldsymbol{\varphi}(\boldsymbol{Y})\boldsymbol{A}, \tag{9.5}
$$

where $\boldsymbol{A} \in \mathbb{R}^{N\times n}$.

Proof Let $D^* = \tilde{D} + D_\perp$, where $\tilde{D}$ belongs to the space generated by the columns of $\varphi(Y)$, hence it can be written as in (9.5), and $D_\perp$ is orthogonal to this space, hence

$$\varphi(Y)^T D_\perp = 0, \quad \tilde{D}^T D_\perp = 0.$$

Assuming that X is optimal in (9.4), its objective can be written

$$\begin{aligned}\|\varphi(Y) - DX\|_F^2 &= \mathrm{tr}[(\varphi(Y)^T - X^T D^T)(\varphi(Y) - DX)] \\ &= \mathrm{tr}[\varphi(Y)^T \varphi(Y) - X^T \tilde{D}^T \varphi(Y) - \varphi(Y)^T \tilde{D} X + X^T D_\perp^T D_\perp X]\end{aligned}$$

The last matrix in the above expression is positive semidefinite, so the minimum can be attained only if $D_\perp = 0$. ■

The problem becomes more manageable now, as the matrix A from (9.5), which replaces D as unknown, has size $N \times n$, which depends on the data, not on the feature space dimension. Still, the size is much larger than for the standard DL problem, since we typically have $N \gg m$.

The specific form (9.5) of the dictionary allows an expression of the optimization objective that does not depend on the feature space dimension:

$$\begin{aligned}f(A, X) = \|\varphi(Y) - DX\|_F^2 &= \|\varphi(Y)(I - AX)\|_F^2 \\ &= \mathrm{tr}[(I - X^T A^T) K (I - AX)]\end{aligned} \tag{9.6}$$

where

$$K = \varphi(Y)^T \varphi(Y), \quad K \in \mathbb{R}^{N \times N}, \tag{9.7}$$

is a symmetric positive semidefinite matrix (called also Gram matrix, like all matrices of this form) whose elements are

$$k_{ij} = \varphi(y_i)^T \varphi(y_j) = \kappa(y_i, y_j)$$

and hence can be computed by direct application of the kernel function to all pairs of signals. Computing K is not much more complex than computing $Y^T Y$, since a kernel computation involves operations with vectors of size m. However, this means $O(mN^2)$ operations, which is clearly more than the cost of a single iteration of our usual DL algorithms. Since K must be computed only once, we can for the moment accept the increase in complexity for the promise of introducing nonlinearity in the DL setup.

9.3 Kernel OMP

Let us start with the sparse coding stage looking at the OMP Algorithm 1.1. To adapt it to the kernel approach, we need to look at all scalar products and replace them with expressions that include the kernel matrix (9.7).

It is clear from (9.5) that an atom has the form

$$\boldsymbol{d}_j = \varphi(\boldsymbol{Y})\boldsymbol{a}_j, \tag{9.8}$$

hence the atoms are obtained by a transformation of the columns of $\boldsymbol{A}$. The kernel version of OMP results from this particular form of the atoms. We make clear from the beginning that the atoms $\boldsymbol{d}_j$ must not appear explicitly in the algorithm.

Assume that we are given the dictionary (9.5) through the matrix $\boldsymbol{A}$ and a signal $\boldsymbol{y}$, which is a column of $\boldsymbol{Y}$, and we want to compute the sparse representation $\boldsymbol{x}$ of the feature vector $\varphi(\boldsymbol{y})$ via OMP. So, denoting as usual the sparsity level by s, the problem we attempt to solve is

$$\begin{array}{ll} \min\limits_{\boldsymbol{x}} & \|\varphi(\boldsymbol{y}) - \varphi(\boldsymbol{Y})\boldsymbol{A}\boldsymbol{x}\|^2 \\ \text{s.t.} & \|\boldsymbol{x}\|_0 \le s \end{array} \tag{9.9}$$

We denote $\mathcal{S}$ the support computed by the OMP algorithm at some point. Let us look first at how the optimal representation given by the least squares solution (3.42) is transformed by our particular form of the dictionary. Since $\boldsymbol{D}_{\mathcal{S}} = \varphi(\boldsymbol{Y})\boldsymbol{A}_{\mathcal{S}}$, this representation is

$$\boldsymbol{x}_{\mathcal{S}} = \left(\boldsymbol{D}_{\mathcal{S}}^T\boldsymbol{D}_{\mathcal{S}}\right)^{-1}\boldsymbol{D}_{\mathcal{S}}^T\varphi(\boldsymbol{y}) = \left(\boldsymbol{A}_{\mathcal{S}}^T\boldsymbol{K}\boldsymbol{A}_{\mathcal{S}}\right)^{-1}\boldsymbol{A}_{\mathcal{S}}^T\boldsymbol{k}, \tag{9.10}$$

where

$$\boldsymbol{k} = \varphi(\boldsymbol{Y})^T\varphi(\boldsymbol{y}) \tag{9.11}$$

is the column of $\boldsymbol{K}$ in the same position as $\boldsymbol{y}$ in $\boldsymbol{Y}$. Note that the inner products that appear in the least squares formula eliminate the feature vectors, replaced by the kernel matrix $\boldsymbol{K}$.

We now examine how the next atom is selected. The representation error (residual) corresponding to the support $\mathcal{S}$ is

$$\boldsymbol{e} = \varphi(\boldsymbol{y}) - \boldsymbol{D}_{\mathcal{S}}\boldsymbol{x}_{\mathcal{S}} = \varphi(\boldsymbol{y}) - \varphi(\boldsymbol{Y})\boldsymbol{A}_{\mathcal{S}}\boldsymbol{x}_{\mathcal{S}}. \tag{9.12}$$

We do not have to explicitly compute this vector, since it belongs to the feature vector space. To select the next atom that enters the representation, we need to compute the inner product of the atoms with the residual. Using (9.8) and (9.12), we obtain

Algorithm 9.1: Kernel OMP

Data: dictionary $\boldsymbol{A} \in \mathbb{R}^{N \times n}$
kernel matrix $\boldsymbol{K} \in \mathbb{R}^{N \times N}$ as defined in (9.7)
vector $\boldsymbol{k} \in \mathbb{R}^{N}$ as defined in (9.11) or (9.14)
sparsity level s
stopping error ε
Result: representation support $\mathcal{S}$, solution $\boldsymbol{x}$

Initialize $\mathcal{S} = \emptyset$, $\boldsymbol{e} = \boldsymbol{y}$
while $|\mathcal{S}| < s$ and $\|\boldsymbol{e}\| > \varepsilon$ **do**
  Find new index: $i = \arg\max_{j \notin \mathcal{S}} |\boldsymbol{a}_j^T (\boldsymbol{k} - \boldsymbol{K}\boldsymbol{A}_{\mathcal{S}}\boldsymbol{x}_{\mathcal{S}})|$
  Build new support: $\mathcal{S} \leftarrow \mathcal{S} \cup \{i\}$
  Compute new solution: $\boldsymbol{x}_{\mathcal{S}} = \left(\boldsymbol{A}_{\mathcal{S}}^T \boldsymbol{K} \boldsymbol{A}_{\mathcal{S}}\right)^{-1} \boldsymbol{A}_{\mathcal{S}}^T \boldsymbol{k}$

$$\boldsymbol{d}_j^T \boldsymbol{e} = \boldsymbol{a}_j^T (\boldsymbol{k} - \boldsymbol{K}\boldsymbol{A}_{\mathcal{S}}\boldsymbol{x}_{\mathcal{S}}). \tag{9.13}$$

The selected atom is the one for which the absolute value of the above product is maximum.

The above derivation of the kernel OMP algorithm is valid when the signal $\boldsymbol{y}$ belongs to the set of training signals, which is true in the DL process. However, the algorithm must also work for any other signal, say $\boldsymbol{z}$, like when the dictionary is used after learning. Revisiting relations (9.10) and (9.13) we conclude that they hold by simply redefining

$$\boldsymbol{k} = \varphi(\boldsymbol{Y})^T \varphi(\boldsymbol{z}) = \kappa(\boldsymbol{Y}, \boldsymbol{z}). \tag{9.14}$$

So, we just need to compute the kernel values for all the pairs formed by a training signal and the new signal.

Kernel OMP is summarized as Algorithm 9.1; it was first proposed in [140]; some precursory ideas appeared in [195]. Even if the kernel matrix $\boldsymbol{K}$ is precomputed, the complexity is much higher than that of standard OMP. First of all, the atoms and the vector $\boldsymbol{k}$, which is a substitute of the signal to be represented, have size N instead of m; this is already an increase of at least one order of magnitude, if not more. The multiplication $\boldsymbol{K}\boldsymbol{A}_{\mathcal{S}}$ needs $O(sN)$ operations for the whole algorithm (the kernel matrix must multiply an atom only once, the result being reused in subsequent iterations), but this is negligible. The overall complexity is (see Sect. 1.3) $O(Ns(n + s^2))$, N/m times than that of the standard algorithm.

Of course, in the DL context one may design a kernel Batch OMP, inspired from Algorithm 3.3. The main effort is to precompute the matrix $\boldsymbol{G} = \boldsymbol{D}^T \boldsymbol{D} = \boldsymbol{A}^T \boldsymbol{K} \boldsymbol{A}$, which means $O(n^2 N)$ operations. Note that the size of $\boldsymbol{G}$ is still $n \times n$, like in the standard Batch OMP; hence, only the precomputation increases the cost of the kernel version. The reader is invited to fill all the details in problem **P** 9.1.

Algorithm 9.1 is not the only possibility of solving the sparse representation problem with feature vectors. For example, if the error norm is combined with an ℓ_1 sparsity enforcer like in basis pursuit (1.9), the problem becomes (we denote z the signal to be represented)

$$\min_{x} \quad \|\varphi(z) - \varphi(Y)Ax\|^2 + \lambda\|x\|_1. \tag{9.15}$$

The objective is immediately transformed into

$$\kappa(z, z) - 2k^T Ax + x^T A^T K Ax + \lambda\|x\|_1.$$

The error term is thus transformed into a regular quadratic function and the overall objective is a convex function that can be minimized with various methods, see, for example, [76].

9.4 Kernel DL Algorithms

We can give now kernel versions of the dictionary update stage in several standard DL algorithms, paralleling the developments from Chap. 3. The form (9.6) of the optimization objective suggests formally replacing D with A and searching the inner products that replace the feature vectors with the kernel. For example, we replace the representation error (2.4) with

$$E = I - AX. \tag{9.16}$$

Besides sparse coding, another common operation is atom normalization. Most DL algorithms normalize the atoms at the end of each iteration if they are not already normalized in the dictionary update stage. Due to (9.8), a normalized atom satisfies

$$a_j^T K a_j = 1 \Leftrightarrow \left\| K^{\frac{1}{2}} a_j \right\| = 1, \tag{9.17}$$

and not $\|a_j\| = 1$ as one may hastily conclude. So, atom normalization is done by

$$a_j \leftarrow a_j / \sqrt{a_j^T K a_j}. \tag{9.18}$$

Since it requires $O(N^2)$ operations, it is not negligible in the total cost of an algorithm.

9.4.1 Kernel MOD

Using the trace differentiation rules (3.14) for the objective function (9.6), we obtain the gradient

$$\frac{\partial f}{\partial \boldsymbol{A}} = 2\boldsymbol{K}\boldsymbol{A}\boldsymbol{X}\boldsymbol{X}^T - 2\boldsymbol{K}\boldsymbol{X}^T = -2\boldsymbol{K}\boldsymbol{E}\boldsymbol{X}^T. \tag{9.19}$$

Note that, with the redefined error (9.16), this is the gradient from (3.11) multiplied with $\boldsymbol{K}$; this kind of modification will often appear. For fixed representations $\boldsymbol{X}$, the optimal matrix $\boldsymbol{A}$ is obtained by setting the above gradient to zero (and assuming that $\boldsymbol{K}$ is nonsingular, i.e., positive definite, which is the generic situation), and thus getting

$$\boldsymbol{A} = \boldsymbol{X}^T(\boldsymbol{X}\boldsymbol{X}^T)^{-1}. \tag{9.20}$$

This relation defines the kernel MOD algorithm. Comparing with the optimal dictionary (3.31) of the standard MOD, we note the absence of any signal information in (9.20); this should not be too surprising, since the form (9.5) ensures by default that the dictionary covers the subspace of the signals (via the feature vectors). Combining (9.5) and (9.20), the dictionary update of the kernel MOD is

$$\boldsymbol{D} = \boldsymbol{\varphi}(\boldsymbol{Y})\boldsymbol{X}^T(\boldsymbol{X}\boldsymbol{X}^T)^{-1},$$

which is (3.31) with $\boldsymbol{Y}$ replaced by $\boldsymbol{\varphi}(\boldsymbol{Y})$, exactly as we should expect.

9.4.2 Kernel SGK

We optimize atom $\boldsymbol{d}_j$ while all the other atoms and the representations are fixed. Like in Sect. 3.3.2, $\mathcal{I}_j$ denotes the indices of the signals that use $\boldsymbol{d}_j$ in their representation. Similarly with (3.19) and in harmony with (9.16), we define

$$\boldsymbol{F} = \left[\boldsymbol{I} - \sum_{i \neq j} \boldsymbol{a}_i \boldsymbol{x}_i^T\right]_{\mathcal{I}_j}. \tag{9.21}$$

With only $\boldsymbol{a}_j$ as variable, the representation error is

$$f_j(\boldsymbol{a}_j, \boldsymbol{X}_{j,\mathcal{I}_j}) = \|\boldsymbol{\varphi}(\boldsymbol{Y})(\boldsymbol{F} - \boldsymbol{a}_j\boldsymbol{X}_{j,\mathcal{I}_j})\|_F^2. \tag{9.22}$$

We lighten the notation like in Sect. 3.3.2: $\boldsymbol{a}$ is the atom and $\boldsymbol{x}^T$ the corresponding compressed row from $\boldsymbol{X}$. Proposition 3.3 is transformed as follows:

Proposition 9.5 *The solution to the problem*

$$\min_{\boldsymbol{a}} \quad \left\| \varphi(\boldsymbol{Y})(\boldsymbol{F} - \boldsymbol{a}\boldsymbol{x}^T) \right\|_F^2 \tag{9.23}$$

is

$$\boldsymbol{a} = \frac{\boldsymbol{F}\boldsymbol{x}}{\|\boldsymbol{x}\|^2}. \tag{9.24}$$

Proof Expressing the objective as

$$\left\| \varphi(\boldsymbol{Y})(\boldsymbol{F} - \boldsymbol{a}\boldsymbol{x}^T) \right\|_F^2 = \operatorname{tr}[\boldsymbol{F}^T \boldsymbol{K} \boldsymbol{F}] - 2\boldsymbol{x}^T \boldsymbol{F}^T \boldsymbol{K} \boldsymbol{a} + \|\boldsymbol{x}\|^2 \boldsymbol{a}^T \boldsymbol{K} \boldsymbol{a} \tag{9.25}$$

and equating its derivative with respect to $\boldsymbol{a}$ with zero leads to (9.24) ($\boldsymbol{K}$ is nonsingular). ■

Like in (9.20), the matrix $\boldsymbol{K}$ has vanished; the result is formally identical to (3.22), but with a new meaning for $\boldsymbol{F}$; note also that $\boldsymbol{F}$ has N rows, which leads to a higher complexity than standard SGK. However, the matrix $\boldsymbol{K}$ has not completely disappeared; it is still present in the normalization (9.17) that is performed after all atoms are updated (and, of course, in the sparse coding step). Modifying the SGK dictionary update from Algorithm 3.5 according to (9.24) is a simple task that is left to the reader.

9.4.3 Kernel AK-SVD

The atom update specific to AK-SVD (Sect. 3.5) adds atom normalization to the optimization problem (9.23). The kernel version of Proposition 3.4 is as follows:

Proposition 9.6 *The solution to the problem*

$$\begin{aligned} \min_{\boldsymbol{a}} \quad & \left\| \varphi(\boldsymbol{Y})(\boldsymbol{F} - \boldsymbol{a}\boldsymbol{x}^T) \right\|_F^2 \\ s.t. \quad & \left\| \boldsymbol{K}^{\frac{1}{2}} \boldsymbol{a} \right\| = 1 \end{aligned} \tag{9.26}$$

is

$$\boldsymbol{a} = \frac{\boldsymbol{F}\boldsymbol{x}}{\left\| \boldsymbol{K}^{\frac{1}{2}} \boldsymbol{F}\boldsymbol{x} \right\|} = \frac{\boldsymbol{F}\boldsymbol{x}}{(\boldsymbol{x}^T \boldsymbol{F}^T \boldsymbol{K} \boldsymbol{F}\boldsymbol{x})^{\frac{1}{2}}}. \tag{9.27}$$

Proof Reminding that atom normalization means (9.17), we denote $\tilde{a} = K^{\frac{1}{2}}a$. Inserting the norm constraint $a^T Ka = 1$ into (9.25) leaves us with the task of maximizing $x^T F^T Ka = x^T F^T K^{\frac{1}{2}}\tilde{a}$ under the constraint $\|\tilde{a}\| = 1$. The solution is

$$\tilde{a} = \frac{K^{\frac{1}{2}}Fx}{\left\|K^{\frac{1}{2}}Fx\right\|},$$

which gives (9.27). ■

Now, the kernel matrix appears in the optimal atom formula (9.27). The second relation is more practical, since it works directly with the kernel matrix. In fact, in view of (9.18), relation (9.27) can be simply seen as the computation of the vector Fx followed by normalization. We need $O(N^2)$ operations for computing the atom with (9.27). (The computation of $K^{\frac{1}{2}}$ would be painful, since it requests $O(N^3)$ operations.)

We go now to the optimization of the representation when the atom is fixed, i.e., we present the kernel version of Proposition 3.8.

Proposition 9.7 *The solution to the problem*

$$\min_{x} \left\|\varphi(Y)(F - ax^T)\right\|_F^2 \tag{9.28}$$

is

$$x = F^T Ka. \tag{9.29}$$

Proof Putting $a^T Ka = 1$ into (9.25), taking the derivative with respect to x, and equating it with zero gives (9.29) immediately. ■

Due to the presence of K, the complexity of computing (9.29) is $O(N^2)$.

The dictionary update stage of AK-SVD is summarized in Algorithm 9.2, where relations (9.27) and (9.29) are written with the expression of the current atom and its corresponding compressed representations. Otherwise, it is very similar with AK-SVD; compare with Algorithm 3.7. The cost of the dictionary update is $O(N^2n)$, compared with $O(smN)$ for standard AK-SVD. We cannot take full advantage anymore of the sparsity of X. Even if the size of F reflects the sparsity and so multiplications with F are overall efficient, costing $O(sN^2)$, there is no shortcut for the multiplications with K.

Algorithm 9.2: Kernel AK-SVD Dictionary Update

Data: kernel matrix $\boldsymbol{K} \in \mathbb{R}^{N \times N}$
current dictionary $\boldsymbol{A} \in \mathbb{R}^{N \times n}$
representation matrix $\boldsymbol{X} \in \mathbb{R}^{n \times N}$
Result: updated dictionary $\boldsymbol{A}$

Compute error $\boldsymbol{E} = \boldsymbol{I} - \boldsymbol{A}\boldsymbol{X}$
for $j = 1 : n$ **do**
  Modify error: $\boldsymbol{F} = \boldsymbol{E}_{\mathcal{I}_j} + \boldsymbol{a}_j \boldsymbol{X}_{j,\mathcal{I}_j}$
  Update atom: $\boldsymbol{a}_j = \boldsymbol{F}\boldsymbol{X}_{j,\mathcal{I}_j}^T$
  Normalize atom: $\boldsymbol{a}_j \leftarrow (\boldsymbol{a}_j^T \boldsymbol{K} \boldsymbol{a}_j)^{\frac{1}{2}}$
  Update representation: $\boldsymbol{X}_{j,\mathcal{I}_j}^T = \boldsymbol{F}^T \boldsymbol{K} \boldsymbol{a}_j$
  Recompute error: $\boldsymbol{E}_{\mathcal{I}_j} = \boldsymbol{F} - \boldsymbol{a}_j \boldsymbol{X}_{j,\mathcal{I}_j}$

9.4.4 *Kernel K-SVD*

For the kernel version of K-SVD, we need to optimize the atom and the representation simultaneously, providing an analogue of Proposition 3.7. However, for reasons that will be clear below, the problem cannot be cast directly as a rank-1 approximation with obvious solution.

Proposition 9.8 *Consider the problem*

$$\begin{aligned} \min_{\boldsymbol{a},\boldsymbol{x}} \quad & \left\| \varphi(\boldsymbol{Y})(\boldsymbol{F} - \boldsymbol{a}\boldsymbol{x}^T) \right\|_F^2 \\ s.t. \quad & \left\| \boldsymbol{K}^{\frac{1}{2}} \boldsymbol{a} \right\| = 1 \end{aligned} \tag{9.30}$$

In the eigenvalue decomposition

$$\boldsymbol{F}^T \boldsymbol{K} \boldsymbol{F} = \boldsymbol{U} \boldsymbol{\Lambda} \boldsymbol{U}^T, \tag{9.31}$$

with orthogonal $\boldsymbol{U}$ and diagonal $\boldsymbol{\Lambda}$, assume that the first diagonal element of $\boldsymbol{\Lambda}$, denoted σ_1^2, is the largest eigenvalue; denote $\boldsymbol{u}_1$ the first column of $\boldsymbol{U}$. Then, the solution to problem (9.30) is

$$\boldsymbol{a} = \sigma_1^{-1} \boldsymbol{F} \boldsymbol{u}_1, \quad \boldsymbol{x} = \sigma_1 \boldsymbol{u}_1. \tag{9.32}$$

Proof Inserting the optimal representations (9.29) into (9.25) and denoting again $\tilde{\boldsymbol{a}} = \boldsymbol{K}^{\frac{1}{2}} \boldsymbol{a}$, we obtain

$$\left\| \varphi(\boldsymbol{Y})(\boldsymbol{F} - \boldsymbol{a}\boldsymbol{x}^T) \right\|_F^2 = \operatorname{tr}[\boldsymbol{F}^T \boldsymbol{K} \boldsymbol{F}] - \tilde{\boldsymbol{a}}^T \boldsymbol{K}^{\frac{1}{2}} \boldsymbol{F} \boldsymbol{F}^T \boldsymbol{K}^{\frac{1}{2}} \tilde{\boldsymbol{a}}. \tag{9.33}$$

Using the SVD decomposition $\boldsymbol{F}^T \boldsymbol{K}^{\frac{1}{2}} = \boldsymbol{U\Sigma V}^T$, the optimum is attained for

$$\tilde{\boldsymbol{a}} = \boldsymbol{v}_1 \Leftrightarrow \boldsymbol{a} = \boldsymbol{K}^{-\frac{1}{2}} \boldsymbol{v}_1.$$

The expression (9.29) of the optimal representation leads to

$$\boldsymbol{x} = \boldsymbol{F}^T \boldsymbol{K a} = \boldsymbol{U\Sigma V}^T \boldsymbol{v}_1 = \sigma_1 \boldsymbol{u}_1.$$

So, we have obtained the second relation from (9.32).

To obtain the first relation and to avoid the computation of $\boldsymbol{K}^{\frac{1}{2}}$, we note that

$$\boldsymbol{F}^T \boldsymbol{K F} = \boldsymbol{U\Sigma\Sigma}^T \boldsymbol{U}^T,$$

hence $\boldsymbol{U}$ from (9.31) indeed appears in the SVD of $\boldsymbol{F}^T \boldsymbol{K}^{\frac{1}{2}}$ and the first diagonal element of $\boldsymbol{\Lambda} = \boldsymbol{\Sigma\Sigma}^T$ is indeed σ_1^2. The optimal atom results from (9.27) and is

$$\boldsymbol{a} = \frac{\boldsymbol{F x}}{\left\| \boldsymbol{K}^{\frac{1}{2}} \boldsymbol{F x} \right\|} = \frac{\sigma_1 \boldsymbol{F u}_1}{\left\| \boldsymbol{V \Sigma}^T \boldsymbol{U}^T \cdot \sigma_1 \boldsymbol{u}_1 \right\|} = \sigma_1^{-1} \boldsymbol{F u}_1,$$

which completely proves (9.32). ■

The fact that we need only an eigenvector instead of two singular vectors does not make the solution easier to compute, since these are practically equivalent problems. The only benefit is that we can work directly with $\boldsymbol{K}$ and there is no need for $\boldsymbol{K}^{\frac{1}{2}}$. In any case, kernel K-SVD is considerably more complex than kernel AK-SVD, hence it should be used only if the latter (or other algorithms) are not satisfactory.

Remark 9.9 Another way of proving Proposition 9.8 is by writing its objective as

$$\|\varphi(\boldsymbol{Y})\boldsymbol{F} - \varphi(\boldsymbol{Y})\boldsymbol{a} \cdot \boldsymbol{x}^T\|_F^2 \tag{9.34}$$

and seeing it as a rank-1 approximation problem, like in the standard K-SVD approach. This is possible because $\varphi(\boldsymbol{Y})\boldsymbol{a}$ is a vector that, although not free, belongs to the range of the matrix $\varphi(\boldsymbol{Y})\boldsymbol{F}$. So, ignoring the huge (possibly infinite) size, we could use the SVD decomposition

$$\varphi(\boldsymbol{Y})\boldsymbol{F} = \tilde{\boldsymbol{U}}\tilde{\boldsymbol{\Sigma}}\tilde{\boldsymbol{V}}^T$$

to obtain the optimal rank-1 approximation

$$\varphi(\boldsymbol{Y})\boldsymbol{a} = \tilde{\boldsymbol{u}}_1, \quad \boldsymbol{x} = \tilde{\sigma}_1 \tilde{\boldsymbol{v}}_1.$$

Noticing that

$$F^T K F = \tilde{V}\tilde{\Sigma}^T \tilde{\Sigma} \tilde{V}^T$$

and identifying with (9.31), it results that $\tilde{V} = U$ and $\tilde{\sigma}_1 = \sigma_1$. Hence the optimal representation x is the same as in (9.32). The optimal atom results as shown at the end of the proof of Proposition 9.8. ■

9.5 Size Reduction

The only drawback of the above kernel algorithms is the dimension of matrix K, which is $N \times N$. So, already for N in the thousands, which is a rather common number of signals, the complexity of the operations involving K and even its storing space may reach an unacceptable level.

9.5.1 *Nyström Sampling*

A cure was proposed in [79], namely to approximate (9.4) by approximating $\varphi(Y)$ with a matrix $\hat{Y}$ of size $\hat{m} \times N$, with $m < \hat{m} \ll \tilde{m}$. We can see $\hat{Y}$ as a compressed version of the feature vectors $\varphi(Y)$; at the same time, it is still an extension of the signal matrix Y. Since the compressed feature vectors have a reasonable size, the problem can be solved in the usual DL framework, with one of the standard algorithms acting with $\hat{Y}$ as input.

Inspired by (9.7), the immediate algorithm is to compute a rank $\hat{m}$ approximation $\hat{K}$ of K and then factorize it as $\hat{K} = \hat{Y}^T \hat{Y}$. Since K is a positive semidefinite matrix, the optimal way to compute this approximation starts by computing the eigenvalue decomposition $K = U \Sigma U^T$, with the eigenvalues ordered decreasingly on the diagonal of Σ. Denoting $\hat{\Sigma}$ the $\hat{m} \times \hat{m}$ upper left block of Σ and $\hat{U}$ the first $\hat{m}$ columns of U, the optimal rank-$\hat{m}$ approximation of K is $\hat{K} = \hat{U}\hat{\Sigma}\hat{U}^T$, thus obtaining

$$\hat{Y} = \hat{\Sigma}^{\frac{1}{2}} \hat{U}^T. \tag{9.35}$$

This direct approach is hardly feasible, because it implies the eigenvalue decomposition of K, which is a large matrix.

A simpler solution is given by the Nyström method, which amounts to sampling the matrix K, with the purpose of cheaply computing a low rank approximation. We present only the basic idea; for more details and alternatives, see [114]. Essentially, the method is composed of the following operations. First, p columns of K are chosen randomly, with $\hat{m} \le p \ll N$. Then, with no loss of generality, the matrix is rearranged such that these columns are the first and thus we have

$$K = \begin{bmatrix} W & S^T \\ S & B \end{bmatrix}, \quad C \triangleq \begin{bmatrix} W \\ S \end{bmatrix}.$$

A rank-p approximation $\check{K}$ of K is

$$\check{K} = CW^{-1}C^T. \tag{9.36}$$

The first p columns of $\check{K}$ coincide with those of K; so, we could say that the approximation is optimal in matching the selected columns. We assume that W is invertible; if not, the pseudo-inverse can be used instead of W^{-1}, but this complicates the description. Using the eigenvalue decomposition $W = Q\Lambda Q^T$ (note that this is a $p \times p$ matrix, hence "small"), the matrix (9.36) becomes

$$\check{K} = CQ\Lambda^{-1}Q^TC^T.$$

Assuming that the eigenvalues of Λ are ordered decreasingly, we denote $\hat{\Lambda}$ the $\hat{m} \times \hat{m}$ upper left block of Λ and by $\hat{Q}$ the first $\hat{m}$ columns of Q. The approximation of $\varphi(Y)$ is then

$$\hat{Y} = \hat{\Lambda}^{-\frac{1}{2}}\hat{Q}^TC^T. \tag{9.37}$$

The role of p in the above construction is to control the trade-off between the quality of approximation and complexity. When $p = \hat{m}$, the complexity is the lowest possible, but the approximation may highly depend on the selected columns of K. At the other end, when $p = N$, the complexity is the highest but the solution is optimal, since in this case (9.37) coincides with the SVD-based approximation (9.35).

A more complex alternative to the random selection of the p columns of K (and implicitly of $\varphi(z)$) is to try to choose representative columns. The problem is known as Column Subset Selection and can be approximately solved via different deterministic algorithms. One of them is the greedy solution proposed in [71]; the bibliography of the cited work points to other interesting algorithms.

We summarize the Nyström sampling approach in Algorithm 9.3. The loops 2 and 3 compute the matrix C, which is an $N \times p$ block of the kernel matrix K; its computation requires $O(mpN)$ operations. The eigenvalue decomposition in line 5 needs $O(p^3)$ operations and the matrix multiplication from line 6 needs $O(\hat{m}pN)$. Since $\hat{m} > m$, the DL algorithm has also larger complexity than the standard linear approach using directly the signals Y.

In Algorithm 9.3, we have denoted the dictionary with $\hat{D}$ to stress that it matches the compressed feature vectors $\hat{Y}$, with length $\hat{m}$ signals. We are still left with at least a question regarding this dictionary: how do we use it to represent a new signal z? The answer is that we use $\hat{D}$ as we always did for sparse representations, the target being not z, but the approximation $\hat{z} \in \mathbb{R}^{\hat{m}}$ of the feature vector $\varphi(z)$.

Algorithm 9.3: Kernel DL with Reduced Feature Vectors

Data: signals set $\boldsymbol{Y} \in \mathbb{R}^{m \times N}$
kernel function $\kappa(\cdot, \cdot)$
reduced feature vectors size $\hat{m}$
Nyström sampling size p, with $p \geq \hat{m}$
number of atoms n
Result: dictionary $\hat{\boldsymbol{D}} \in \mathbb{R}^{\hat{m} \times n}$

Reorder randomly the columns of $\boldsymbol{Y}$
for $i = 1 : N$ **do**
  for $j = 1 : p$ **do**
    Compute $c_{ij} = \kappa(\boldsymbol{y}_i, \boldsymbol{y}_j)$
Compute sorted eigenvalue decomposition $\boldsymbol{C}(1 : p, 1 : p) = \boldsymbol{Q}\boldsymbol{\Lambda}\boldsymbol{Q}^T$
Compute reduced feature vectors $\hat{\boldsymbol{Y}} = \boldsymbol{\Lambda}(1 : \hat{m}, 1 : \hat{m})^{-\frac{1}{2}}\boldsymbol{Q}(1 : p, 1 : \hat{m})^T \boldsymbol{C}^T$
Design dictionary $\hat{\boldsymbol{D}}$ by calling preferred DL algorithm with inputs $\hat{\boldsymbol{Y}}$ and n

To compute it, let us look at the expression of the available training signals. Since $\boldsymbol{C}$ represents the first p columns of $\boldsymbol{K}$, the column ℓ of the compressed feature matrix (9.37), which is an approximation of $\boldsymbol{\varphi}(\boldsymbol{y}_\ell)$, is

$$\hat{\boldsymbol{y}}_\ell = \hat{\boldsymbol{\Lambda}}^{-\frac{1}{2}} \hat{\boldsymbol{Q}}^T [\kappa(\boldsymbol{y}_1, \boldsymbol{y}_\ell) \ \ldots \ \kappa(\boldsymbol{y}_p, \boldsymbol{y}_\ell)]^T.$$

So, we approximate the feature vector $\boldsymbol{\varphi}(\boldsymbol{z})$ of the new signal with

$$\hat{\boldsymbol{z}} = \hat{\boldsymbol{\Lambda}}^{-\frac{1}{2}} \hat{\boldsymbol{Q}}^T [\kappa(\boldsymbol{y}_1, \boldsymbol{z}) \ \ldots \ \kappa(\boldsymbol{y}_p, \boldsymbol{z})]^T.$$

So, when representing the signal $\boldsymbol{z}$ via this approach, the extra computation required by $\hat{\boldsymbol{z}}$ is only $O(mp)$ operations, both for kernel values computation and for the matrix vector multiplication (note that p has still an effect on complexity!). This is better than the computation (9.14) appearing in the kernel algorithms, which requires $O(mN)$ operations. If OMP is used for representation, the size $\hat{m}$ of $\hat{\boldsymbol{z}}$ makes the representation clearly faster than the kernel OMP Algorithm 9.1. The factor N disappears from the complexity expression.

9.5.2 *Changing the Target*

Another complexity reduction was proposed in [1], this time by avoiding many computations involving $\boldsymbol{K}$, moving the problem from the signal space into the coefficient space. By denoting $\boldsymbol{C} = \boldsymbol{A}\boldsymbol{X} \in \mathbb{R}^{N \times N}$, the objective (9.6) of kernel DL becomes

$$\|\boldsymbol{\varphi}(\boldsymbol{Y}) - \boldsymbol{\varphi}(\boldsymbol{Y})\boldsymbol{C}\|_F^2. \tag{9.38}$$

The idea is to compute $\boldsymbol{C}$ in the beginning and then optimize $\boldsymbol{A}$ and $\boldsymbol{X}$ that satisfy $\boldsymbol{AX} \approx \boldsymbol{C}$, thus partially avoiding further use of $\boldsymbol{Y}$ or $\boldsymbol{K}$. The trivial minimizer $\boldsymbol{C} = \boldsymbol{I}$ of (9.38) is not acceptable since the matrix $\boldsymbol{C}$ should actually have rank n, by construction. Oppositely, one can see $\boldsymbol{C}$ as a matrix of coefficients that linearly produce each signal as a combination of *other* signals, thus being the solution of

$$\begin{aligned} \min_{\boldsymbol{C}} \quad & \|\boldsymbol{\varphi}(\boldsymbol{Y}) - \boldsymbol{\varphi}(\boldsymbol{Y})\boldsymbol{C}\|_F^2 \\ \text{s.t.} \quad & c_{\ell\ell} = 0, \ \ell = 1 : N \end{aligned} \tag{9.39}$$

This problem can be solved approximately by applying the Kernel OMP Algorithm 9.1 for the representation of each feature vector $\boldsymbol{\varphi}(\boldsymbol{y}_\ell)$ using all other feature vectors, with an appropriate error threshold. Thus, for signal $\boldsymbol{y}_\ell$, the problem is to find the $(N-1)$-length vector $\tilde{\boldsymbol{c}}_\ell$ such that $\left\|\boldsymbol{\varphi}(\boldsymbol{y}_\ell) - \boldsymbol{\varphi}(\tilde{\boldsymbol{Y}}_\ell)\tilde{\boldsymbol{c}}_\ell\right\| \le \epsilon$, where $\tilde{\boldsymbol{Y}}_\ell$ is the signal matrix $\boldsymbol{Y}$ with the ℓth column removed and ϵ is the signal error threshold. The ℓth column of $\boldsymbol{C}$ is obtained by inserting a zero before the ℓth position of $\tilde{\boldsymbol{c}}_\ell$. Although the resulting matrix $\boldsymbol{C}$ will not have a rank exactly equal to n, it will have a distribution of singular values that is far from that of the unit matrix.

Once in possession of $\boldsymbol{C}$, the DL problem becomes standard, since it has the form

$$\begin{aligned} \min_{\boldsymbol{A},\boldsymbol{X}} \quad & \|\boldsymbol{C} - \boldsymbol{AX}\|_F^2 \\ \text{s.t.} \quad & \|\boldsymbol{x}_\ell\|_0 \le s, \ \ell = 1 : N \\ & \|\boldsymbol{a}_j\| = 1, \ j = 1 : n \end{aligned} \tag{9.40}$$

The typical alternating optimization can be used. However, when the dictionary $\boldsymbol{A}$ is fixed, the representation matrix $\boldsymbol{X}$ can be found using kernel OMP for solving problems like (9.9), similarly to the standard kernel DL algorithms. Only when $\boldsymbol{X}$ is fixed, the problem (9.40) is solved for $\boldsymbol{A}$. This can be done atom by atom, in AK-SVD style. The method from [1] orders the atoms in order of the size of the residual matrix. The complexity of updating an atom is thus at most $O(Nn)$ instead of the $O(N^2)$ needed by the kernel AK-SVD; moreover, since $\boldsymbol{X}$ is sparse, the complexity is actually much lower.

9.6 Classification with Kernel DL

The use of kernel DL methods for classification is relatively straightforward. Posing the classification problem exactly like in Chap. 8, the classification algorithms presented there have to be modified by replacing DL with kernel DL for learning the dictionaries and OMP with kernel OMP for computing representations. We present here only few methods, to illustrate the necessary transformations.

9.6.1 Kernel SRC

We start the discussion with the SRC method from Sect. 8.3; remind that DL is not involved here, but only sparse representation. Extending SRC to feature vectors was proposed in [76, 217, 222]; the dictionary is

$$\boldsymbol{D} = [\varphi(\boldsymbol{Y}_1)\ \varphi(\boldsymbol{Y}_2)\ \dots\ \varphi(\boldsymbol{Y}_c)]\boldsymbol{A} \triangleq \varphi(\boldsymbol{Y})\boldsymbol{A}, \tag{9.41}$$

where the diagonal matrix $\boldsymbol{A}$ has the role of normalizing the atoms. Since

$$\|\varphi(\boldsymbol{y}_j)\| = \sqrt{\varphi(\boldsymbol{y}_j)^T\varphi(\boldsymbol{y}_j)} = \sqrt{k_{j,j}},$$

it results that $\boldsymbol{A} = \text{diag}(1/\sqrt{k_{j,j}})_{j=1:n}$. Given a test signal $\boldsymbol{z}$, we find its sparse representations $\boldsymbol{x}$ using kernel OMP, called with the vector $\boldsymbol{k}$ from (9.14), or by solving the kernel Basis Pursuit problem (9.15). Then, SRC proceeds as in the standard case. The reader is invited to write the computable form of the expression of the error (8.3), depending on the kernel matrix $\boldsymbol{K}$ and the vector $\boldsymbol{k}$, see problem **P** 9.5.

The SRC version with DL trains dictionaries $\boldsymbol{D}_i$ for each class, working then with the concatenated dictionary (8.5) instead of (9.41). A class dictionary has the form $\varphi(\boldsymbol{Y}_i)\boldsymbol{A}_i$, where the dictionary $\boldsymbol{A}_i$ with n_i atoms is designed using the kernel AK-SVD Algorithm 9.2 applied to the training data $\boldsymbol{Y}_i$. The overall dictionary is

$$\boldsymbol{D} = [\varphi(\boldsymbol{Y}_1)\boldsymbol{A}_1\ \varphi(\boldsymbol{Y}_2)\boldsymbol{A}_2\ \dots\ \varphi(\boldsymbol{Y}_c)\boldsymbol{A}_c] \triangleq \varphi(\boldsymbol{Y})\boldsymbol{A}, \tag{9.42}$$

where

$$\boldsymbol{A} = \text{diag}(\boldsymbol{A}_1, \dots, \boldsymbol{A}_c) \in \mathbb{R}^{N \times n}.$$

For classifying a test signal $\boldsymbol{z}$, we can run kernel OMP with this dictionary. However, such an approach would not take advantage of the block diagonal structure of $\boldsymbol{A}$. A more efficient kernel OMP version can be derived, see problem **P** 9.6.

Remark 9.10 A modification is proposed in [222] to reduce the size of the problem. Since the feature vectors are large (possibly infinite), we can try to compress them by a simple multiplication with a matrix $\boldsymbol{P}$ with $\hat{m}$ rows, thus working with $\boldsymbol{P}\varphi(\boldsymbol{Y})$ instead of $\varphi(\boldsymbol{Y})$. This technique is somewhat akin to compressed sensing [24], remind Sect. 2.4.4. However, a general matrix would not be satisfactory, by the simple reason that we cannot efficiently compute $\boldsymbol{P}\varphi(\boldsymbol{Y})$; the solution is to take

$\boldsymbol{P} = \boldsymbol{B}\varphi(\boldsymbol{Y})^T$, with $\boldsymbol{B} \in \mathbb{R}^{\hat{m} \times N}$. So, instead of minimizing

$$\|\varphi(\boldsymbol{y}) - \varphi(\boldsymbol{Y})\boldsymbol{x}\|,$$

we minimize

$$\|\boldsymbol{P}\varphi(\boldsymbol{y}) - \boldsymbol{P}\varphi(\boldsymbol{Y})\boldsymbol{x}\| = \|\boldsymbol{B}\kappa(\boldsymbol{Y}, \boldsymbol{y}) - \boldsymbol{B}\boldsymbol{K}\boldsymbol{x}\| \tag{9.43}$$

with sparse $\boldsymbol{x}$. The two problems are not equivalent, but the second is a standard sparse representation problem with an $\hat{m} \times N$ matrix; hence, kernel OMP or other specialized algorithm is not necessary. Assuming that the matrix $\boldsymbol{B}\boldsymbol{K}$ is precomputed, then the complexity of solving (9.43) is clearly lower. ■

9.6.2 An Example

To illustrate the ability of kernel DL methods to solve classification problems, we consider the two-class problem from the lower right of Fig. 9.1, which cannot be solved by classifiers based on sparse representations. We generate $N_1 = N_2 = 100$ signals for each class, an example being shown in the upper left of Fig. 9.2. The inner (red) class is obtained with random vectors following a Gaussian distribution with zero mean and standard deviation equal to 0.1. The outer (green) class is formed with vectors with random direction; their magnitude obeys to a Gaussian distribution with mean equal to 1 and standard deviation equal to 0.1.

We trained two classifiers for this problem, using an RBF kernel (9.3) with $\sigma = 1$. The sparsity level is $s = 1$, which is actually the only choice, since the signals have length $m = 2$.

The first classifier is SRC with kernel DL, described in Sect. 9.6.1. The class dictionaries are designed with the kernel AK-SVD Algorithm 9.2. The classification results for many test points are shown in Fig. 9.2, for different number of atoms: $n_1 = n_2 = 3, 5, 8$ in the upper right, lower left, and lower right panels. We see that the actual boundary between the two classes appears to be a polygon with n sides, a reminiscence of the linear character of the sparse representation. (In any case, we would not expect the boundary to be circular, due to the peculiarities of the training signals.) However, besides this geometric characteristic, the points with small magnitude are in the "red" class, while those with large magnitude are in the "green" class, which shows that the classifier does the right job. We also note that the classification results depend also on the initial dictionaries used in kernel AK-SVD. Hence, the boundary between classes is different for different initializations; however, a larger number of atoms diminishes the variability.

The second classifier uses the Nyström method for dimensional reduction described in Sect. 9.5.1. The first size parameter is $p = 200$, which is the maximum possible value (the total number of training signals). We try two values for $\hat{m}$,

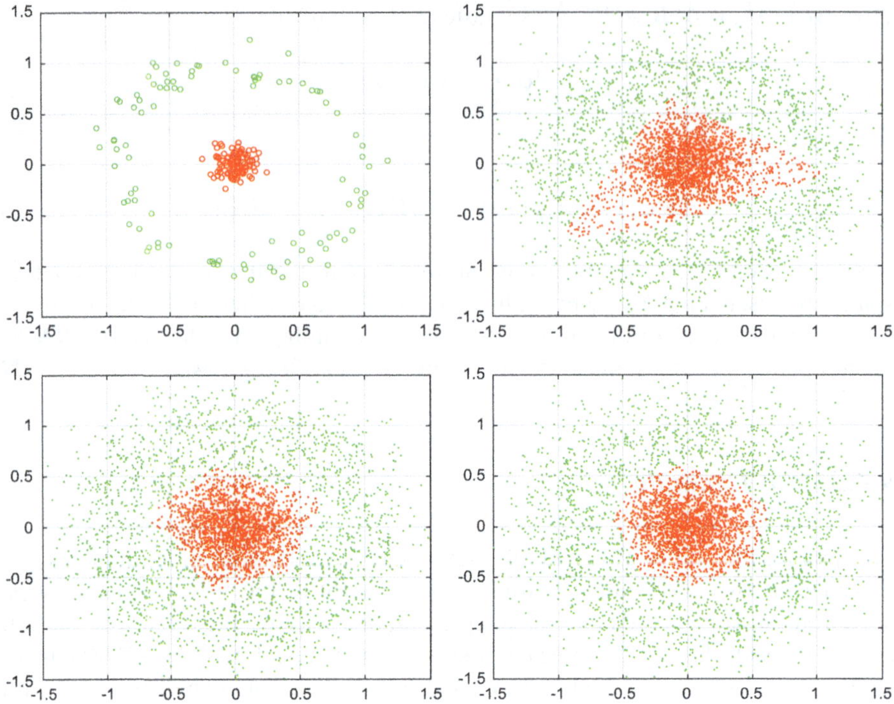

Fig. 9.2 Classification problem and results with SRC, dictionaries trained with kernel AK-SVD. Up left: training signals. Results with 3 atoms/class (up right), 5 atoms/class (down left), 8 atoms/class (down right)

namely 20 and 30. The reduced matrix $\hat{\boldsymbol{Y}}$ is fed to the discriminative DL algorithm that minimizes (8.26), with $\alpha = 0.1$; the number of atoms is $n = 2\hat{m}$. Figure 9.3 shows the results. As expected, larger $\hat{m}$ and n lead to better approximation and classification, respectively. In the left figure, where $\hat{m} = 20$, besides the ellipsoidal form of the inner class, there are a few points with large magnitude that are wrongly put in the "red" class (three of them are in the lower part of the figure). This is a typical event for small $\hat{m}$. When $\hat{m}$ is large, the classification is very good, comparable to that provided by the kernel methods.

9.6.3 Kernel Discriminative DL

As discussed in Sect. 8.5.2, discriminative DL produces a classifier matrix $\boldsymbol{W}$ that aims to reconstruct the label matrix $\boldsymbol{H}$ from the sparse representations. The optimization problem (8.26) combines the sparse representation error with the label reconstruction error. Its adaptation to the kernel idea has the form

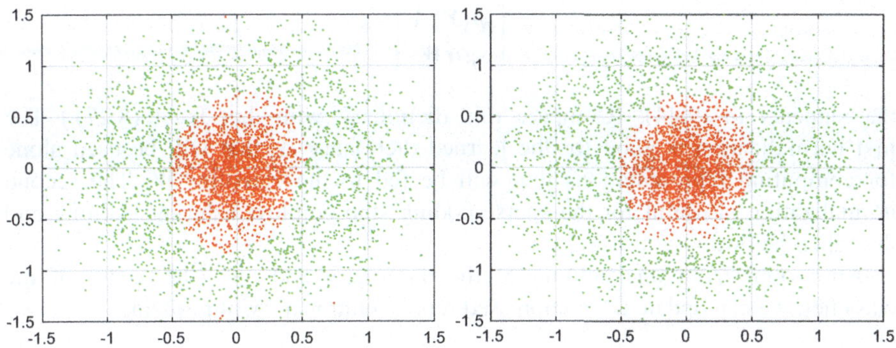

Fig. 9.3 Classification results using Nyström method. Left: $\hat{m} = 20$, $n = 40$. Right: $\hat{m} = 30$, $n = 60$

$$\min_{\boldsymbol{A},\boldsymbol{W},\boldsymbol{X}} \|\varphi(\boldsymbol{Y}) - \varphi(\boldsymbol{Y})\boldsymbol{A}\boldsymbol{X}\|_F^2 + \alpha \|\boldsymbol{H} - \boldsymbol{W}\boldsymbol{X}\|_F^2 \tag{9.44}$$

Although it seems no longer possible to build an objective like (8.27) and immediately call a kernel DL algorithm, the derivation of a specific algorithm is not complicated. In particular, the AK-SVD algorithm can be easily adapted.

Let us look first at the *dictionary update* step. We note that when $\boldsymbol{X}$ is fixed, the variables $\boldsymbol{A}$ and $\boldsymbol{W}$ are decoupled. So, to optimize atom $\boldsymbol{a}$ of $\boldsymbol{A}$, we use the notations from Sect. 9.4.2 and simply conclude that the result is given by Proposition 9.5. To optimize column $\boldsymbol{w}_j$ of the classifier matrix $\boldsymbol{W}$, we proceed like in Sect. 3.3.2; although we work now with the label matrix $\boldsymbol{H}$ instead of the signals matrix and with the classifier $\boldsymbol{W}$ instead of the dictionary, the functional form of the objective is the same, hence the solution is the same. So, denoting again $\mathcal{I}_j$ the indices of the nonzero elements on row j of the representation matrix $\boldsymbol{X}$ and defining

$$\boldsymbol{G} = \left[\boldsymbol{H} - \sum_{i \neq j} \boldsymbol{w}_i \boldsymbol{x}_i^T \right]_{\mathcal{I}_j}, \tag{9.45}$$

we end up with a formulation like in Proposition 3.3. Re-denoting $\boldsymbol{w}$ the current column of the classifier, its optimal expression is

$$\boldsymbol{w} = \frac{\boldsymbol{G}\boldsymbol{x}}{\|\boldsymbol{x}^2\|}, \tag{9.46}$$

which is a virtual copy of (3.22).

These updates are made in SGK style. If we aim to an AK-SVD version, normalization can be made in two meaningful ways. The first is to divide both $\boldsymbol{a}$ and $\boldsymbol{w}$ with $\sqrt{\boldsymbol{a}^T \boldsymbol{K}\boldsymbol{a}}$, such that the dictionary atoms have unit norm, like in Algorithm 9.2. The second way is to normalize by $\sqrt{\boldsymbol{a}^T \boldsymbol{K}\boldsymbol{a} + \alpha \boldsymbol{w}^T \boldsymbol{w}}$, considering the norm of the atoms of the extended dictionary

$$\begin{bmatrix} \varphi(Y)A \\ \sqrt{\alpha}W \end{bmatrix}$$

that appears in (9.44). (Compare this dictionary with that from (8.27).) The first normalization is good for the learned dictionary $\boldsymbol{A}$, that will be used alone for computing representations that will be fed to the classifier $\boldsymbol{W}$. The second normalization is useful for the sparse coding step employed during learning and described below.

After optimizing one column of the dictionary and one (the same) of the classifier, we can update the compressed representation $\boldsymbol{x}$ by minimizing

$$\left\| \varphi(\boldsymbol{Y})(\boldsymbol{F} - \boldsymbol{a}\boldsymbol{x}^T) \right\|_F^2 + \alpha \left\| \boldsymbol{G} - \boldsymbol{w}\boldsymbol{x}^T \right\|_F^2. \tag{9.47}$$

The solution is

$$\boldsymbol{x} = \frac{\boldsymbol{F}^T \boldsymbol{K}\boldsymbol{a} + \alpha \boldsymbol{G}^T \boldsymbol{w}}{1+\alpha} \tag{9.48}$$

and is in fact the weighted combination of (9.29) and (3.37).

The *sparse coding* step can be solved by building a special OMP. Denoting $\boldsymbol{y}$ a training signal and $\boldsymbol{h}$ its associated label vector, the sparse representation $\boldsymbol{x}$ (note that now $\boldsymbol{x}$ denotes a column of the representation matrix $\boldsymbol{X}$) results from minimizing

$$\|\varphi(\boldsymbol{y}) - \varphi(\boldsymbol{Y})\boldsymbol{A}\boldsymbol{x}\|^2 + \alpha \|\boldsymbol{h} - \boldsymbol{W}\boldsymbol{x}\|^2, \tag{9.49}$$

constrained to $\|\boldsymbol{x}\|_0 \le s$, where s is the given sparsity level. We follow the developments from Sect. 9.3, modifying as appropriate to (9.49).

Denoting $\mathcal{S}$ the current support computed by the OMP algorithm, the optimal representation with that support is, similarly with (9.10),

$$\boldsymbol{x}_\mathcal{S} = \left(\boldsymbol{A}_\mathcal{S}^T \boldsymbol{K} \boldsymbol{A}_\mathcal{S} + \alpha \boldsymbol{W}_\mathcal{S}^T \boldsymbol{W}_\mathcal{S} \right)^{-1} (\boldsymbol{A}_\mathcal{S}^T \boldsymbol{k} + \alpha \boldsymbol{W}_\mathcal{S}^T \boldsymbol{h}), \tag{9.50}$$

where $\boldsymbol{k}$ is given in (9.11).

The selection criterion (9.13) for atom j becomes

$$\boldsymbol{a}_j^T (\boldsymbol{k} - \boldsymbol{K}\boldsymbol{A}_\mathcal{S}\boldsymbol{x}_\mathcal{S}) + \alpha \boldsymbol{w}_j^T (\boldsymbol{h} - \boldsymbol{W}_\mathcal{S}\boldsymbol{x}_\mathcal{S}), \tag{9.51}$$

the maximum absolute value deciding which column of the pair dictionary-classifier will join the current support. The expression from (9.51) is the weighted combination of the criteria of kernel OMP and standard OMP. The reader is invited to fill in the details in problem **P** 9.7.

Using the above kernel discriminative DL algorithm on the example from Section 9.6.2, we obtain results similar to those in Figure 9.2, but typically with a smaller number of atoms.

Remark 9.11 An alternative approach to kernel discriminative DL is inspired from the ℓ_1 penalty methods discussed in Chap. 5. The problem is

$$\min_{\boldsymbol{A},\boldsymbol{W},\boldsymbol{X}} \|\varphi(\boldsymbol{Y}) - \varphi(\boldsymbol{Y})\boldsymbol{A}\boldsymbol{X}\|_F^2 + \alpha \|\boldsymbol{H} - \boldsymbol{W}\boldsymbol{X}\|_F^2 + \lambda \|\boldsymbol{X}\|_{1,1} \tag{9.52}$$

the advantage being the lack of explicit sparsity constraints; however, there are now two trade-off parameters to be chosen. Several methods can be adapted to this kind of problem. For example, an online algorithm was described in [109].

Another kernel discriminative DL approach [75] moves the problem into a space where signals and labels become more related. ■

9.6.4 Multiple Kernels

Choosing the right kernel for a classification application is usually based on experience and on insights on the properties of the classes; in the worst case, it is a matter of trial and error. A way to relieve this burden is to use a linear combination of kernels, which is the most popular technique from the family of multiple kernel learning [81]. The kernel function is defined by

$$\kappa(\boldsymbol{y}, \boldsymbol{z}) = \sum_{i=1}^{M} \beta_i \kappa_i(\boldsymbol{y}, \boldsymbol{z}), \tag{9.53}$$

where $\kappa_i(\cdot, \cdot)$ are standard kernels (like the polynomial or Gaussian) and β_i are weights, satisfying $\beta_i \geq 0$, $\sum_{i=1}^{M} \beta_i = 1$. Note that the different kernels may have the same functional form, but different parameters; for example, distinct values (c, d) in the polynomial kernel (9.2) would give distinct kernels in (9.53). It is clear that (9.53) implies a similar relation of the kernel matrices:

$$\boldsymbol{K} = \sum_{i=1}^{M} \beta_i \boldsymbol{K}_i. \tag{9.54}$$

Using these expressions in the algorithms from this chapter is the only required modification in order to work with multiple kernels.

The only remaining issue is how to choose the weights. A possibility is to compute the weights outside the classification process; for example, one can use the classification results with each kernel or the distance from an ideal kernel. The latter approach is based on the alignment score between kernel matrices $\boldsymbol{K}$ and $\boldsymbol{J}$

$$A(\boldsymbol{K}, \boldsymbol{J}) = \frac{\mathrm{tr}[\boldsymbol{K}\boldsymbol{J}]}{\|\boldsymbol{K}\|_F \, \|\boldsymbol{J}\|_F}. \tag{9.55}$$

The alignment score (9.55) is the cosine of the angle between the two kernel matrices, when they are vectorized. We define the ideal kernel matrix $\boldsymbol{K}_{\text{ideal}}$, having element (ℓ, j) equal to one if the signals $\boldsymbol{y}_\ell$ and $\boldsymbol{y}_j$ belong to the same class and zero otherwise. (Note that such a kernel matrix does not have a kernel function behind.) Based on the alignment score, the weights are taken as

$$\beta_i = \frac{A(\boldsymbol{K}_i, \boldsymbol{K}_{\text{ideal}})}{\sum_{j=1}^{M} A(\boldsymbol{K}_j, \boldsymbol{K}_{\text{ideal}})}. \tag{9.56}$$

Intuitively, kernel matrices that are closer to the ideal get a larger weight.

Another possibility is to consider that the weights are variables to be optimized such that the outcome of classification is better. In our case, the weights may be optimized during the DL process. Inserting (9.54) into the representation error expression (9.6) gives a linear form in the weight vector $\boldsymbol{\beta}$. Denoting $\boldsymbol{c}$ the vector of coefficients of the linear form, the weight optimization problem becomes

$$\begin{aligned} \min_{\boldsymbol{\beta}} \quad & \boldsymbol{c}^T \boldsymbol{\beta} \\ \text{s.t.} \quad & \sum_{i=1}^{M} \beta_i = 1 \\ & \beta_i \geq 0, \; i = 1 : M \end{aligned} \tag{9.57}$$

Such a problem has an immediate solution: if c_i is the smallest element of the vector $\boldsymbol{c}$, then $\beta_i = 1$ and all the other weights are zero. This means that a single kernel is actually selected; this is likely to make the DL process unstable, since a different kernel may be selected in each iteration. A single kernel may also increase the danger of overfitting the classifier to the available training signals. So, a gentler way of modifying the weights is necessary; such a solution is given in [175]. Other approaches can be found in [186], [225]; the latter work uses the dimensionality reduction trick that leads to (9.43) and so the weight optimization problem no longer has a trivial solution.

Problems

P 9.1 Write a batch version of the kernel OMP Algorithm 9.1, inspired by the Batch OMP Algorithm 3.3. Analyze its complexity and compare with Batch OMP.

P 9.2 Consider the linear kernel $\kappa(\boldsymbol{y}, \boldsymbol{z}) = \boldsymbol{y}^T \boldsymbol{z}$. Show that the kernel OMP Algorithm 9.1 coincides with standard OMP.

P 9.3 Reminding that $\boldsymbol{D}^T\boldsymbol{D} = \boldsymbol{A}^T\boldsymbol{K}\boldsymbol{A}$, revisit the algorithms from Sect. 4.4 that jointly optimize representation error and coherence and adapt them to kernel DL. (The relation above says that the scalar product between two atoms is $\boldsymbol{d}_i^T\boldsymbol{d}_j = \boldsymbol{a}_i^T\boldsymbol{K}\boldsymbol{a}_j$; otherwise, mutual coherence is defined as usual, as the maximum of $|\boldsymbol{d}_i^T\boldsymbol{d}_j|$ over all $i \neq j$.)

P 9.4 The APrU Algorithm 5.1 is based on solving the DL problem (5.4). If we want to find its kernel version, we need to solve the dictionary update problem

$$\begin{array}{ll} \min\limits_{\boldsymbol{a},\boldsymbol{x}} & \left\| \varphi(\boldsymbol{Y})(\boldsymbol{F} - \boldsymbol{a}\boldsymbol{x}^T) \right\|_F^2 + \lambda \|\boldsymbol{x}\|_1 \\ \text{s.t.} & \left\| \boldsymbol{K}^{\frac{1}{2}}\boldsymbol{a} \right\| = 1 \end{array}$$

which is the adaptation of (5.8). Show that, if $\boldsymbol{x}$ is fixed, then the optimal atom $\boldsymbol{a}$ is given by (9.27). Also, when $\boldsymbol{a}$ is fixed (and normalized), then the optimal representation is

$$\boldsymbol{x} = \text{Soft\_th}\left(\boldsymbol{F}^T\boldsymbol{K}\boldsymbol{a}, \frac{\lambda}{2}\right).$$

This is the adaptation of relation (5.11). Using the above relations, rewrite Algorithm 5.1 and its One-Stage version described in Sect. 5.2.

P 9.5 The kernel SRC method works with the dictionary (9.41). Given a test signal $\boldsymbol{z}$, we compute its sparse representation $\boldsymbol{x}$ with that dictionary. We see $\boldsymbol{x}$ as a concatenation of vectors $\boldsymbol{x}_i$ of length N_i, each corresponding to a class. Similarly, we see the diagonal normalization matrix $\boldsymbol{A}$ as formed by diagonal blocks $\boldsymbol{A}_i$ of size $N_i \times N_i$. The class where $\boldsymbol{z}$ belongs is that for which $\|\varphi(\boldsymbol{z}) - \varphi(\boldsymbol{Y}_i)\boldsymbol{A}_i\boldsymbol{x}_i\|$, $i = 1 : c$, is minimum. Show how these quantities are computed efficiently.

P 9.6 Write an efficient version of the kernel OMP Algorithm 9.1 for the dictionary (9.42) resulted from SRC classification with DL, taking advantage of the simplifications that the block diagonal structure of the matrix $\boldsymbol{A}$ brings into the basic operations (9.10) and (9.13).

P 9.7 Write a complete form of the AK-SVD algorithm for solving the kernel discriminative DL problem from Sect. 9.6.3, including the details of the adapted OMP. Additionally, prove relations (9.48), which gives the optimal representations (with given sparsity pattern), and (9.50), (9.51), which define the OMP adapted to the problem.

P 9.8 Modify the AK-SVD algorithm for kernel discriminative DL from Sect. 9.6.3 to obtain an algorithm for solving the kernel version of the label consistent DL problem (8.28).

Chapter 10
Cosparse Representations

Abstract So far we have approached dictionary learning from the perspective of Chap. 2, where we are interested in the few nonzero entries of the representations. In literature this process is also called the synthesis-based sparse representation model. Recent years have shown approximation improvements when instead we analyze the set of atoms that do not participate in signal representation. If the sparse DL quest is to learn a dictionary able to identify the low-dimensional space that is the true origin of a given class of signals, in this new analysis-based cosparse representation model we are interested in finding its null-space complement. Throughout this chapter we look at representation and learning challenges posed by the cosparse domain and compare them to similar obstacles encountered by its sparse sibling.

10.1 The Cosparse Model

In the synthesis model [158], presented in Chap. 1 and used throughout the book until now, we were interested in finding the basis of a low dimensional subspace $\mathcal{Y}$ where lies a given signal $\boldsymbol{y} \in \mathbb{R}^m$. The basis was formed by few atoms chosen from an overcomplete dictionary. By varying the atoms (i.e., the support), the model comprised a union of subspaces.

In the cosparse analysis model [137], we apply the analysis dictionary $\boldsymbol{\Omega} \in \mathbb{R}^{n \times m}$ to the signal $\boldsymbol{y}$ and obtain the sparse vector

$$\boldsymbol{\gamma} = \boldsymbol{\Omega} \boldsymbol{y}. \tag{10.1}$$

The dictionary is overcomplete, with $n > m$, and its atoms are the rows of the dictionary matrix $\boldsymbol{\Omega}$. We are interested in the set Λ consisting of the rows orthogonal to the signal, which generate the subspace $\mathcal{Y}^{\perp}$; so, the analysis approach is dual to the synthesis one: we work with the orthogonal complement of the low dimensional subspace containing the signal. For example, in Fig. 10.1, the signal $\boldsymbol{y}$ is orthogonal to 11 atoms leading to the sparse representation $\boldsymbol{\gamma}$ whose nonzero entries determine that the true signal subspace $\mathcal{Y}$ is in fact at most tridimensional. Like in the synthesis model, a union of subspaces is obtained by varying the set Λ.

B. Dumitrescu, P. Irofti, *Dictionary Learning Algorithms and Applications*,
https://doi.org/10.1007/978-3-319-78674-2_10

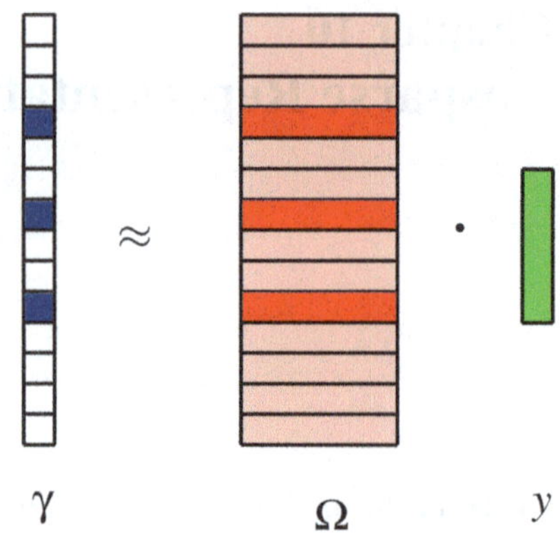

Fig. 10.1 Cosparse representation of a signal. The atoms orthogonal to the signal are pink. The remaining red atoms produce the blue nonzero coefficients

Given a known dictionary $\mathbf{\Omega}$, computing the sparse coefficients is just a matter of computing the multiplication $\mathbf{\Omega} \boldsymbol{y}$, a much easier task compared to its synthesis correspondent that is only approximately solved by calling OMP or other methods, e.g., based on a convex relaxation. The assumption here is that $\boldsymbol{y}$ is a signal whose associated null-space $\mathcal{Y}^{\perp}$ contains most of the dictionary atoms. In order to generate such a signal, we can start from an arbitrary vector $\boldsymbol{u} \in \mathbb{R}^m$ that we project onto the subspace orthogonal to the randomly chosen atoms set Λ. This subspace is $\text{Ker}\mathbf{\Omega}_{\Lambda}$, where $\mathbf{\Omega}_{\Lambda}$ is the dictionary restriction to the rows in Λ. The projection is

$$z = (\boldsymbol{I} - \mathbf{\Omega}_{\Lambda}^{\dagger} \mathbf{\Omega}_{\Lambda})\boldsymbol{u}, \tag{10.2}$$

where $\mathbf{\Omega}_{\Lambda}^{\dagger}$ is the pseudo-inverse of $\mathbf{\Omega}_{\Lambda}$; if $\mathbf{\Omega}_{\Lambda}$ has linearly independent rows, then $\mathbf{\Omega}_{\Lambda}^{\dagger} = \mathbf{\Omega}_{\Lambda}^{T}(\mathbf{\Omega}_{\Lambda}\mathbf{\Omega}_{\Lambda}^{T})^{-1}$; however, this is hardly the case, since typically $|\Lambda| > m$. This operation is the cosparse equivalent of the representation $\boldsymbol{D}_{\text{true}}\boldsymbol{x}$ in the synthesis model when generating artificial data in (2.11). The sparsity target s is replaced by the cosparsity level $|\Lambda|$ and the nonzero positions are implicitly given by the complement set Λ^c; the associated representation coefficients result from the multiplication $\mathbf{\Omega}_{\Lambda^c} z$.

Limiting the study to ideal signals, such as the ones generated in (10.2), is of course impractical for real-world applications. Instead, we assume that the true signal is perturbed by noise. The dictionary atoms might no longer be truly orthogonal to the signal $\boldsymbol{y}$, but we expect that most of them have a negligible contribution which allows us to treat them as if they were. Following the signal model of (2.11), we assume that our data is perturbed by additive noise $\boldsymbol{v}$ drawn from a zero mean Gaussian distribution

$$\boldsymbol{y} = z + \boldsymbol{v}. \tag{10.3}$$

Thus, once we decide on an atom set Λ (a matter discussed in the next section), we can recover z directly through subspace projection (10.2).

Table 10.1 Sparse vs cosparse: the basic notions

	Synthesis (sparse)	Analysis (cosparse)
Signal	$\boldsymbol{y} \in \mathbb{R}^m$	$\boldsymbol{y} \in \mathbb{R}^m$
Dictionary	$\boldsymbol{D} \in \mathbb{R}^{m \times n}$	$\boldsymbol{\Omega} \in \mathbb{R}^{n \times m}$
Atoms	Columns of $\boldsymbol{D}$	Rows of $\boldsymbol{\Omega}$
Exact representation	$\boldsymbol{y} = \boldsymbol{D}\boldsymbol{x}$, $\boldsymbol{x}$ sparse	$\boldsymbol{\gamma} = \boldsymbol{\Omega}\boldsymbol{y}$, $\boldsymbol{\gamma}$ sparse
(Co)support	$\mathcal{S} = \{j \mid x_j \neq 0\}$	$\Lambda = \{i \mid \gamma_i = 0\}$
(Co)sparsity level	$\lvert\mathcal{S}\rvert$	$\lvert\Lambda\rvert$
Noisy representation	$\boldsymbol{y} = \boldsymbol{z} + \boldsymbol{v}$, $\boldsymbol{z} = \boldsymbol{D}\boldsymbol{x}$	$\boldsymbol{y} = \boldsymbol{z} + \boldsymbol{v}$, $\boldsymbol{\Omega}_\Lambda \boldsymbol{z} = 0$

For convenience, Table 10.1 gives a condensed view on the similarities and differences between the synthesis and analysis approaches to sparse representation, and also the basic nomenclature.

10.2 Representation

Given the fixed cosparse dictionary $\boldsymbol{\Omega} \in \mathbb{R}^{n \times m}$ and a noisy signal $\boldsymbol{y} \in \mathbb{R}^m$, the representation problem consists of recovering the vector $\boldsymbol{z} \in \mathbb{R}^m$, which is close to $\boldsymbol{y}$ and belongs to a low-dimensional subspace, by finding the atoms subset Λ (named cosupport) orthogonal to $\boldsymbol{z}$.

If Λ consists of $n - s$ atoms, which usually implies that $\boldsymbol{z}$ lies in an s-dimensional subspace, the representation problem is

$$\begin{aligned} \min_{\boldsymbol{z},\Lambda} \quad & \|\boldsymbol{y} - \boldsymbol{z}\|^2 \\ \text{s.t.} \quad & \boldsymbol{\Omega}_\Lambda \boldsymbol{z} = 0 \\ & \text{rank}(\boldsymbol{\Omega}_\Lambda) = m - s \end{aligned} \tag{10.4}$$

where the optimization goal is naturally derived from (10.3) in the same way the problem (1.6), which we copy here,

$$\begin{aligned} \min_{\boldsymbol{x}} \quad & \|\boldsymbol{y} - \boldsymbol{D}\boldsymbol{x}\|^2 \\ \text{s.t.} \quad & \|\boldsymbol{x}\|_0 \leq s \end{aligned}$$

was derived from the synthesis model (2.11). We can also see that by rewriting the sparsity inducing constraint $\boldsymbol{\Omega}_\Lambda \boldsymbol{z} = 0$ as $\|\boldsymbol{\Omega}\boldsymbol{z}\|_0 \leq s$, the problem (10.4) turns into the synthesis dual. The extra rank constraint ensures that the subspace where $\boldsymbol{z}$ lies is indeed s-dimensional, since this condition is not necessarily imposed by the first constraint of (10.4) and by the cardinality of Λ.

To better understand this rank constraint (also called co-rank) and its associated subspace pursuit, let us start with the simple case where the dictionary is square and invertible ($n = m$). It is obvious that the sparse and cosparse models coincide and that $\boldsymbol{D} = \boldsymbol{\Omega}^{-1}$. If the signal z belongs to the subspace $\mathcal{Y}$ of dimension $s \ll m$, then the cosparse model seeks to find its orthogonal complement $\mathcal{Y}^{\perp}$ whose dimension, the co-rank, is $m - s$. Thus the true sparse support $\mathcal{S}$ and the cosupport Λ are complementary sets. As a side note, it is also known [66] that in the undercomplete denoising case, where $\boldsymbol{D} = \boldsymbol{\Omega}^{\dagger}$, the two models are nearly equivalent.

In the overcomplete case ($n > m$), if every subset with m atoms is linearly independent, implying that $\boldsymbol{\Omega}$ has full-spark (see Definition 1.13), then the null-space associated with the true signal z always satisfies the inequality $|\Lambda| \leq m$. Accordingly, the sparse representation $\boldsymbol{y}$ is in fact quite dense when $n \gg m$. Note that $|\Lambda| = m$ means that $z = 0$, as the unique solution to $\boldsymbol{\Omega}_{\Lambda} z = 0$. Moreover, the conditions $|\Lambda| = m - s$ and $\text{rank}(\boldsymbol{\Omega}_{\Lambda}) = m - s$ are equivalent.

For non-full spark overcomplete analysis dictionaries, finding $\mathcal{Y}^{\perp}$ is more complicated because the cardinality of the cosupport can go beyond the signal dimension m. This means that we can build a cosupport with $|\Lambda| = m$ such that $\boldsymbol{\Omega}_{\Lambda} z = 0$. Even more, when we decide to add an atom to the set we might find that $\boldsymbol{y}$ gains more than one zero entry due to the other existing collinear atoms (or linearly dependent on the atoms already in Λ) in $\boldsymbol{\Omega}$. If $\boldsymbol{\Omega}_{\Lambda}$ has not full rank, then the dimension of $\mathcal{Y}^{\perp}$ is smaller than $m - s$, which means the dimension of $\mathcal{Y}$ is larger than s and hence the (co)sparsity goal is in fact not achieved. In this case, the second constraint of (10.4) cannot be directly related with the cardinality of Λ and is the only possible formulation.

Remark 10.1 Both the synthesis and analysis models cover a union of subspaces in $\mathbb{R}^m$. One may wonder which model is richer. It was shown in Remark 1.3 that the synthesis model comprises $\binom{n}{s}$ subspaces of dimension s. The above discussion of the analysis model in the full-spark case shows that an s-dimensional subspace is described (via its orthogonal complement) by $m - s$ atoms. Hence, there are $\binom{n}{m-s}$ such subspaces. Since typically $s \ll m$, it results that the analysis model is significantly richer.

The problems (10.4) and Sect. 10.2 have a clear denoising significance, the vectors z and $\boldsymbol{Dx}$, respectively, being approximations of the underlying clean signal manifested through the noisy $\boldsymbol{y}$. In light of the number of subspaces for the analysis and synthesis models, we should expect that z is closer to $\boldsymbol{y}$ than $\boldsymbol{Dx}$. However, we should not conclude that the cosparse denoising is better. ■

As we increase the number of atoms present in overcomplete dictionaries, sparsity driven results start to drift away from their cosparse counterparts. In fact, by representing the dictionaries and their associated solutions as polytopes, it has been geometrically shown in [66] that the synthesis model is a generalization of the analysis model for the l_1 formulation. Unfortunately we cannot state that for l_0, which is our main interest.

Algorithm 10.1: Backward Greedy

Data: dictionary $\boldsymbol{\Omega} \in \mathbb{R}^{n \times m}$
signal $\boldsymbol{y} \in \mathbb{R}^m$
orthogonal threshold ε_0
co-rank $m - s$
stopping error ε

Result: cosupport Λ, recovered signal $\boldsymbol{z} \in \mathbb{R}^m$

1 Initialize $\Lambda = \emptyset$, $\boldsymbol{z} = \boldsymbol{y}$, $\boldsymbol{e} = 0$
2 **while** $rank(\boldsymbol{\Omega}_\Lambda) < m - s$ *and* $\|\boldsymbol{e}\| < \varepsilon$ **do**
3 Find new index: $k = \arg\min_{j \notin \Lambda} \left|\boldsymbol{\omega}_j^T \boldsymbol{z}\right|$
4 Build new cosupport: $\Lambda \leftarrow \Lambda \cup \{k\}$
5 Compute new solution: $\boldsymbol{z} = (\boldsymbol{I} - \boldsymbol{\Omega}_\Lambda^\dagger \boldsymbol{\Omega}_\Lambda)\boldsymbol{y}$
6 Look for extra orthogonal atoms: $\Lambda \leftarrow \Lambda \cup \{j \mid j \notin \Lambda \text{ and } \left|\boldsymbol{\omega}_j^T \boldsymbol{z}\right| < \varepsilon_0\}$
7 Compute new residual: $\boldsymbol{e} = \boldsymbol{y} - \boldsymbol{z}$

10.2.1 Backward Greedy

In this section we present a first OMP-like attempt for solving (10.4). Indeed, the Backward Greedy (BG) [159] method selects atoms based on their correlation with the current approximation of the true signal and iteratively increases the cosupport until the co-rank criteria or the target error is met. The differences are in the details: we are looking for the most decorrelated (orthogonal) atoms such that the null space of the recovered signal $\boldsymbol{z}$ is at least $(m - s)$-dimensional.

We summarize the Backward Greedy instructions in Algorithm 10.1. BG starts with an empty cosupport set Λ, which leads to the initial approximation $\boldsymbol{z} = \boldsymbol{y}$ of the true signal. We stop in step 2 when either a null space of dimension $m - s$ is found or when the target error ε is met; note that initially the residual is zero and the error increases as the cosupport grows; hence, if we want a final error less than ε, we need to backtrack to the last but one iteration; this is not explicitly written in the algorithm. In step 3, we mimic the greedy column selection from the synthesis approach by selecting atoms from $\boldsymbol{\Omega}$ on a row-by-row basis. On each iteration the row $\boldsymbol{\omega}_k^T$ with the smallest correlation with the current $\boldsymbol{z}$ is selected and the cosupport Λ is updated (step 4). For reasons described around (10.2), the new solution $\boldsymbol{z}$ computed in step 5 is

$$\boldsymbol{z} = \boldsymbol{P}_\Lambda \boldsymbol{y}, \tag{10.5}$$

where

$$\boldsymbol{P}_\Lambda = \boldsymbol{I} - \boldsymbol{\Omega}_\Lambda^\dagger \boldsymbol{\Omega}_\Lambda \tag{10.6}$$

is the projection matrix on $\text{Ker}\boldsymbol{\Omega}_\Lambda$. In the general case where we deal with redundant dictionaries, the fresh coefficients of $\boldsymbol{\Omega z}$ might reveal some new (almost) orthogonal atoms corresponding to (almost) zero entries. As such, BG performs an additional cosupport update in step 6, where it expands the set Λ to contain all atoms that have become (almost) orthogonal on the current z because they are (almost) linear combinations of the existing atoms; they correspond to coefficients of $\boldsymbol{\Omega z}$ that are less than the given threshold ε_0. This operation is specific to the analysis model and has no counterpart in the synthesis case because it would be redundant in the context of OMP, which guarantees the selection of linearly independent atoms. Step 7 updates the residual and starts a new iteration. In principle, each iteration adds a single independent atom to the cosupport, hence exactly $m - s$ iterations are necessary to attain the rank target.

Let us analyze the complexity of a single iteration from Algorithm 10.1. Step 3 requires the multiplication $\boldsymbol{\Omega z}$, which amounts to $O(mn)$ operations. The cosupport update in step 6 has a similar complexity and the result can in fact be used at the next step 3. The most demanding task is computing the new solution z in step 5.

The computation of the pseudo inverse required by $\boldsymbol{P}_\Lambda$ from (10.6) can be avoided through orthogonalization of the new atom on the existing selected set. Let $(\boldsymbol{\omega}_k^\perp)^T$ be the orthogonalized atom at step k; note that the atoms chosen in step 6 have very little contribution and their orthogonalized versions should be zero or very small, as those atoms are almost linear combinations of other selected atoms; the number of independent orthogonalized atoms may also serve as value for the rank of $\boldsymbol{\Omega}_\Lambda$, although the number of iterations is a relevant enough information. Then, we can rewrite (10.6) as

$$\begin{aligned} \boldsymbol{P}_\Lambda &= \boldsymbol{I} - \sum_{i=1}^{k} \boldsymbol{\omega}_i^\perp (\boldsymbol{\omega}_i^\perp)^T \\ &= \boldsymbol{I} - \sum_{i=1}^{k-1} \boldsymbol{\omega}_i^\perp (\boldsymbol{\omega}_i^\perp)^T - \boldsymbol{\omega}_k^\perp (\boldsymbol{\omega}_k^\perp)^T. \end{aligned} \tag{10.7}$$

(Note that $(\boldsymbol{\omega}_i^\perp)^T \boldsymbol{P}_\Lambda \boldsymbol{y} = 0$ and $\boldsymbol{P}_\Lambda z = z$, which justifies the first equality; see also problem **P** 10.2.) The second equality allows the simplification of (10.5) to

$$z^{(k)} = z^{(k-1)} - \boldsymbol{\omega}_k^\perp (\boldsymbol{\omega}_k^\perp)^T \boldsymbol{y}, \tag{10.8}$$

where $z^{(k)}$ represents the solution at iteration k. In this new form the matrix inversion is avoided and step 5 requires only $O(m^2)$ operations, mostly for the orthogonalization process, which can be done by a modified Gram-Schmidt algorithm or by updating a QR factorization. Ignoring the $O(m)$ norm (step 2) and the substraction (step 7) operations, BG performs around $O(m(m - s)(n + m))$ operations for cosparsity level $m - s$, which for small values of s can be rounded

Algorithm 10.2: Optimized Backward Greedy Atom Selection

Data: dictionary $\boldsymbol{\Omega} \in \mathbb{R}^{n \times m}$
signal $\boldsymbol{y} \in \mathbb{R}^m$
current solution $\boldsymbol{z} \in \mathbb{R}^m$
current cosupport Λ
Result: new index k

1 **for** $j \notin \Lambda$ **do**
2 Build new cosupport: $\bar{\Lambda} \leftarrow \Lambda \cup \{j\}$
3 Compute new solution: $\bar{\boldsymbol{z}} = (\boldsymbol{I} - \boldsymbol{\Omega}_{\bar{\Lambda}}^{\dagger} \boldsymbol{\Omega}_{\bar{\Lambda}})\boldsymbol{y}$
4 Compute new residual: $\boldsymbol{e}_j = \bar{\boldsymbol{z}} - \boldsymbol{z}$
5 Find new index: $k = \arg\min_{j \notin \Lambda} \|\boldsymbol{e}_j\|$

up to $O(m^2(n + m))$. As expected, this requires more computations than OMP's $O(ms(n + s^2))$ due to the larger cardinality of the cosupport compared with that of the regular support.

10.2.2 Optimized Backward Greedy

A simple but expensive way to improve BG results is to extend the atom selection stage. The Optimized Backward Greedy (OBG) [159] algorithm modifies BG atoms selection by sweeping through all remaining row candidates in dictionary $\boldsymbol{\Omega}$ and selecting the one that changes $\boldsymbol{z}$ the least. This is the cosparse equivalent of Orthogonal Least Squares (see Remark 1.7).

We present only the atom selection strategy in Algorithm 10.2 because the rest of OBG instructions are identical to those of BG. The main loop iterates over all atoms j currently not included in cosupport Λ (step 1) and proceeds to temporarily include each in the extended set $\bar{\Lambda}$ (step 2) for computing a tentative solution $\bar{\boldsymbol{z}}$ (step 3) whose difference to the existing solution is stored in $\boldsymbol{e}_j$ (step 4). In the end, atom k with the smallest signal energy impact is chosen (step 5).

The selection process can be further improved, at the cost of complexity, by adapting look-ahead strategies from the sparse model. For example, whereas OBG looks at the impact on the solution between two iterations, LAOLS (Algorithm 3.12) investigates what happens to the final solution if atom j were to be chosen. For a compromise between complexity and performance, one might investigate adapting the sparse coding algorithms Projection-Based OMP [34] and Projection-Based OLS [99] that are not discussed in this book.

In terms of complexity, we can simplify the necessary instructions by using the orthogonalization process described around equation (10.7). Indeed, we can rewrite the residual from step 4 as

$$\begin{aligned} e_j &= \bar{z} - z \\ &= z - (\omega_j{}^{\perp})^T \omega_j{}^{\perp} y - z \\ &= (\omega_j{}^{\perp})^T \omega_j{}^{\perp} y, \end{aligned} \tag{10.9}$$

which means that we can reduce steps 2–4 to just computing $(\omega_j{}^{\perp})^T$. Thus we need to find the smallest atom-solution correlation and step 5 becomes

$$k = \arg\min_{j \notin \Lambda} \left| (\omega_j^{\perp})^T z \right| . \tag{10.10}$$

The equality $(\omega_j^{\perp})^T z = (\omega_j^{\perp})^T y$ and other details of OBG are discussed in problem **P** 10.3.

In conclusion, the complexity overhead of the OBG algorithm is given by the orthogonalization process which is performed on average in $O(n^2)$ operations at each iteration.

10.2.3 Other Algorithms

There are many other approaches besides the BG and OBG algorithms described above. From the same greedy family, we mention Greedy Analysis Pursuit (GAP) [137], which starts from the other end, by considering a cosupport containing all the atoms. Then, the cosupport is shrunk until it reaches the desired $m - s$ cardinality, by removing atoms that have large projections on the current solution. We note that GAP solves a slightly different problem, assuming that some measurements of the signal are available, in compressed sensing style. An improvement of GAP is presented [121].

On different directions of attack, there are analysis algorithmic adaptations for most sparse representation algorithms such as Subspace Pursuit (SP), Compressive Sampling Matching Pursuit (CoSaMP), Iterative Hard Thresholding (IHT), and Hard Thresholding Pursuit (HTP) [78].

10.3 Cosparse DL

We are now interested in finding a proper analysis dictionary $\boldsymbol{\Omega}$ for a class of m-dimensional signals that are known (or assumed) to lie on a union of much smaller subspaces of dimension s, such that applying the dictionary to any given signal from this class would reveal its low-dimensional true-self z.

Given a training signal set $\boldsymbol{Y} \in \mathbb{R}^{m \times N}$, our goal is to design the dictionary atoms such that there always exists an almost orthogonal subset describing the $m - s$ null-space associated with a signal from $\boldsymbol{Y}$. Formally, the cosparse DL problem is

$$
\begin{aligned}
\min_{\boldsymbol{\Omega},\boldsymbol{Z},\{\Lambda_\ell\}_{\ell=1:N}} \quad & \|\boldsymbol{Y}-\boldsymbol{Z}\|_F^2 \\
\text{s.t.} \quad & \boldsymbol{\Omega}_{\Lambda_\ell}\boldsymbol{z}_\ell = 0,\ \ \ell = 1:N \\
& \text{rank}(\boldsymbol{\Omega}_{\Lambda_\ell}) = m - s,\ \ \ell = 1:N \\
& \left\|\boldsymbol{\omega}_j^T\right\| = 1,\ \ j = 1:n
\end{aligned}
\tag{10.11}
$$

where the variables are the analysis dictionary $\boldsymbol{\Omega}$, with normalized rows $\boldsymbol{\omega}_j^T$, and the recovered signals matrix $\boldsymbol{Z} \in \mathbb{R}^{m\times N}$. Recall the synthesis dictionary learning problem

$$
\begin{aligned}
\min_{\boldsymbol{D},\boldsymbol{X}} \quad & \|\boldsymbol{Y}-\boldsymbol{D}\boldsymbol{X}\|_F^2 \\
\text{s.t.} \quad & \|\boldsymbol{x}_\ell\|_0 \le s,\ \ \ell = 1:N \\
& \|\boldsymbol{d}_j\| = 1,\ \ j = 1:n
\end{aligned}
$$

and note that the sparsity and atom normalization constraints are similar, but in the analysis case we need to do extra work to ensure proper atom selection via co-rank constraints for reasons described in Sect. 10.2.

The algorithms for solving the cosparse DL problem (10.11) follow the standard alternate optimization structure of most DL algorithms. The coding or representation step assumes that the dictionary $\boldsymbol{\Omega}$ is fixed and computes the recovered signals $\boldsymbol{Z}$ using one of the methods discussed in the previous section. The dictionary update step assumes that the recovered signals $\boldsymbol{Z}$ are fixed or, at least, the cosupports Λ_ℓ, $\ell = 1:N$, are fixed, and optimizes the dictionary $\boldsymbol{\Omega}$. In what follows, we describe two algorithms for solving (10.11) or a close relative of it, insisting mostly on the dictionary update step.

10.3.1 Analysis K-SVD

Following the one atom at a time atom update approach of K-SVD, Analysis K-SVD [159] sequentially updates each row $\boldsymbol{\omega}_j^T$ of the cosparse dictionary $\boldsymbol{\Omega}$ by keeping the other atoms fixed. Let $\mathcal{J}$ be the indices of the signals that are (nearly) orthogonal to the current atom, i.e., the signals whose cosupport contains $\boldsymbol{\omega}_j^T$. Allowing also the representations to change, the adaptation of problem (10.11) to the optimization of $\boldsymbol{\omega}_j^T$ leads to

Algorithm 10.3: Analysis K-SVD Dictionary Update

Data: current dictionary $\boldsymbol{\Omega} \in \mathbb{R}^{n \times m}$
signals set $\boldsymbol{Y} \in \mathbb{R}^{m \times N}$
recovered signals matrix $\boldsymbol{Z} \in \mathbb{R}^{m \times N}$
Result: updated dictionary $\boldsymbol{\Omega}$

for $j = 1 : n$ **do**
  Find set $\mathcal{J}$ of signals whose cosupport contains the current atom
  Update atom: $\boldsymbol{\omega}_j = \arg\min_{\boldsymbol{\omega}} \|\boldsymbol{\omega}^T \boldsymbol{Y}_{\mathcal{J}}\|$ s.t. $\|\boldsymbol{\omega}\| = 1$

$$\begin{aligned} \min_{\boldsymbol{\omega}_j, \boldsymbol{Z}_{\mathcal{J}}} \quad & \|\boldsymbol{Y}_{\mathcal{J}} - \boldsymbol{Z}_{\mathcal{J}}\|_F^2 \\ \text{s.t.} \quad & \boldsymbol{\Omega}_{\bar{\Lambda}_\ell} z_\ell = 0, \ \bar{\Lambda}_\ell = \Lambda_\ell \setminus \{j\}, \ \forall \ell \in \mathcal{J} \\ & \boldsymbol{\omega}_j^T \boldsymbol{Z}_{\mathcal{J}} = 0 \\ & \left\|\boldsymbol{\omega}_j^T\right\| = 1 \end{aligned} \tag{10.12}$$

where $\bar{\Lambda}_\ell$ represents the cosupport of signal z_ℓ without atom $\boldsymbol{\omega}_j^T$. The constraints ensure that the new normalized atom does not interfere with fellow cosupport atoms and that the updated representations $\boldsymbol{Z}$ remain orthogonal to their existing cosupport.

However, solving (10.12) turns out to be difficult, due to the large number of variables and the hard constraints. So, the optimization problem (10.12) is further simplified to

$$\begin{aligned} \min_{\boldsymbol{\omega}_j{}^T} \quad & \left\|\boldsymbol{\omega}_j^T \boldsymbol{Y}_{\mathcal{J}}\right\|^2 \\ \text{s.t.} \quad & \left\|\boldsymbol{\omega}_j^T\right\| = 1 \end{aligned} \tag{10.13}$$

which expresses the simple desire to have an atom as (nearly) orthogonal as possible on the signals in whose cosupport the atom is present, disregarding the effect of (and on) the other atoms. The solution of (10.13) is the singular vector corresponding to the smallest singular value of $\boldsymbol{Y}_{\mathcal{J}}$. This neatly corresponds to the optimal K-SVD solution from (3.35), where we were interested in the dominant singular vector, with the observation that now the representations update is skipped. In fact, Analysis K-SVD is closer to being the cosparse equivalent of SGK from Algorithm 3.5 than that of K-SVD (Algorithm 3.6). Even though the dependence on formerly updated atoms is lost, thus making room for parallelism, computing the singular vector is a much more intensive task than in the synthesis case.

Efficiency-wise, there is no definitive solution for this job and in practice there exist multiple competing algorithms [165], such as Lanczos, Jacobi–Davidson, JDCG, LOBPCG or the inverse-free preconditioned Krylov subspace method [119], and often the winner is decided based on eigen specific properties of the input matrix. However, the straightforward and popular method of computing the smallest singular value and its associated vectors is through inverse iterative methods. The singular vectors of matrix $\boldsymbol{Y}$ are identical to the eigenvectors of $\boldsymbol{Y}^T\boldsymbol{Y}$, thus the smallest singular value corresponds to the largest eigenvalue of $(\boldsymbol{Y}^T\boldsymbol{Y})^{-1}$ and its associated eigenvector is computed via the inverse power method. The main computations are the Gram matrix factorization $O(mN^2)$ and the least squares solution needed by the power method $O(m^2N)$. In the standard scenario where $N \gg m$, the total number of operations is asymptotically $O(mN^2)$. We note that computing the full singular value decomposition of matrix $\boldsymbol{Y}$ costs $O(4mN^2)$ instructions and that above we assumed that it is sufficient to perform just a few iterations of the inverse power method.

Finally, the Analysis K-SVD dictionary update recipe is summarized in Algorithm 10.3. Step 1 iterates the dictionary rows and for each atom j we find the recovered signals that include $\boldsymbol{\omega}_j^T$ in their cosupport (step 2). The column indices are stored in set $\mathcal{J}$ helping step 3 to update the atom through the SVD decomposition of the input signals $\boldsymbol{Y}_{\mathcal{J}}$ as earlier described.

10.3.2 Analysis SimCO

The cosparse adaptation [50] of SimCO [44], Algorithm 3.9 from Chap. 3, trains the analysis dictionary $\boldsymbol{\Omega}$ so that given the signals $\boldsymbol{Y}$ the product $\boldsymbol{\Omega Y}$ leads to the sparse representations $\boldsymbol{\Gamma}$. The DL problem is

$$\begin{aligned} \min_{\boldsymbol{\Omega},\boldsymbol{\Gamma}} \quad & \|\boldsymbol{\Gamma} - \boldsymbol{\Omega Y}\|_F^2 \\ \text{s.t.} \quad & \|\boldsymbol{\gamma}_\ell\|_0 = s,\ \ell = 1:N \\ & \left\|\boldsymbol{\omega}_j^T\right\| = 1,\ j = 1:n \end{aligned} \tag{10.14}$$

and is a simple reformulation of the original objective from (10.11) where we no longer use the null-space based projection $\boldsymbol{z}$ from (10.2) and focus instead on its sparse representation $\boldsymbol{\gamma}$ from (10.1).

A first consequence of this rewrite is that it allows Analysis SimCO to compute the representations directly by performing hard-thresholding on the smallest $m - s$ entries of each column of the product $\boldsymbol{\Omega Y}$, which is obviously faster than using the BG algorithm.

Considering the nonzero pattern of the representations fixed, as in the synthesis formulation, the dictionary error function (3.40) becomes

$$\begin{aligned} f(\boldsymbol{\Omega}) = \min_{\boldsymbol{\Gamma}} \quad & \|\boldsymbol{\Gamma} - \boldsymbol{\Omega}\boldsymbol{Y}\|_F^2 \\ \text{s.t.} \quad & \boldsymbol{\Gamma}_\Lambda = 0 \end{aligned} \tag{10.15}$$

where Λ is the known cosupport. The dictionary update problem can then be formulated similarly to (3.43)

$$\begin{aligned} \min_{\boldsymbol{\Omega}} \quad & f(\boldsymbol{\Omega}) \\ \text{s.t.} \quad & \|\boldsymbol{\omega}_\ell^T\| = 1, \ \ell = 1:n \end{aligned} \tag{10.16}$$

The gradient of (10.15) is

$$\boldsymbol{G} = 2\boldsymbol{\Gamma}\boldsymbol{Y}^T - 2\boldsymbol{\Omega}\boldsymbol{Y}\boldsymbol{Y}^T, \tag{10.17}$$

which allows us to solve (10.16) via the same descent methods as described in Sect. 3.7.

Problems

P 10.1 Prove that the analysis model is equivalent to the synthesis model when the dictionary is square and has full rank.

P 10.2 Prove that the matrix $\boldsymbol{P}_\Lambda$ from (10.7) is indeed a projection onto $\mathrm{Ker}\boldsymbol{\Omega}_\Lambda$. Show that i) $\boldsymbol{P}_\Lambda$ is idempotent ($\boldsymbol{P}_\Lambda^2 = \boldsymbol{P}_\Lambda$), ii) $\boldsymbol{P}_\Lambda \boldsymbol{z} = \boldsymbol{z}$, $\forall \boldsymbol{z} \in \mathrm{Ker}\boldsymbol{\Omega}_\Lambda$, and iii) $(\boldsymbol{I} - \boldsymbol{P}_\Lambda)\boldsymbol{y}$ is orthogonal on $\mathrm{Ker}\boldsymbol{\Omega}_\Lambda$, for all vectors $\boldsymbol{y}$.

P 10.3 Let $\boldsymbol{y}$ be a signal and $\boldsymbol{z}$ its approximation at some iteration of the OBG Algorithm 10.2, where the cosupport is Λ. Let $\bar{\boldsymbol{z}}$ be the approximation when adding atom $\boldsymbol{\omega}_j^T$ to the cosupport and $(\boldsymbol{\omega}_j^\perp)^T$ its orthogonalized version with respect to the already chosen atoms.

i) Show that $(\boldsymbol{\omega}_j^\perp)^T \boldsymbol{z} = (\boldsymbol{\omega}_j^\perp)^T \boldsymbol{y}$. Hint: write $\boldsymbol{y} = \boldsymbol{z} + \boldsymbol{u}$, where $\boldsymbol{u} \perp \mathrm{Ker}\boldsymbol{\Omega}_\Lambda$. It results that $\boldsymbol{u}$ is in the span of $\boldsymbol{\Omega}_\Lambda^T$, hence $(\boldsymbol{\omega}_j^\perp)^T \boldsymbol{u} = 0$.
ii) The error minimized in OBG is $\|\boldsymbol{z} - \bar{\boldsymbol{z}}\|$. Show that the same result is obtained by minimizing $\|\boldsymbol{y} - \bar{\boldsymbol{z}}\|$, i.e., minimizing the residual with respect to the initial signal (which is more intuitive). Hint: write $\|\boldsymbol{y} - \bar{\boldsymbol{z}}\|^2 = \|(\boldsymbol{y} - \boldsymbol{z}) + (\boldsymbol{z} - \bar{\boldsymbol{z}})\|^2$ and develop without breaking the parentheses; then use i).

P 10.4 Modify the Optimized Backward Greedy algorithm such that it follows the LAOLS strategy and compares the impact of choosing atom $\boldsymbol{\omega}_j^T$ on the final solution $\boldsymbol{x}$.

P 10.5 Section 5.7 presents a selection method for DL, that greedily chooses the atoms from a large set $\boldsymbol{A}$ of candidates. Adapt the method to cosparse DL, with the problem (10.14) as target optimization goal. What would be the best strategy: to start with a void dictionary and add atoms or to start with a dictionary made of all candidates and remove atoms?

References

1. V. Abrol, P. Sharma, A.K. Sao, Greedy dictionary learning for kernel sparse representation based classifier. Pattern Recogn. Lett. **78**, 64–69 (2016)
2. P.A. Absil, R. Mahony, R. Sepulchre, *Optimization Algorithms on Matrix Manifolds* (Princeton University Press, Princeton, 2009)
3. A. Agarwal, A. Anandkumar, P. Jain, P. Netrapalli, R. Tandon, Learning sparsely used overcomplete dictionaries, in *Conference on Learning Theory* (2014), pp. 123–137
4. M. Aharon, M. Elad, A. Bruckstein, K-SVD: an algorithm for designing overcomplete dictionaries for sparse representation. IEEE Trans. Signal Process. **54**(11), 4311–4322 (2006)
5. M. Aharon, M. Elad, A. Bruckstein, On the uniqueness of overcomplete dictionaries, a practical way to retrieve them. Linear Algebra Appl. **416**(1), 48–67 (2006)
6. S. Arora, R. Ge, A. Moitra, New algorithms for learning incoherent overcomplete dictionaries, in *Conference on Learning Theory* (2014), pp. 779–806
7. F. Bach, R. Jenatton, J. Mairal, G. Obozinski, Optimization with sparsity-inducing penalties. Found. Trends Mach. Learn. **4**(1), 1–106 (2011)
8. M. Bahri, Y. Panagakis, S. Zafeiriou, Robust Kronecker-decomposable component analysis for low rank modeling. Preprint (2017). arXiv:1703.07886
9. C. Bao, H. Ji, Y. Quan, Z. Shen, Dictionary learning for sparse coding: algorithms and convergence analysis. IEEE Trans. Pattern Anal. Mach. Intell. **38**(7), 1356–1369 (2016)
10. D. Barchiesi, M.D. Plumbley, Learning incoherent dictionaries for sparse approximation using iterative projections and rotations. IEEE Trans. Signal Process. **61**(8), 2055–2065 (2013)
11. A. Beck, M. Teboulle, Fast iterative shrinkage-thresholding algorithm for linear inverse problems. SIAM J. Imag. Sci. **2**(1), 183–202 (2009)
12. S. Becker, J. Bobin, E.J. Candès, NESTA: a fast and accurate first-order method for sparse recovery. SIAM J. Imag. Sci. **4**(1), 1–39 (2011)
13. T. Blumensath, M.E. Davies, Iterative hard thresholding for compressed sensing. Appl. Comput. Harm. Anal. **27**(3), 265–274 (2009)
14. S. Boyd, L. Vandenberghe, *Convex Optimization* (Cambridge University Press, Cambridge, 2004)
15. S. Boyd, N. Parikh, E. Chu, B. Peleato, J. Eckstein, Distributed optimization and statistical learning via the alternating direction method of multipliers. Found. Trends® Mach. Learn. **3**(1), 1–122 (2011)
16. H. Bristow, A. Eriksson, S. Lucey, Fast convolutional sparse coding, in *Proceedings of IEEE Conference on Computer Vision and Pattern Recognition* (2013), pp. 391–398

B. Dumitrescu, P. Irofti, *Dictionary Learning Algorithms and Applications*,
https://doi.org/10.1007/978-3-319-78674-2

17. A.M. Bruckstein, D.L. Donoho, M. Elad, From sparse solutions of systems of equations to sparse modeling of signals and images. SIAM Rev. **51**(1), 34–81 (2009)
18. O. Bryt, M. Elad, Compression of facial images using the K-SVD algorithm. J. Vis. Commun. Image Represent. **19**(4), 270–282 (2008)
19. T. Cai L. Wang, Orthogonal matching pursuit for sparse signal recovery with noise. IEEE Trans. Inf. Theory **57**(7), 4680–4688 (2011)
20. T. Cai, L. Wang, G. Xu, New bounds for restricted isometry constants. IEEE Trans. Inf. Theory **56**(9), 4388–4394 (2010)
21. T. Cai, L. Wang, G. Xu, Shifting inequality and recovery of sparse signals. IEEE Trans. Signal Proc. **58**(3), 1300–1308 (2010)
22. E.J. Candès, The restricted isometry property and its implications for compressed sensing. C. R. Math. **346**(9–10), 589–592 (2008)
23. E.J. Candès, T. Tao, Decoding by linear programming. IEEE Trans. Inf. Theory **51**(12), 4203–4215 (2005)
24. E.J. Candès, L. Wakin, An introduction to compressive sampling. IEEE Signal Process. Mag. **25**(2), 21–30 (2008)
25. E.J. Candès, J. Romberg, T. Tao, Robust uncertainty principles: exact signal reconstruction from highly incomplete frequency information. IEEE Trans. Inf. Theory **52**(2), 489–509 (2006)
26. J.D. Carroll, J.J. Chang, Analysis of individual differences in multidimensional scaling via an n-way generalization of "Eckart-Young" decomposition. Psychometrika **35**(3), 283–319 (1970)
27. P.G. Casazza, R.G. Lynch, A brief introduction to Hilbert space frame theory and its applications, in *Proceedings of Symposia in Applied Mathematics*, ed. by K.A. Okoudjou, vol. 73 (American Mathematical Society, Providence, 2016), pp. 1–52
28. M.V. Casillas, V. Puig, L.E. Garza-Castanón, A. Rosich, Optimal sensor placement for leak location in water distribution networks using genetic algorithms. Sensors **13**(11), 14984–15005 (2013)
29. V. Cevher, A. Krause, Greedy dictionary selection for sparse representation. IEEE J. Sel. Topics Signal Process. **5**(5), 979–988 (2011)
30. J.M. Chambers, Algorithm 410: partial sorting. Commun. ACM **14**(5), 357–358 (1971)
31. L.H. Chang, J.Y. Wu, An improved rip-based performance guarantee for sparse signal recovery via orthogonal matching pursuit. IEEE Trans. Inf. Theory **60**(9), 5702–5715 (2014)
32. H. Chang, M.K. Ng, T. Zeng, Reducing artifacts in JPEG decompression via a learned dictionary. IEEE Trans Signal Process. **62**(3), 718–728 (2014)
33. R. Chartrand, Shrinkage mappings and their induced penalty functions, in *ICASSP*, Florence (2014), pp. 1026–1029
34. S. Chatterjee, M. Vehkapera, M. Skoglund, Projection-based and look-ahead strategies for atom selection. IEEE Trans. Signal Process. **60**(2), 634–647 (2012)
35. Y. Chen, C. Caramanis, Noisy and missing data regression: distribution-oblivious support recovery, in *International Conference on Machine Learning*, Atlanta, GA (2013)
36. S. Chen, S.A. Billings, W. Luo, Orthogonal least squares methods and their application to non-linear system identification. Int. J. Control **50**(5), 1873–1896 (1989)
37. S.S. Chen, D.L. Donoho, M.A. Saunders, Atomic decomposition by basis pursuit. SIAM J. Sci. Comput. **20**(1), 33–61 (1998)
38. W. Chen, M.R.D. Rodrigues, I.J. Wassell, On the use of unit-norm tight frames to improve the average MSE performance in compressive sensing applications. IEEE Signal Process. Lett. **19**(1), 8–11 (2012)
39. X. Chen, Z. Du, J. Li, X. Li, H. Zhang, Compressed sensing based on dictionary learning for extracting impulse components. Signal Process. **96**, 94–109 (2014)
40. Y. Cong, J. Yuan, J. Luo, Towards scalable summarization of consumer videos via sparse dictionary selection. IEEE Trans. Multimed. **14**(1), 66–75 (2012)
41. Y. Cong, J. Liu, G. Sun, Q. You, Y. Li, J. Luo, Adaptive greedy dictionary selection for web media summarization. IEEE Trans. Image Process. **26**(1), 185–195 (2017)

42. K. Dabov, A. Foi, V. Katkovnik, K. Egiazarian, Image denoising by sparse 3-D transform-domain collaborative filtering. IEEE Trans. Image Process. **16**(8), 2080–2095 (2007)
43. W. Dai, O. Milenkovic, Subspace pursuit for compressive sensing signal reconstruction. IEEE Trans. Inf. Theory **55**(5), 2230–2249 (2009)
44. W. Dai, T. Xu, W. Wang, Simultaneous codeword optimization (SimCO) for dictionary update and learning. IEEE Trans. Signal Process. **60**(12), 6340–6353 (2012)
45. H.P. Dang, P. Chainais, Towards dictionaries of optimal size: a Bayesian non parametric approach. J. Signal Process. Syst. 1–12 (2016). https://doi.org/10.1007/s11265-016-1154-1
46. C.F. Dantas, M.N. da Costa, R. da Rocha Lopes, Learning dictionaries as a sum of Kronecker products. IEEE Signal Process. Lett. **24**(5), 559–563 (2017)
47. I. Daubechies, M. Defrise, C. De Mol, An iterative thresholding algorithm for linear inverse problems with a sparsity constraint. Commun. Pure Appl. Math. **57**(11), 1413–1457 (2004)
48. M.A. Davenport, M.B. Wakin, Analysis of orthogonal matching pursuit using the restricted isometry property. IEEE Trans. Inf. Theory **56**(9), 4395–4401 (2010)
49. P. Davies, N.J. Higham, Numerically stable generation of correlation matrices and their factors. BIT Numer. Math. **40**(4), 640–651 (2000)
50. J. Dong, W. Wang, W. Dai, Analysis SimCo: a new algorithm for analysis dictionary learning, in *2014 IEEE International Conference on Acoustics, Speech and Signal Processing (ICASSP)* (IEEE, New York, 2014), pp. 7193–7197
51. D.L. Donoho, Compressed sensing. IEEE Trans. Inf. Theory **52**(4), 1289–1306 (2006)
52. D.L. Donoho, M. Elad, Optimally sparse representation in general (non-orthogonal) dictionaries via ℓ_1 minimization. Proc. Natl. Acad. Sci. **100**, 2197–2202 (2003)
53. D.L. Donoho, X. Huo, Uncertainty principles and ideal atomic decomposition. IEEE Trans. Inf. Theory **47**(7), 2845–2862 (2001)
54. D.L. Donoho, J.M. Johnstone, Ideal spatial adaptation by wavelet shrinkage. Biometrika **81**(3), 425–455 (1994)
55. D.L. Donoho, A. Maleki, A. Montanari, Message-passing algorithms for compressed sensing. Proc. Natl. Acad. Sci. **106**(45), 18914–18919 (2009)
56. M.F. Duarte, M.A. Davenport, D. Takbar, J.N. Laska, T. Sun, K.F. Kelly, R.G. Baraniuk, Single-pixel imaging via compressive sampling. IEEE Signal Process. Mag. **25**(2), 83–91 (2008)
57. J.M. Duarte-Carvajalino, G. Sapiro, Learning to sense sparse signals: simultaneous sensing matrix and sparsifying dictionary optimization. IEEE Trans. Image Process. **18**(7), 1395–1408 (2009)
58. B. Dumitrescu, Designing incoherent frames with only matrix-vector multiplications. IEEE Signal Process. Lett. **24**(9), 1265–1269 (2017)
59. B. Dumitrescu, P. Irofti, Low dimensional subspace finding via size-reducing dictionary learning, in *International Workshop Machine Learning for Signal Processing*, Vietri sul Mare (2016)
60. B. Dumitrescu, P. Irofti, Regularized K-SVD. IEEE Signal Process. Lett. **24**(3), 309–313 (2017)
61. M. Elad, *Sparse and Redundant Representations: From Theory to Applications in Signal Processing* (Springer, Berlin, 2010)
62. M. Elad, M. Aharon, Image denoising via sparse and redundant representations over learned dictionaries. IEEE Trans. Image Process. **15**(12), 3736–3745 (2006)
63. M. Elad, A.M. Bruckstein, A generalized uncertainty principle and sparse representation in pairs of bases. IEEE Trans. Inf. Theory **48**(9), 2558–2567 (2002)
64. M. Elad, I. Yavneh, A plurality of sparse representations is better than the sparsest one alone. IEEE Trans. Inf. Theory **55**(10), 4701–4714 (2009)
65. M. Elad, J.L. Starck, P. Querre, D.L. Donoho, Simultaneous cartoon and texture image inpainting using morphological component analysis (MCA). Appl. Comput. Harmon. Anal. **19**(3), 340–358 (2005)
66. M. Elad, P. Milanfar, R. Rubinstein, Analysis versus synthesis in signal priors. Inverse Prob. **23**(3), 947 (2007)

67. K. Engan, S.O. Aase, J.H. Husoy, Method of optimal directions for frame design, in *IEEE International Conference on Acoustics Speech Signal Proceedings*, vol. 5 (1999), pp. 2443–2446
68. O.D. Escoda, L. Granai, P. Vandergheynst, On the use of a priori information for sparse signal approximations. IEEE Trans. Signal Process. **54**(9), 3468–3482 (2006)
69. M.J. Fadili, J.L. Starck, F. Murtagh, Inpainting and zooming using sparse representations. Comput. J. **52**(1), 64–79 (2007)
70. Y. Fang, J. Wu, B. Huang, 2D sparse signal recovery via 2D orthogonal matching pursuit. Sci. China Inf. Sci. **55**(4), 889–897 (2012)
71. A.K. Farahat, A. Elgohary, A. Ghodsi, M.S. Kamel, Greedy column subset selection for large-scale data sets. Springer Knowl. Inf. Syst. **45**(1), 1–34 (2015)
72. J. Feng, L. Song, X. Yang, W. Zhang, Sub clustering K-SVD: size variable dictionary learning for sparse representations, in *16th IEEE International Conference on Image Processing* (2009), pp. 2149–2152
73. J. Feng, L. Song, X. Yang, W. Zhang, Learning dictionary via subspace segmentation for sparse representation, in *18th IEEE International Conference on Image Processing* (2011), pp. 1245–1248
74. Y. Fu, J. Gao, Y. Sun, X. Hong, Joint multiple dictionary learning for tensor sparse coding, in *2014 International Joint Conference on Neural Networks (IJCNN)* (IEEE, New York, 2014), pp. 2957–2964
75. M.J. Gangeh, A. Ghodsi, M.S. Kamel, Kernelized supervised dictionary learning. IEEE Trans. Signal Process. **61**(19), 4753–4767 (2013)
76. S. Gao, I.W.H. Tsang, L.T. Chia, Sparse representation with kernels. IEEE Trans. Image Process. **22**(2), 423–434 (2013)
77. A. Gersho, R.M. Gray, *Vector Quantization and Signal Compression* (Springer Science & Business Media, Berlin, 1992)
78. R. Giryes, S. Nam, M. Elad, R. Gribonval, M. Davies, Greedy-like algorithms for the cosparse analysis model. Linear Algebra Appl. **441**, 22–60 (2014)
79. A. Golts, M. Elad, Linearized kernel dictionary learning. IEEE J. Sel. Top. Signal Process. **10**(4), 726–739 (2016)
80. G.H. Golub, C. Van Loan, *Matrix Computations*, 4th edn. (Johns Hopkins University Press, Baltimore, 2013)
81. M. Gönen, E. Alpaydin, Multiple kernel learning algorithms. J. Mach. Learn. Res. **12**, 2211–2268 (2011)
82. I.F. Gorodnitsky, B.D. Rao, Sparse signal reconstruction from limited data using FOCUSS: a re-weighted minimum norm algorithm. IEEE Trans. Signal Process. **45**(3), 600–616 (1997)
83. J.C. Gower, G.B. Dijksterhuis, *Procrustes Problems*, vol. 3 (Oxford University Press, Oxford, 2004)
84. R. Gribonval, M. Nielsen, Sparse decompositions in "incoherent" dictionaries, in *2003 Proceedings of the International Conference on Image Processing, 2003 (ICIP 2003)*, vol. 1 (IEEE, New York, 2003), pp. I–33
85. R. Gribonval, K. Schnass, Dictionary identification-sparse matrix-factorization via ℓ_1-minimization. IEEE Trans. Inf. Theory **56**(7), 3523–3539 (2010)
86. R. Grosse, R. Raina, H. Kwong, A.Y. Ng, Shift-invariant sparse coding for audio classification, in *Proceedings of 23rd Conference on Uncertainty in Artificial Intelligence* (AUAI Press, Corvallis, 2007), pp. 149–158
87. C. Guichaoua, Dictionary learning for audio inpainting, 2012. HAL Robotics [cs.RO]. 2012. http://dumas.ccsd.cnrs.fr/dumas-00725263
88. C. Guillemot, O. Le Meur, Image Inpainting. IEEE Signal Process. Mag. **31**(1), 127–144 (2014)
89. O.G. Guleryuz, Nonlinear approximation based image recovery using adaptive sparse reconstructions and iterated denoising-Part II: adaptive algorithms. IEEE Trans. Image Process. **15**(3), 555–571 (2006)

90. J. Han, J. Pei, Y. Yin, R. Mao, Mining frequent patterns without candidate generation: a frequent-pattern tree approach. Data Min. Knowl. Disc. **8**(1), 53–87 (2004)
91. R.A. Harshman, Foundations of the parafac procedure: models and conditions for an "explanatory" multimodal factor analysis. UCLA Working Papers in Phonetics, vol. 16, pp. 1–84 (1970)
92. S. Hawe, M. Seibert, M. Kleinsteuber, Separable dictionary learning, in *Proceedings of the IEEE Conference on Computer Vision and Pattern Recognition* (2013), pp. 438–445
93. F.L. Hitchcock, The expression of a tensor or a polyadic as a sum of products. Stud. Appl. Math. **6**(1–4), 164–189 (1927)
94. T. Hofmann, B. Schölkopf, A.J. Smola, Kernel methods in machine learning. Ann. Stat. **36**(3), 1171–1220 (2008)
95. S.H. Hsieh, C.S. Lu, S.C. Pei, 2D sparse dictionary learning via tensor decomposition, in *2014 IEEE Global Conference on Signal and Information Processing (GlobalSIP)* (IEEE, New York, 2014), pp. 492–496
96. P. Irofti, Sparse denoising with learned composite structured dictionaries, in *19th International Conference on System Theory, Control and Computing* (2015), pp. 331–336
97. P. Irofti, Efficient parallel implementation for single block orthogonal dictionary learning. J. Control Eng. Appl. Inf. **18**(3), 101–108 (2016)
98. P. Irofti, B. Dumitrescu, GPU parallel implementation of the approximate K-SVD algorithm using OpenCL, in *22nd European Signal Processing Conference* (2014), pp. 271–275
99. P. Irofti, B. Dumitrescu, Overcomplete dictionary design: the impact of the sparse representation algorithm, in *The 20th International Conference on Control Systems and Computer Science* (2015), pp. 901–908
100. P. Irofti, B. Dumitrescu, Regularized algorithms for dictionary learning, in *Proceedings of International Conference Communications*, Bucharest (2016), pp. 439–442
101. P. Irofti, B. Dumitrescu, Overcomplete dictionary learning with Jacobi atom updates, in *39th International Conference on Telecommunications and Signal Processing* (2016), pp. 421–424
102. P. Irofti, F. Stoican, Dictionary learning strategies for sensor placement and leakage isolation in water networks, in *The 20th World Congress of the International Federation of Automatic Control* (2017), pp. 1589–1594
103. M.G. Jafari, M.D. Plumbley, Fast dictionary learning for sparse representations of speech signals. IEEE J. Sel. Topics Signal Process. **5**(5), 1025–1031 (2011)
104. R. Jenatton, R. Gribonval, F. Bach, Local stability and robustness of sparse dictionary learning in the presence of noise. Preprint (2012). arXiv:1210.0685
105. Z. Jiang, Z. Lin, L.S. Davis, Label consistent K-SVD: learning a discriminative dictionary for recognition. IEEE Trans. Pattern Anal. Mach. Intell. **35**(11), 2651–2664 (2013)
106. C. Jung, L. Jiao, H. Qi, T. Sun, Image deblocking via sparse representation. Signal Process. Image Commun. **27**(6), 663–677 (2012)
107. S.P. Kasiviswanathan, H. Wang, A. Banerjee, P. Melville, Online L1-dictionary learning with application to novel document detection, in *Advances in Neural Information Processing Systems*, ed. by F. Pereira, C.J.C. Burges, L. Bottou, K.Q. Weinberger, vol. 25 (Curran Associates, Inc., Red Hook, 2012), pp. 2258–2266
108. M.E. Kilmer, C.D. Martin, Factorization strategies for third-order tensors. Linear Algebra Appl. **435**(3), 641–658 (2011)
109. S.J. Kim, Online kernel dictionary learning, in *GlobalSIP*, Orlando, December 2015
110. T.G. Kolda, B.W. Bader, The tophits model for higher-order web link analysis, in *Workshop on Link Analysis, Counterterrorism and Security*, vol. 7 (2006), pp. 26–29
111. T.G. Kolda, B.W. Bader, Tensor decompositions and applications. SIAM Rev. **51**(3), 455–500 (2009)
112. S. Kong, D. Wang, A dictionary learning approach for classification: separating the particularity and the commonality, in *Computer Vision–ECCV 2012* (2012), pp. 186–199
113. K. Kreutz-Delgado, J.F. Murray, B.D. Rao, K. Engan, T.W. Lee, T.J. Sejnowski. Dictionary learning algorithms for sparse representation. Neural Comput. **15**(2), 349–396 (2003)

114. S. Kumar, M. Mohri, A. Talwalkar, Sampling methods for the Nyström method. J. Mach. Learn. Res. **13**, 981–1006 (2012)
115. L. Le Magoarou, R. Gribonval, Chasing butterflies: in search of efficient dictionaries, in *International Conference on Acoustics Speech Signal Processing (ICASSP)*, Brisbane, April 2015, pp. 3287–3291
116. H. Lee, A. Battle, R. Raina, A.Y. Ng, Efficient sparse coding algorithms, in *Advances in Neural Information Processing Systems 19*, ed. by P.B. Schölkopf, J.C. Platt, T. Hoffman (MIT Press, Cambridge, 2007), pp. 801–808
117. S. Lesage, R. Gribonval, F. Bimbot, L. Benaroya, Learning unions of orthonormal bases with thresholded singular value decomposition, in *Proceedings. IEEE International Conference on Acoustics, Speech, and Signal Processing (ICASSP '05), 2005*, vol. 5, March 2005, pp. v/293–v/296
118. N. Li, S. Kindermann, C. Navasca, Some convergence results on the regularized alternating least-squares method for tensor decomposition. Linear Algebra Appl. **438**(2), 796–812 (2013)
119. Q. Liang Q. Ye, Computing singular values of large matrices with an inverse-free preconditioned Krylov subspace method. Electron. Trans. Numer. Anal. **42**, 197–221 (2014)
120. T. Lin, S. Liu, H. Zha, Incoherent dictionary learning for sparse representation, in *21st International Conference on Pattern Recognition (ICPR)*, Tsukuba, November 2012, pp. 1237–1240
121. Z. Liu, J. Li, W. Li, P. Dai, A modified greedy analysis pursuit algorithm for the cosparse analysis model. Numer. Algorithms **74**(3), 867–887 (2017)
122. S. Lloyd, Least squares quantization in PCM. IEEE Trans. Info. Theory **28**(2), 129–137 (1982)
123. S. Mahdizadehaghdam, L. Dai, H. Krim, E. Skau, H. Wang, Image classification: a hierarchical dictionary learning approach, in *IEEE International Conference on Acoustics Speech Signal Processing (ICASSP)*, New Orleans (2017), pp. 2597–2601
124. B. Mailhé, M. Plumbley, Fixed points of dictionary learning algorithms for sparse representations. IEEE Trans. Inf. Theory (2013, submitted) https://hal.inria.fr/file/index/docid/807545/filename/locOpt.pdf
125. B. Mailhé, S. Lesage, R. Gribonval, F. Bimbot, P. Vandergheynst, Shift-invariant dictionary learning for sparse representations: extending K-SVD, in *European Signal Processing Conference (EUSIPCO)*, Lausanne, August 2008
126. J. Mairal, F. Bach, J. Ponce, G. Sapiro, A. Zisserman, Discriminative learned dictionaries for local image analysis, in *Proceedings of IEEE Conference on Computer Vision and Pattern Recognition* (2008)
127. J. Mairal, M. Elad, G. Sapiro, Sparse representation for color image restoration. IEEE Trans. Image Process. **17**(1), 53–69 (2008)
128. J. Mairal, G. Sapiro, M. Elad, Learning multiscale sparse representations for image and video restoration. SIAM Multiscale Model. Simul. **7**(1), 214–241 (2008)
129. J. Mairal, F. Bach, J. Ponce, G. Sapiro, Online learning for matrix factorization and sparse coding. J. Mach. Learn. Res. **11**, 19–60 (2010)
130. J. Mairal, F. Bach, J. Ponce, Task-driven dictionary learning. IEEE Trans. Pattern Anal. Mach. Intell. **34**(4), 791–804 (2012)
131. M. Marsousi, K. Abhari, P. Babyn, J. Alirezaie, An adaptive approach to learn overcomplete dictionaries with efficient numbers of elements. IEEE Trans. Signal Process. **62**(12), 3272–3283 (2014)
132. K. Marwah, G. Wetzstein, Y. Bando, R. Raskar, Compressive light field photography using overcomplete dictionaries and optimized projections. ACM Trans. Graph. **32**(4), 46 (2013)
133. J.L. Massey, T. Mittelholzer, Welch's bound and sequence sets for code-division multiple-access systems, in *Sequences II*, ed. by R. Capocelli, A. De Santis, U. Vaccaro (Springer, Berlin, 1993), pp. 63–78
134. R. Mazhar, P.D. Gader, EK-SVD: optimized dictionary design for sparse representations, in *19th International Conference on Pattern Recognition* (2008), pp. 1–4

135. Q. Mo, Y. Shen, A remark on the restricted isometry property in orthogonal matching pursuit. IEEE Trans. Inf. Theory **58**(6), 3654–3656 (2012)
136. J.F. Murray, K. Kreutz-Delgado, Sparse image coding using learned overcomplete dictionaries, in *International Workshop on Machine Learning for Signal Processing* (2004), pp. 579–588
137. S. Nam, M.E. Davies, M. Elad, R. Gribonval, The cosparse analysis model and algorithms. Appl. Comput. Harmon. Anal. **34**(1), 30–56 (2013)
138. D. Needell R. Vershynin, Uniform uncertainty principle and signal recovery via regularized orthogonal matching pursuit. Found. Comput. Math. **9**(3), 317–334 (2009)
139. M. Nejati, S. Samavi, N. Karimi, S.M.R. Soroushmehr, K. Najarian, Boosted dictionary learning for image compression. IEEE Trans. Image Process. **25**(10), 4900–4915 (2016)
140. V.H. Nguyen, V.M. Patel, N.M. Nasrabadi, R. Chellappa, Design of non-linear kernel dictionaries for object recognition. IEEE Trans. Image Process. **22**(12), 5123–5135 (2013)
141. B.A. Olshausen, D.J. Field, Emergence of simple-cell receptive field properties by learning a sparse code for natural images. Nature **381**, 607–609 (1996)
142. B.A. Olshausen, D.J. Field, Sparse coding with an overcomplete basis set: a strategy employed by V1? Vision Res. **37**(23), 3311–3325 (1997)
143. N. Parikh, S. Boyd, Proximal algorithms. Found. Trends Optim. **1**(3), 123–231 (2014)
144. Y.C. Pati, R. Rezaiifar, P.S. Krishnaprasad, Orthogonal matching pursuit: recursive function approximation with applications to wavelet decomposition, in *27th Asilomar Conference on Signals Systems Computers*, vol. 1, November 1993, pp. 40–44
145. D.S. Pham, S. Venkatesh, Joint learning and dictionary construction for pattern recognition, in *Proceedings of IEEE Conference on Computer Vision and Pattern Recognition* (2008)
146. G. Pope, C. Aubel, C. Studer, Learning phase-invariant dictionaries, in *International Conference on Acoustics Speech Signal Processing (ICASSP)*, Vancouver, May 2013, pp. 5979–5983
147. J. Portilla, V. Strela, M.J. Wainwright, E.P. Simoncelli, Image denoising using scale mixtures of Gaussians in the wavelet domain. IEEE Trans. Image Process. **12**(11), 1338–1351 (2003)
148. A. Rakotomamonjy, Direct optimization of the dictionary learning problem. IEEE Trans. Signal Process. **61**(22), 5495–5506 (2013)
149. I. Ramirez, P. Sprechmann, G. Sapiro, Classification and clustering via dictionary learning with structured incoherence and shared features, in *IEEE Conference on Computer Vision Pattern Recognition (CVPR)*, San Francisco, June 2010
150. I. Ramirez, G. Sapiro, An MDL framework for sparse coding and dictionary learning, *IEEE Trans. Signal Proc.*, **60**(6), 2913–2927 (2012)
151. J. Ranieri, A. Chebira, M. Vetterli, Near-optimal sensor placement for linear inverse problems. IEEE Trans. Signal Process. **62**(5), 1135–1146 (2014)
152. N. Rao, F. Porikli, A clustering approach to optimize online dictionary learning, in *International Conference on Acoustics Speech Signal Processing (ICASSP)*, Kyoto, March 2012, pp. 1293–1296
153. S. Ravishankar, Y. Bresler, MR image reconstruction from highly undersampled k-space data by dictionary learning. IEEE Trans. Med. Imag. **30**(5), 1028–1041 (2011)
154. L. Rebollo-Neira, Dictionary redundancy elimination. IEE Process. Vis. Image Signal Process. **151**(1), 31–34 (2004)
155. L.A. Rossman, EPANET 2: users manual. US Environmental Protection Agency. Office of Research and Development. National Risk Management Research Laboratory (2000)
156. R. Rubinstein, M. Zibulevsky, M. Elad, Efficient implementation of the K-SVD algorithm using batch orthogonal matching pursuit. Technical Report CS-2008-08, Technion University, Haifa (2008)
157. R. Rubinstein, M. Zibulevsky, M. Elad, Double sparsity: learning sparse dictionaries for sparse signal approximation. IEEE Trans. Signal Process. **58**(3), 1553–1564 (2010)
158. R. Rubinstein, A.M. Bruckstein, M. Elad, Dictionaries for sparse representations modeling. Proc. IEEE **98**(6), 1045–1057 (2010)

159. R. Rubinstein, T. Peleg, M. Elad. Analysis K-SVD: a dictionary-learning algorithm for the analysis sparse model. IEEE Trans. Signal Process. **61**(3), 661–677 (2013)
160. C. Rusu, B. Dumitrescu, Stagewise K-SVD to design efficient dictionaries for sparse representations. IEEE Signal Process. Lett. **19**(10), 631–634 (2012)
161. C. Rusu, B. Dumitrescu, Block orthonormal overcomplete dictionary learning, in *21st European Signal Processing Conference (EUSIPCO 2013)*, September 2013, pp. 1–5
162. C. Rusu, B. Dumitrescu, An initialization strategy for the dictionary learning problem, in *International Conference Acoustics Speech Signal Processing*, Florence (2014), pp. 6731–6735
163. C. Rusu, N. Gonzalez-Prelcic, Designing incoherent frames through convex techniques for optimized compressed sensing. IEEE Trans. Signal Process. **64**(9), 2334–2344 (2016)
164. C. Rusu, B. Dumitrescu, S.A. Tsaftaris, Explicit shift-invariant dictionary learning. IEEE Signal Process. Lett. **24**(1), 6–9 (2014)
165. Y. Saad, *Numerical Methods for Large Eigenvalue Problems*, Revised edn. (SIAM, Providence, 2011)
166. M. Sadeghi, M. Babaie-Zadeh, Incoherent unit-norm frame design via an alternating minimization penalty method. IEEE Signal Process. Lett. **24**(1), 32–36 (2017)
167. M. Sadeghi, M. Babaie-Zadeh, C. Jutten, Dictionary learning for sparse representation: a novel approach. IEEE Signal Process. Lett. **20**(12), 1195–1198 (2013)
168. M. Sadeghi, M. Babaie-Zadeh, C. Jutten, Learning overcomplete dictionaries based on atom-by-atom updating. IEEE Trans. Signal Process. **62**(4), 883–891 (2014)
169. S. K. Sahoo, A. Makur, Dictionary training for sparse representation as generalization of K-means clustering. IEEE Signal Process. Lett. **20**(6), 587–590 (2013)
170. J. Scarlett, J.S. Evans, S. Dey, Compressed sensing with prior information: information-theoretic limits and practical decoders. IEEE Trans. Signal Process. **61**(2), 427–439 (2013)
171. O.G. Sezer, O. Harmanci, O.G. Guleryuz, Sparse orthonormal transforms for image compression, in *2008 15th IEEE International Conference on Image Processing*, October 2008, pp. 149–152
172. L. Shao, R. Yan, X. Li, Y. Liu, From heuristic optimization to dictionary learning: a review and comprehensive comparison of image denoising algorithms. IEEE Trans. Cybern. **44**(7), 1001–1013 (2014)
173. L. Shen, G. Sun, Q. Huang, S. Wang, Z. Lin, E. Wu, Multi-level discriminative dictionary learning with application to large scale image classification. IEEE Trans. Image Process. **24**(10), 3109–3123 (2015)
174. B. Shen, B.D. Liu, Q. Wang, Elastic net regularized dictionary learning for image classification. Multimed. Tools Appl. **75**, 8861–8874 (2016)
175. A. Shrivastava, V.M. Patel, R. Chellappa, Multiple kernel learning for sparse representation-based classification. IEEE Trans. Image Process. **23**(7), 3013–3024 (2014)
176. C.D. Sigg, T. Dikk, J.D. Buhmann, Learning dictionaries with bounded self-coherence. IEEE Signal Process. Lett. **19**(19), 861–865 (2012)
177. K. Skretting, K. Engan, Recursive least squares dictionary learning. IEEE Trans. Signal Process. **58**(4), 2121–2130 (2010)
178. K. Skretting, K. Engan, Image compression using learned dictionaries by RLS-DLA and compared with K-SVD, in *International Conference on Acoustics Speech Signal Processing (ICASSP)*, Prague (2011), pp. 1517–1520
179. K. Skretting, S.A. Husoy, Texture classification using sparse frame based representations. EURASIP J. Appl. Signal Process. (2006). Article ID 52561
180. L.N. Smith, M. Elad, Improving dictionary learning: multiple dictionary updates and coefficient reuse. IEEE Signal Process. Lett. **20**(1), 79–82 (2013)
181. J.L. Starck, E.J. Candès, D.L. Donoho, The curvelet transform for image denoising. IEEE Trans. Image Process. **11**(6), 670–684 (2002)
182. P. Stoica, P. Ahgren, Exact initialization of the recursive least-squares algorithm. Int. J. Adapt. Control Signal Process. **16**, 219–230 (2002)

183. B.L. Sturm, M.G. Christensen, Comparison of orthogonal matching pursuit implementations, in *Proceedings of European Signal Processing Conference (EUSIPCO)*, Bucharest (2012), pp. 220–224
184. J. Sun, Q. Qu, J. Wright, Complete dictionary recovery over the sphere, in *International Conference on Sampling Theory and Applications (SampTA)* (2015), pp. 407–410
185. J.J. Thiagarajan, K.N. Ramamurthy, A. Spanias, Shift-invariant sparse representation of images using learned dictionaries, in *IEEE Workshop Machine Learning Signal Processing*, Cancun, October 2008, pp. 145–150
186. J.J. Thiagarajan, K.N. Ramamurthy, A. Spanias, Multiple kernel sparse representations for supervised and unsupervised learning. IEEE Trans. Image Process. **23**(7), 2905–2915 (2014)
187. R. Tibshirani, Regression Shrinkage and Selection via the Lasso. J. R. Stat. Soc. B **58**(1), 267–288 (1996)
188. M.E. Tipping, Sparse Bayesian learning and the relevance vector machine. J. Mach. Learn. Res. **1**, 211–244 (2001)
189. I. Tosic, P. Frossard, Dictionary learning. IEEE Signal Process. Mag. **28**(2), 27–38 (2011)
190. J.A. Tropp, Greed is good: algorithmic results for sparse approximation. IEEE Trans. Inf. Theory **50**(10), 2231–2242 (2004)
191. J.A. Tropp, I.S. Dhillon, R.W. Heath Jr., T. Strohmer, Designing structured tight frames via an alternating projection method. IEEE Trans. Inf. Theory **51**(1), 188–209 (2005)
192. E.V. Tsiligianni, L.P. Kondi, A.K. Katsaggelos, Construction of incoherent unit norm tight frames with application to compressed sensing. IEEE Trans. Info. Theory **60**(4), 2319–2330 (2014)
193. D. Vainsencher, S. Mannor, A.M. Bruckstein, The sample complexity of dictionary learning. J. Mach. Learn. Res. **12**, 3259–3281 (2011)
194. C.F. Van Loan, N. Pitsianis, Approximation with Kronecker products, in *Linear Algebra for Large Scale and Real-Time Applications* (Springer, Berlin, 1993), pp. 293–314
195. P. Vincent, Y. Bengio, Kernel matching pursuit. Mach. Learn. **48**, 165–187 (2002)
196. T.H. Vu, V. Monga, Learning a low-rank shared dictionary for object classification, in *2016 IEEE International Conference on Image Processing (ICIP)* (IEEE, New York, 2016), pp. 4428–4432
197. S. Waldron, Generalized Welch bound equality sequences are tight frames. IEEE Trans. Inf. Theory **49**(9), 2307–2309 (2003)
198. Z. Wang, A.C. Bovik, H.R. Sheikh, E.P. Simoncelli, Image quality assessment: from error visibility to structural similarity. IEEE Trans. Image Process. **13**(4), 600–612 (2004)
199. S. Wang, L. Zhang, Y. Liang, Q. Pan, Semi-coupled dictionary learning with applications to image super-resolution and photo-sketch synthesis, in *2012 IEEE Conference on Computer Vision and Pattern Recognition (CVPR)*, Providence (2012), pp. 2216–2223
200. H. Wang, J. Wright, D. Spielman, A batchwise monotone algorithm for dictionary learning (2015). arXiv:1502.00064
201. Z. Wang, Y. Yang, J. Yang, T. Huang, Designing a composite dictionary adaptively from joint examples, in *IEEE Visual Communication Image Processing (VCIP)*, Singapore, December 2015
202. A.G. Weber, The USC-SIPI image database (1997)
203. L. Welch, Lower bounds on the maximum cross correlation of signals. IEEE Trans. Inf. Theory **20**(3), 397–399 (1974)
204. H. Wersing, J. Eggert, E. Körner, Sparse coding with invariance constraints. Lect. Notes Comput. Sci. **2714**, 385–392 (2003)
205. D.P. Wipf, B.D. Rao, Sparse Bayesian learning for basis selection. IEEE Trans. Signal Process. **52**(8), 2153–2164 (2004)
206. B. Wohlberg, Efficient algorithms for convolutional sparse representations. IEEE Trans. Image Process. **15**(1), 301–315 (2016)
207. J. Wright, A.Y. Yang, S. Shankar Sastry, Y. Ma, Robust face recognition via sparse representation. IEEE Trans. Pattern Anal. Mach. Intell. **31**(2), 210–227 (2009)

208. Z. Xing, M. Zhou, A. Castrodad, G. Sapiro, L. Carin, Dictionary learning for noisy and incomplete hyperspectral images. SIAM J. Imag. Sci. **5**(1), 33–56 (2012)
209. M. Xu, X. Jia, M. Pickering, A.J. Plaza, Cloud removal based on sparse representation via multitemporal dictionary learning. IEEE Trans. Geosci. Remote Sens. **54**(5), 2998–3006 (2016)
210. M. Yaghoobi, T. Blumensath, M.E. Davies, Dictionary learning for sparse approximations with the majorization method. IEEE Trans. Signal Process. **57**(6), 2178–2191 (2009)
211. M. Yaghoobi, L. Daudet, M.E. Davies, Parametric dictionary design for sparse coding. IEEE Trans. Signal Process. **57**(12), 4800–4810 (2009)
212. J. Yang, J. Wright, T.S. Huang, Y. Ma, Image super-resolution via sparse representation. IEEE Trans. Image Process. **19**(11), 2861–2873 (2010)
213. M. Yang, L. Zhang, J. Yang, D. Zhang, Metaface learning for sparse representation based face recognition, in *International Conference on Image Processing (ICIP)*, Hong Kong, September 2010
214. M. Yang, L. Zhang, X. Feng, D. Zhang, Fisher discrimination dictionary learning for sparse representation, in *IEEE International Conference on Computer Vision* (2011), pp. 543–550
215. J. Yang, Z. Wang, Z. Lin, S. Cohen, T. Huang, Coupled dictionary training for image super-resolution. IEEE Trans. Image Process. **21**(8), 3467–3478 (2012)
216. M. Yang, D. Dai, L. Shen, L. Van Gool, Latent dictionary learning for sparse representation based classification, in *IEEE Conference on Computer Vision Pattern Recognition* (2014), pp. 4138–4145
217. J. Yin, Z. Liu, Z. Jin, W. Yang, Kernel sparse representation based classification. Neurocomputing **77**, 120–128 (2012)
218. C.H. Zhang, Nearly unbiased variable selection under minimax concave penalty. Ann. Stat. **38**(2), 894–942 (2010)
219. Z. Zhang, S. Aeron, Denoising and completion of 3d data via multidimensional dictionary learning, in *Proceedings of the Twenty-Fifth International Joint Conference on Artificial Intelligence (IJCAI-16)* (2016), pp. 2371–2377
220. Q. Zhang, B. Li, Discriminative K-SVD for dictionary learning in face recognition, in *Proceedings of IEEE Conf. Computer Vision and Pattern Recognition* (2010), pp. 2691–2698
221. Q. Zhang, B. Li, *Dictionary Learning in Visual Computing* (Morgan & Claypool, San Rafael, 2015)
222. L. Zhang, W.D. Zhou, P.C. Chang, J. Liu, Z. Yan, T. Wang, F.Z. Li, Kernel sparse representation-based classifier. IEEE Trans. Signal Process. **60**(4), 1684–1695 (2012)
223. H. Zhang, Y. Zhang, T.S. Huang, Simultaneous discriminative projection and dictionary learning for sparse representation based classification. Pattern Recogn. **46**, 346–354 (2013)
224. Z. Zhang, Y. Xu, J. Yang, X. Li, D. Zhang. A survey of sparse representation: algorithms and applications. IEEE Access **3**, 490–530 (2015)
225. G. Zhang, H. Sun, G. Xia, Q. Sun, Multiple kernel sparse representation-based orthogonal discriminative projection and its cost-sensitive extension. IEEE Trans. Image Process. **25**(9), 4271–4285 (2016)
226. M. Zhou, H. Chen, J. Paisley, L. Ren, L. Li, Z. Xing, D. Dunson, G. Sapiro, L. Carin, Nonparametric Bayesian dictionary learning for analysis of noisy and incomplete images. IEEE Trans. Image Process. **21**(1), 130–144 (2012)
227. H. Zou, T. Hastie, Regularization and variable selection via the elastic net. J. R. Stat. Soc. Ser. B Stat. Methodol. **67**(2), 301–320 (2005)

Index

Symbols
0-norm, 2
1ONB, 174, 206
2D signal, 187
2D-OMP, 188

A
AK-SVD, 59, 63, 78, 105, 141, 169, 183, 186, 196
AK-SVDc, 105
AK-SVDr, 94
alternating least squares, 194
analysis dictionary, 257
analysis K-SVD, 265
analysis model, 257
analysis SimCO, 267
Approximate K-SVD, 59
APrU – Atom Profile Updating, 119
atom replacement, 72, 148, 150, 155, 160
atom splitting, 155
average sparsity level, 116

B
backward greedy, 261
basis pursuit, 8, 12, 185, 204
Batch OMP, 50
BFGS, 104
BG, 261
block coordinate descent, 45
blocking effect, 37
BM3D, 83

C
CANDECOMP/PARAFAC, 193
cell splitting, 156
Cholesky factorization, 47, 58
Cholesky factorization update, 139
circulant matrices, 181
classification, 247
    class scatter, 218
    discrimination, 217, 218, 220
    shared dictionary, 220
    supervised, 209
    unsupervised, 210, 216
classifier matrix, 222
clustering, 157
co-rank, 260
coherence
    cumulative, 16
    mutual, 16, 173
    penalty, 126, 215, 221
    root mean square, 99
column subset selection, 245
composite denoising, 205, 206
composite dictionary, 204
compressive sampling, 41
condition number, 17, 32
conditioning, 31
convex relaxation, 8
convolutional sparse coding, 184
coordinate descent, 54, 118
cosparse representation, 259
cosparsity level, 258
cosupport, 259
coupled DL, 130

B. Dumitrescu, P. Irofti, *Dictionary Learning Algorithms and Applications*,
https://doi.org/10.1007/978-3-319-78674-2

CP, 193
critical point, 75

D
data
  artificial, 34
  images, 34
  overlapping patches, 37
deblurring, 130
DFT, 2
dictionary initialization, 70, 153
dictionary learning, 26
dictionary update, 31
  cycles, 68
  inner iterations, 60, 62, 66
discriminative atoms, 211
discriminative DL, 130, 223, 250
DL-COPAR, 221
DLENE, 155
double sparsity, 169
  AK-SVD, 170
  generalization, 172
dynamic range, 38

E
elastic net regularization, 126
Enhanced K-SVD, 160
EPANET, 225
error
  denoising, 37
  expectation, 28
  norm, 27
  relative representation, 72
  representation, 27, 54
external dictionary, 204

F
factorization, 30
fault detection and isolation, 212
feature vector, 231
Fisher discrimination, 218
FISTA, 12, 119
fixed point, 75
frame
  equiangular, 97
  Grassmannian, 98, 111
  potential, 99
  tight, 97
  unit norm, tight, 97

G
Gauss–Seidel, 62
Gram matrix, 49, 107, 235
greedy, 8
greedy adaptive dictionary, 171
greedy analysis pursuit, 264

H
Hadamard product, 192
halftoning, 129
Hanoi water network, 211
hard thresholding, 124

I
ICD, 112
IDL, 104
impulsive noise, 127
incomplete data, 140
indetermination, 29
indian buffet process, 162
internal dictionary, 204
IPR, 109
ISTA, 13

J
Jacobi, 62
JMDL, 198

K
K-SVD, 58, 76, 78, 92, 105, 223, 224
K-SVDr, 93
K-TSVD, 201
KDRSDL, 198
kernel, 232
  Epanechnikov, 158
  Gaussian, 158
  Mercer, 233
  multiple, 253
  polynomial, 233
  RBF, 234
  statistics, 158
  trick, 233
kernel AK-SVD, 241
kernel DL, 234
kernel K-SVD, 242
kernel MOD, 239
kernel OMP, 237
kernel SGK, 240

kernel SRC, 248
Khatri–Rao product, 192

L
label consistency matrix, 224
label matrix, 209
LAOLS, 84
lasso, 8
latent DL, 227
LC-KSVD, 224
leak magnitude, 213
local minima, 29, 31
long signal, 184

M
Mahalanobis distance, 205
mask, 140
masked OMP, 40, 141
matrix inversion lemma, 138
mean shift, 158
MOD, 57, 75, 78, 92, 118
mode-*k* product, 193
multiscale dictionary, 172

N
N-BOMP, 199
normalization, 238
  atom, 26, 46, 56
  indetermination, 29
  signal, 28
NP-hard, 5
NSGK, 68, 78
nuclear norm, 198, 221
Nyquist rate, 41
Nyström method, 244

O
OBG, 263
OLS, 8, 11, 263
OMP, 8, 10, 185, 188, 236
  group-level, 127
  shift invariant, 182
One Stage DL, 119
operator
  $\mathcal{R}$, 197
  vec, 34, 52
optimized backward greedy, 263
orthogonal base
  union, 174
orthogonal bases
  training, 174
overall sparsity measure, 191
overcomplete DCT, 3
overcompleteness factor, 4

P
PAK-SVD, 78
partial sorting, 175, 177
peak signal to noise ratio, 38
pixel averaging, 38, 40
PNSGK, 68, 78
power method, 61
pressure residual, 212
principal component analysis, 28
principal directions, 151
Procrustes, 105
proximal gradient method, 124
proximal operator, 123
PSGK, 62
PSNR, 38

R
recovery ratio, 35, 82
recursive least squares, 137
regularized AK-SVD, 94
regularized ALS, 196
regularized K-SVD, 93
regularized OMP, 92
regularized SimCO, 95
representation power, 149
restricted isometry property, 17
Riemannian submanifold, 190
RMSE, 33, 79
root mean square error, 33
rotation, 106, 181

S
SBO, 177, 206
  classification, 180
  parallel, 179
SeDiL, 190
sensor placement, 225
separable dictionary, 188
SGK, 56, 58, 63, 78, 92, 141, 182, 186
shared dictionary, 220
shift invariance, 172, 181
sigmoid function, 219
signal to noise ratio, 34, 38
SimCO, 64, 92, 94

SimCOr, 95
similarity measure, 232
SNR, 34
soft thresholding, 13, 119
softmax function, 217
source separation, 216
spark, 17, 30
sparse coding, 31
sparse representation, 1
  approximate, 6
  convolutional, 184
  exact, 5
Sparse Representation-based Classifier, 214
Sparsenet, 53, 75, 92
sparsity level, 2
SRC, 214, 248
SSIM, 38
Stagewise K-SVD, 151
Stiefel manifold, 64
structural similarity, 38
SuKro, 197
super-resolution, 130
support, 2, 26

T
t-product, 199
t-SVD, 200
tensor, 192
  ALS, 194
  CP decomposition, 193
  dictionary update, 195
  f-diagonal, 200
  fiber, 193
  identity, 200
  matricization, 193
  orthogonal, 200
  SVD, 199
  t-linear, 201
  transpose, 200
tight frame
  unit norm, 100
tree hierarchy, 228
TS norm, 201

U
uniqueness, 30
UNTF, 97
UONB, 176

V
vector quantization, 28
video summarization, 134

W
water distribution network, 211
weighted OMP, 163
Welch bound, 98

The manufacturer's authorised representative in the EU is Springer Nature Customer Service Centre GmbH, Europaplatz 3, 69115 Heidelberg, Germany. If you have any concerns regarding our products, please contact ProductSafety@springernature.com

Printed and bound by CPI Group (UK) Ltd, Croydon, CR0 4YY
30/09/2026
02225509-0008